MANAGEMENT OF TOXIC AND HAZARDOUS WASTES

MANAGEMENT OF TOXIC AND HAZARDOUS WASTES

Edited by

Harasiddhiprasad G. Bhatt
Robert M. Sykes
Thomas L. Sweeney

Library of Congress Cataloging in Publication Data
Main entry under title:

Management of toxic and hazardous wastes.

Revision of papers presented at the third Ohio Environmental Engineering Conference held in Columbus, Ohio, in Mar. 1983.
Includes bibliographies and index
1. Hazardous wastes—Congresses. 2. Water, Underground—Pollution—Congresses.
3. Hazardous waste sites—Congresses.
I. Bhatt, H.G. (Harasiddhiprasad G.) II. Ohio Environmental Engineering Conference (3rd : 1983 : Columbus, Ohio)
TD811.5.M35 1985 628.4'4 85-10180
ISBN 0-87371-023-1

LEWIS PUBLISHERS, INC.
121 South Main Street, Chelsea, Michigan 48118

PRINTED IN THE UNITED STATES OF AMERICA

PREFACE

This book is a product of the Third Ohio Environmental Engineering Conference held in Columbus, Ohio in 1983. The conference was sponsored by the Central Ohio Section of the American Society of Civil Engineers and the Ohio State University Department of Civil Engineering.

Chapters presented here have been updated to reflect present conditions. This book, therefore, is a current reference work on the management of toxic and hazardous wastes.

Increasing attention is now being focused on the problem of groundwater pollution in this country. The demand for cleaning of hazardous waste disposal sites has also grown stronger since the passage of the Comprehensive Environmental Response Compensation and Liability Act of 1980 (commonly known as Superfund). In sections on the impact of groundwater and disposal site cleanup, this book presents twelve chapters on these important aspects of hazardous waste management. Attention has also been focused on waste treatment and recycle, risk assessment, public participation and land disposal. The section on legal considerations provides valuable pointers on the precautions to be taken and pitfalls to be avoided to minimize legal liabilities.

The editors wish to thank the sponsoring organizations and all members of the Conference Committee. Ralph Cox, Tom Davis, Doug Uhren, Dave Pritchard, Bob Smith, Lorey Rogenkemp, Marcia Gibson, and Peggy Vince put in a lot of effort to make the conference a success. Thanks are due to Jerry Jones of Malcolm Pirnie, Inc. who prepared the index. Without the tremendous assistance of Shirley Marlowe, Teresa Grimmett, Marilyn Hunter and Carolyn Strutner of Malcolm Pirnie's word processing staff, this book could not have been completed on schedule. Anne M. Roosen, of the publisher's staff, is responsible for the final form of the book, for which she deserves thanks from all of us.

Finally, we thank all contributors to this book. Their understanding, patience and readiness to update their chapters has enabled us to present to you a truly up-to-date work on this important subject.

Harasiddhiprasad G. Bhatt
Robert M. Sykes
Thomas L. Sweeny

Harasiddhiprasad G. Bhatt is a Senior Project Engineer with Malcolm Pirnie, Inc. He holds BE and ME degrees in Civil Engineering from M.S. University of Baroda in India and an MSCE from Michigan State University. Mr. Bhatt has had more than 28 years of varied experience in undergraduate and graduate teaching, project management, process design for municipal and industrial wastewater treatment, facilities planning and design, and construction administration of water and wastewater treatment facilities. He is a Registered Professional Engineer and has authored several technical articles on solid and hazardous waste management, land application of wastewater and sludge, industrial pretreatments, acid mine drainage abatement, and operation of wastewater management facilities. He is actively associated with several national and international professional societies and is a member of the American Society of Civil Engineers Committee on Hazardous Waste Disposal.

Robert M. Sykes is Professor of Civil Engineering at The Ohio State University in Columbus, Ohio, where he teaches environmental engineering. Before joining the OSU faculty in 1972, he pursued undergraduate studies in civil engineering at Northeastern University in Boston, Massachusetts, and graduate studies in environmental engineering at Purdue University in West Lafayette, Indiana. He is responsible for the development of the product/maintenance theory of biological treatment processes, which is a competitor of the standard biokinetic theory for the design and analysis of these processes. Dr. Sykes is a member of the American Society of Civil Engineers, the American Water Works Association, the Association of Environmental Engineering Professors, the International Association on Water Pollution Research and Control and the Water Pollution Control Federation.

Thomas L. Sweeney received his BS, MS and PhD degrees from Case Western Reserve University. After working for The Standard Oil Company in Research and Development and Commercial Development, he joined the faculty of The Ohio State University, where he is now Professor of Chemical Engineering and Associate Vice President for Research Administration. In addition to teaching conventional Chemical Engineering courses, Dr. Sweeney has taught courses in environmental regulation, environmental science and technology, and environmental pollution abatement at Ohio State. He has presented or published a number of papers and has been a consultant to more than 50 industrial organizations, governmental agencies and legal firms. He is a Registered Professional Engineer in Ohio and is admitted to the practice of law in Ohio and United States courts. He is a member of the Ohio Hazardous Waste Facility Approval Board. Dr. Sweeney has been active in various professional and technical organizations and is a member of the American Institute of Chemical Engineers, the American Chemical Society, the American Society for Engineering Education, and the Ohio State Bar Association.

CONTENTS

SECTION 1: GENERAL CONSIDERATIONS

1. Implementation of RCRA and Superfund by the U.S. EPA—
The State's Perspective 1
Richard A. Valentinetti

2. Conflicts and Hazardous Waste Management—
The Environmentalist's Viewpoint 9
W. B. Clapham, Jr.

3. Public Participation in Ohio EPA's Solid and
Hazardous Waste Program 19
Michael Lewis Greenberg

4. Health and Safety Considerations for Hazardous Waste Workers ... 25
Lynn P. Wallace, Ph.D., P.E., William F. Martin, P.E.

5. Hazardous Waste Management—An Industry Perspective 35
William L. West

6. The Partnership Approach to Hazardous Waste Management 41
Stephen H. Sedam

7. Solid Waste Facility Siting—Community Aspects and Incentives 45
Harry E. Smail, AICP

SECTION 2: IMPACT ON GROUNDWATER

8. Groundwater Contamination Control and Treatment,
Rocky Mountain Arsenal, Colorado 65
Paul MacRoberts, C. B. Hagar, Harry L. Callahan

9. Statistical Evaluation of Hydraulic Conductivity Data for Waste
Disposal Sites 81
Wayne R. Bergstrom, George R. Kunkle

10. Groundwater Monitoring Systems—Only as Good as the Weakest Link 105
David E. Johe, C.P.G.

11. Problems in Assessing Organics Contamination in Groundwater ... 119
Robert A. Saar

12. Private Well Sampling in Vicinity of Re-Solve, Inc., Hazardous Waste Site 129
Thomas E. Tetreault, Paul M. Williams

SECTION 3: TREATMENT

13. Liquid Hazardous Waste Treatment Design 141
T. H. Coughlin, O. A. Clemens, J. Johnson

14. In Situ Stabilization and Closure of an Oily Sludge Lagoon 155
J. W. Thorsen, M. F. Coia, A. A. Metry

15. Hazardous Waste Reduction Through In-Process Controls, Process Substitutions, and Recovery/Recycling Techniques 171
John A. Gurklis

SECTION 4: WASTE RECYCLE

16. New York State Industrial Materials Recycling Program 195
Pickett T. Simpson, P.E.

17. The Role of a Waste Exchange in Industrial Waste Management—Development of the Northeast Industrial Waste Exchange 209
Walker Banning

18. 3P: Pollution Prevention Pays—A 3M Success Story 215
Michael D. Koenigsberger

19. European Network of Waste Exchanges 223
Thomas E. Crepeau, Philip R. Beltz

SECTION 5: LAND DISPOSAL

20. Hazardous Waste Land Disposal Regulations—An Environmentalist Perspective 227
Linda E. Greer, David J. Lennett

21. Influence of Hazardous and Toxic Wastes on the Engineering Behavior of Soils 237
Jeffrey C. Evans, Ph.D, P.E., Hsai-Yang Fang, Ph.D, Irwin Jay Kugelman, Ph.D

22. Site Selection and Design Considerations for Hazardous Waste Land Disposal Facilities 265
Dr. Paul A. Hustad, Dr. John A. Ruf

23. EPA'S Land Disposal Regulations—Waste Disposal Industry's Perspective 283
Reva Rubenstein, Ph.D

SECTION 6: DISPOSAL SITE CLEANUP

24. Waterway Contamination—An Assessment Of Cleanup Priorities .. 289
John C. Henningson

25. Cleanup of a Vinylidene Chloride and Phenol Spill 297
Albert R. Posthuma, John G. Kraus, Julie A. Rutherford

26. Case History—Remedial Investigation Re-Solve, Inc. Hazardous Waste Site 307
Jeffrey A. Cassis, Dana Pedersen

27. Waste Stabilization Basin Discharge Elimination and Remediation—A Case Study 319
William H. Bouck, P.E., Andrew N. Johnson, P.E., Stephen J. Fleischacker

28. Site Safety and Sampling Plans—The First Step in Investigating Abandoned Hazardous Waste Disposal Sites 327
John W. Edwards, Vernon M. Reid, Paul B. MacRoberts

29. Remedial Investigation and Feasibility Study, Tacoma Water Supply Wells, Commencement Bay Area, Tacoma, Washington 335
Mark G. Snyder, Paul B. MacRoberts

30. Soil Investigation at the Re-solve, Inc., Hazardous Waste Site 353
Tom A. Pedersen

SECTION 7: RISK ASSESSMENT

31. Environmental Risk Assessment 367
Lynne M. Miller

SECTION 8: LEGAL CONSIDERATIONS

32. Manufacturer's Warranties on Hazardous Waste Disposal Equipment 373
Stanley A. Reigel

33. Federal and State Enforcement of Hazardous Waste Laws 381
William W. Falsgraf

34. Generator Liability Under Superfund 387
Richard T. Sargeant

35. Environmental Law and Contractor Liability 405
J. Wray Blattner, Edward A. Hogan

INDEX **415**

SECTION I

GENERAL CONSIDERATIONS

CHAPTER 1

IMPLEMENTATION OF RCRA AND SUPERFUND BY THE U.S. EPA - THE STATE'S PERSPECTIVE

Richard A. Valentinetti
Vermont Agency of Environmental Conservation

INTRODUCTION

With the passage of the Clean Air and Water Acts in the early 70s, Congress unwittingly enhanced the problems associated with solid and hazardous waste disposal. Hazardous sludges produced by the treatment of air and water emissions and wastes that would have been discharged to streams and rivers prior to the Clean Water Act were disposed of in landfills, lagoons, underground injection wells, etc. To further complicate this situation, the unforeseen legacy of America's industrialization was beginning to manifest itself in the form of Love Canal, the Kin-buc landfill in New Jersey, and other old waste disposal sites. To close the loop of environmental regulation, in 1976 Congress enacted the Resource Conservation and Recovery Act (RCRA) to deal with the ongoing problems of hazardous waste management. In order to provide for the clean up of closed or inactive hazardous waste disposal sites and emergency spill response, in 1980 Congress passed the Comprehensive Environmental Response, Compensation, and Liability Act (CERCLA), more commonly called Superfund. The following is a discussion of the effect on the states of EPA's implementation of these two Acts.

RCRA

In the interim period between the enactment of RCRA and the promulgation of the first major portion of the RCRA regulations in May 1980, many states took the initiative and developed hazardous waste programs of their own. Once the Federal program was established, however, these programs were pre-empted. In order to remain the sole regulatory authority in the hazardous waste area, these states and other states with no pre-existing programs which wanted primacy had to be authorized by the U.S. EPA. Under RCRA Section 3006, there are two types of authorization, interim and final. Interim authorization is temporary and was created to provide states with enough time to prepare their applications for

Management of Toxic and Hazardous Wastes, H. Bhatt, R. Sykes, T. Sweeney (Editors)

final authorization while maintaining primacy.

There are four criteria by which a state's application for final authorization are judged. The state program must be: equivalent to the Federal program; consistent with the Federal program; no less stringent than Federal program; and provide adequate enforcement. While these criteria in and of themselves may not seem unreasonable, EPA's interpretation and application of the criteria have created a number of problems for the states, some of which are major stumbling blocks in the authorization process. In particular, EPA has taken the equivalency and no less stringent requirements and determined that a state program including its regulations, must be identical to the Federal program. This position will not only force those states with pre-existing regulatory programs to revise extensive portions of their regulations, but it also places the states in a situation where continual regulatory amendments are necessary. This is due to the fact that even at this point in time the EPA regulatory program is not complete. EPA promulgated "Phase I" of its regulations in May 1980 but did not promulgate "Phase II, Component C" until July 26, 1982 and more regulations need to be developed to cover all types of hazardous waste facilities. On top of this, EPA is currently planning potentially major revisions to the regulations it already has on the books. These additions and revisions to the Federal regulations will be putting a strain on both the states and those persons who have to follow and comply with the regulations.

Another difficulty which arises as the result of an identical regulation approach is the inability of state permit writers to allow variances or waivers from the federal permitting requirements on a case by case basis. Both the Clean Air and Clean Water Acts use some form of "best engineering judgment" and other mechanisms which give a permit writer some latitude. Due to the no less stringent criteria of RCRA Section 3009, however, a permit writer cannot consider any site specific conditions which might justify a variance from the technical standards usually applied to a hazardous waste facility unless the regulation involved specifically allows for a variance. This position fails to take into account both the fact that facility sites can be radically different and the fact that technology for dealing with hazardous waste is constantly changing and upgrading. Researchers and engineers are developing new treatment techniques and technologies almost every day. Without the ability to variance, these innovative methods will be more difficult to permit. As the Federal government should not only be encouraging this kind of research but also creating a regulatory program that contains an element of common sense, the states are currently working with EPA to resolve this problem.

The problems the states are experiencing in obtaining final authorization are compounded by the fact that all interim authorizations granted to the states expire on January 26, 1985. If a state with interim authorization has not received final authorization by that date, the primacy for the operation of the program reverts back to EPA. Considering the obstacles to the authorization process that I have described, the potential for program reversion seems to increase each day. If reversions occur, the regulated community will face even further confusion than it is now experiencing as the program shifts back and forth between the states and EPA. If primacy shifts, there will also be double regulation; i.e., by both EPA and the state, in those states which have enacted their own hazardous waste laws. With the proposed cut back on EPA funding, operating duplicative programs is a waste of resources we cannot afford.

Once a state overcomes the difficulties inherent in the authorization process, it then faces the myriad problems associated with the actual implementation of the program. These problems arise in the context of facility permittin, the maintenance of a national data base, Federal funding and resources available to the states, and other areas.

The permitting and proper operation of hazardous waste treatment, storage, and disposal facilities is of vital importance in order to prevent the creation of future Superfund sites. The first step, and perhaps the most difficult, is the siting of a facility. The prevalent "not in my backyard" attitude and mistrust of state and federal regulatory authorities, resulting in part, from past abuses by permitted facilities, have caused companies to go through long, painful, and often unsuccessful attempts to site a facility. To deal with this situation, some states have enacted facility siting laws which establish independent boards with the primary authority for siting decisions. Two state's laws, New Jersey and Massachusetts, illustrate the different approaches that have been taken. New Jersey's siting board is comprised of three industry representatives, three local officials and three environmentalists. The Board either proposes sites itself or acts on the request of industry. If the Board finds that site meets the criteria established by the State Department of Environmental Protection, the site is approved without regard to any local requirements, ordinances or potential objections.

The Massachusetts approach differs from New Jersey's in that its law is predicated on the belief that siting considerations are best addressed through negotiations between developers and the host communities. The Board is comprised of eight state representatives, six representatives of state professional organizations, and seven members of the public. The Board addresses the social and economic impact of proposed facilities and awards technical assistance grants to local assessment committees for the evaluation of the proposed facility and the determination of the compensation a facility developer will have to pay to communities abutting the facility. Siting decisions are reached through compromise and any impasse is resolved through binding arbitration. EPA has not yet promulgated comprehensive federal siting standards, and until it takes that step it is likely that an increasing number of states will enact siting laws. The siting process is not the only cause for delay in facility permitting. Once a site has been finalized, Section 7004 of RCRA requires extensive public participation, including notice of draft permits in major local newspapers and radio stations and a mandatory 45 day comment period. In my opinion, informed and rational public comment on a draft permit is of great assistance to a state agency. Unfortunately, due to past problems and the publicity they received, public comment is often highly emotional and almost impossible to respond to in a manner which helps to allay the concerns that have been expressed. Responding to and interacting with the public can add months to the permitting process.

I have been discussing the delays that occur in the permitting process itself, but for a number of different types of facilities, even the initiation of the permitting process can be delayed. As I mentioned earlier, EPA still needs to promulgate additional technical facility standards. Facilities that are lacking standards include underground storage tanks, boilers that burn hazardous wastes, thermal treatment facilities, and facilities that conduct chemical, physical or biological treatment. Chemical, physical or biological treatment alone covers a significant number of existing facilities. EPA has stated that it intends to promulgate these standards within the next two years, but only time will tell.

In order to effectively manage the RCRA program, a nation wide data base is needed. EPA has established a data base using the information submitted through the RCRA Section 3010 notification requirement for hazardous waste generators, transporters, and facilities and information contained in the preliminary permit application forms ("Part A" of the application) submitted by facilities. Unfortunately this data base is comparatively useless due to the faulty information EPA has received. A number of the notification and application filings were protective; i.e.,persons who were unsure of their status under the RCRA program filed rather than face the penalties associated with non-filing. Some of the filings were totally erroneous, made by persons who were confused by the exceedingly complex RCRA regulations. Some of the filings are no longer accurate as the company involved has gone out of business or started to store its waste for less than 90 days (an activity which has been exempted from the full scope of the permitting requirements). Even though it acknowledges the deficiencies, EPA continues to use this data for statistical purposes, mailing lists, etc. States and the regional EPA offices are starting to do this now.

While facility permitting and inaccuracies in the date base may present difficulties in implementation, the most serious problem facing the states is EPA's cut back in funding of state programs and the proposed elimination of all funding within five to ten years. All environmental programs, including the hazardous waste program, have experienced cut backs in available Federal monies in the past year and further cut backs are proposed for fiscal year 1984. These decreases have a significant impact on state programs as they are taking the place at the same time that states are gearing up for facility permitting, a process which is very resource intensive. It seems paradoxical that at the same time EPA is requiring authorized state programs to be identical to the complex Federal program, it is initiating cut backs. Top EPA officials have claimed that states can replace Federal funding through the imposition of user fees, generator taxes, licensing fees, and other mechanisms. While these measures may help a state to supplement neccesary resources, I know of no state that is currently using any of these mechanisms which could totally support its program with the monies it obtains. EPA should provide funding of state hazardous waste programs in excess of the levels that were available in the fiscal year 1983 and continue that funding for the indefinite future. Without this reversal, there is a good possibility that some states will return or decline to run the RCRA program. Ironically, the recent controversy at EPA may end up helping the states in this area due to increasing Congressional concern over management of the program.

SUPERFUND

From the enactment of Superfund in 1980 to the current controversy over its management by EPA and the firing of top EPA officials, the implementation of the Act has been constantly beset by problems. Some of these involve the actual function of the Act itself at the Federal level, such as the failure to finalize the National Contingency Plan required under Section 105 until July 16, 1982. Others have a more direct impact on the states, such as the development and content of the list of all sites which should have top priority for clean up.

In October 1981, EPA published an interim priority list of 115 hazardous waste sites which was subsequently expanded to 160 sites in July 1982 and to 418 sites in December 1982. One of the problems the states encountered during this process was EPA's failure to provide the states with notification of what sites were to be included on the list until the list was ready for publication in the Federal Register. This is despite the fact that the selection of the sites themselves was based on data gathered primarily by the states. This degree of secrecy on the part of EPA is both irresponsible and violates the spirit of the State/Federal partnership mandated by CERCLA itself. To remedy this situation, future additions to and the subsequent prioritization of sites on the list must be jointly agreed on by the individual states and EPA. In this regard, EPA individual states and EPA. In this regard, EPA should place strong reliance on the state's recommendations as they are usually more familiar with the past history and current status of the sites.

This State/Federal co-operation should also include efforts to correctly categorize the listed sites so that the general public is made more aware of the true status of each site. This categorization should include the enforcement status (is the site the subject of the state, federal or joint litigation) and the remedial status (is the site currently the subject of local, state, federal or private clean up actions and is clean up by parties who disposed of waste at the site in progress or expected in the near future). Although EPA has made an initial effort to provide this data, much of the information was inaccurate and the implications of the categorization were not elaborated upon.

The priority list is not the only listing of hazardous waste sites that EPA has developed. Under Section 103(c) of CERCLA, persons who owned or disposal sites were required to notify EPA by June 1981. This notification program resulted in the listing of a total of 14,000 sites nationwide. While the notifications contained the basic information needed, it was readily apparent that more the situation. Therefore, in 1982, Congress appropriated 10 million dollars to be used by the states for the assessment, inspection and the eventual production of an inventory of sites under Section 3012 of RCRA. The inspection and verification of these sites is of vital importance as there is a question of the validity of the data received under the notification program for hazardous waste generators, transporters, and facilities, there is a good probability that the information received contains protective or erroneous filings. Again, the result of all this is a faulty data base in the EPA computers. Once the states have initiated their investigations, EPA should establish come procedure which would allow states to remove sites from the list when they prove to be no threat to human health or the environment or when they are shown to be outside the scope of the Superfund program.

With the promulgation of the national priority list, the next stop in the process is the investigation of the sites on the list and the development of remedial plans. As is true in all phases of a Superfund clean up, the states must provide matching funds for these activities. In this case, the states must provide 10% of the necessary amount. This requirement has created serious problems. In general, it is difficult to come up with the amount of money needed particularly in those states with a large number of sites and the smaller states. Specifically, as it is often near to impossible for a state to get required money all at once from a state legislature, a vicious cycle is created. It is dufficult to justify the aoppropriation of large sums of money unless a preliminary investigation has taken place. Then it is difficult to obtain money for the final clean up of the site until the results of the remedial studies are in and un approximate cost for closure has been determined. Therefore, to implement the program the state agency responsible for Superfund must first get the money for the preliminary investigation, go back again for the amount needed for the detailed investigation and remedial planning, and go back again for the amount needed for final clean up. These continual requests for money make the appropriation process far more difficult and time consuming than it should be and delay the actual clean up. To avoid this problem, the states should get an initial outlay for investigation and planning without 10% matching funds requirement.

If there is difficulty in obtaining money for the initial work and final clean up, there is even further difficulty in obtaining the money necessary for the post-remedial operation and maintenance of the site by the state which is mandated by Section 103(c) (3) of CERCLA. THe states have a real problem with this provision as there is no clear definition of the time a state must perform post-remedial operation and maintenance. At the present, state obligations could be construed to last indefinitely. Some limiting factors and termination criteria must be established. These could potentially be drawn from the criteria already established under RCRA for the termination of the post-closure maintenance period for hazardous waste land disposal facilities.

Even if limits are set, the states are looking at potentially enormous amounts of money to conduct all the necessary activities. Once again, there are very real problems in obtaining the money from state legislatures. One solution to this dilemma is to amend CERCLA to allow the states to use a portion of the Superfund money for investment purposes on a site specific basis. These could include certificates of deposit and other safe high yielding investments. Ideally the states could then use the interest generated to perform the necessary actions in addition to the invested capital. This scheme is particularly appropriate if EPA decides on a 30 year post-remedial operation and maintenance period; i.e., the time period currently required under RCRA for closed facilities.

I have described a number of specific problems that the states have been experiencing, but there is also an overall problem with the implementation of Superfund caused by EPA's current practice of centralizing the administration of the program in its Washington headquarters. The initial decisions on sites are made by the EPA regional offices working in conjuction with the individual states. Often, however, draft contracts, documents, reports, bid packages, etc., and agreements or decisions mutually approved by a state and the regional office are rejected or unacceptably altered by EPA Headquarters. This causes an inordinate delay in program implementation and an increase in costs. Many of these delays are the result of EPA Headquarters' staff's unfamiliarity and lack of involvement with individual sites. These difficulties are often trivial and could be avoided by some degree of decentralization in administration which would give the regional offices greater authority and discretion in making the necessary decisions.

In hindsight, it is relatively easy to recognize all of these problems. If I had been in charge of the program from its inception, however, it is likely that I would have taken the same route as EPA in a number of these areas. So, for example, while I agree with the concerns of the various Congressional committees regarding the manner in which EPA has conducted the negotiation of clean up agreements with responsible parties, it is my opinion that the prime focus should be on negotiated settlements rather than formal court actions. It is my hope that once the current controversy surrounding the Superfund program is resolved, the implementation of Superfund will proceed in a efficient, orderly manner so that the prime goal of the Act, the clean up of the sites, can be achieved as soon as possible.

I have attempted to outline some of the concerns that are foremost in the mind of state officials. None of the problems are insurmountable and though some may require Congressional action, the basic federal legislation is ounc. Most of the problems remain in the administration of these environmental programs by the federal government. I am hopeful that the new Administration at EPA will take an approach in changing both the actual program and the public perception of that program which results in positive environmental management with maximum delegation of the program to the states. The recent flurry of Congressional activity has certainly started this process in the Superfund. It is my hope that it will continue in the less publicized area of authorization of state programs. EPA's position in this area will have to change in order to fulfill the intent of Congress and the Administration; that is, delegation of the hazardous waste management program to the states. It is my estimate that under current policy only 18-20 states will receive final authorization by January 26, 1985. I do not know how states other than Vermont will specifically react, but it is my feeling that many of the interim authorized states whose final authorization applications are not fully processed by January 26, 1985, thus resulting in program reversion, will not want to go through the process again and will therefore drop any intnetions of reapplying for final authorization.

The Association, as well as myself personally, looks forward to working with EPA in this coming year, a year which I feel is a critical year of decision in the management of our national waste problem.

CHAPTER 2

CONFLICTS AND HAZARDOUS WASTE MANAGEMENT - THE ENVIRONMENTALIST'S VIEWPOINT

W.B. Clapham, Jr.
Cleveland State University

INTRODUCTION

Hazardous wastes have become the most prominent environmental problem of the 1980's. It is often difficult to get through a week without hearing something about another illegal dump, fire, or scandal involving them. But despite their notoriety, it is difficult to put one's finger on the "hazardous waste problem," since it is so multidimensional. People are aware that men and women around the country are trying to do something about "the problem" but it is difficult for them to understand what they are talking about. Meetings can concentrate on the engineering aspects, decision-makers worry about economic or policy questions, regulations set minimum technical standards, and the newspapers compare everything to Love Canal. Many different groups are directly or indirectly involved with hazardous waste management, but few spokespersons for any organization have effectively connected its disparate aspects so that the public at large can understand what is involved and what risks and benefits society faces from the current system of hazardous waste management.

FACILITY SITING AND "THE HAZARDOUS WASTE PROBLEM"

All of the dimensions of hazardous waste management come together with facility siting. It may be possible to regard waste generation or facility management as a technical matter, and one may be content to let regulators and newspapers do their thing, perhaps with a comment for the record or a letter to the editor. But facility siting brings government, the private sector, and the public together in an emotion-charged arena where painful decisions are made. All parties enter the fray loaded for bear, and the battle generally ends when one protagonist runs for cover.

We should not need to be reminded that avoiding a decision that needs to be made is in itself a decision. Not siting a facility may mean that a plant's production costs are much greater than they would otherwise have been or that a company expands production of a

Management of Toxic and Hazardous Wastes, H. Bhatt, R. Sykes, T. Sweeney (Editors)

particular product in some other area. It may mean that hazardous wastes will be handled illegally. These too represent costs for society that can be most painful, as we have seen recently at Times Beach, Missouri.

An industrial society needs to deal with its wastes, just as the human body. Hazardous waste management facilities are the kidneys of an industrial society, and a responsible system for dealing with hazardous wastes is simply essential for the survival of our way of life. We do not yet have an adequate system for dealing with hazardous wates, but it is inconceivable that our economy can continue long without one. It is difficult to tell what kind of system we will end up with, however. Our lurchings from one idea to the other over the last few years make prediction all but impossible.

Conflicts and Vested Interests

Hazardous waste management is an area in which the basic interests of industrial generators and environmentalists overlap almost precisely. Their reasons are very different: industry needs functioning "kidneys" that will let it produce its product at the lowest possible cost, so that it can gain market share and increase profit. Environmentalists need effectively functioning hazardous waste management facilities to minimize the release of hazardous materials into the environment where they can affect public health and ecosystem stability. Regardless of their differences, the similarity of basic interest makes them allies (albeit of convenience) in the matter of hazardous waste management, not adversaries.

Despite this, the most common attitude among the public is the very negative syndrome commonly known as NIMBY (not in my back yard). It has been a remarkably effective tool for organizing successful grass roots resistance to hazardous waste management facilities. It can be very dangerous: Thus far, its success has not crippled hazardous waste management, but many people are assuming that it is both the natural and the appropriate reaction to a proposal for a hazardous waste management facility located in a community. It is a natural reaction; it may or may not be the most appropriate. The distinction is critical.

NIMBY comes from the core of the American character. Our birthright has been one in which we as citizens of this country have the right to make decisions and to have a measure of control over our personal destiny. If we perceive that a corporation intends to trample on our rights, we organize to defend ourselves. The grass roots organizations that have sprouted up around the country to fight secure landfills are the real world analogue to the Hollywood fantasy in which John Wayne pulls together an intrepid band of settlers to fight a raving horde of Indians intent on murdering their wives and children. The right to affect our personal destiny is part of our national culture. It is something that we have been taught from childhood to cherish. That a landfill operator should be able to lower the property values of our homes simply because he can get the right permit smacks of taxation without representation. Goverment and industry must both understand how individual homeowners threatened by the possibility of a secure landfill nearby would view themselves as latter-day minutemen!

Dialogue, Mediation, and Trust in Facility Siting

Ours is also a country where people take social responsibility seriously. A community will recognize when an entrepreneur is making a bona fide effort to treat it fairly. The system will work where the community trusts the operator and the key regulatory agencies. If this statement seems too sweeping, perhaps it is better to say that the system will not work where the community distrusts the operator and the regulatory agency. Trust makes demands on all parties. It means asking the president of LTV Steel to trust an activist whom he sees as a long-hair no-growth environmental extremist wearing gold-rimmed glasses (and vice versa). It means asking the mayor of a community proposed as the site of a secure landfill to trust a company that he (and the voters that keep him in office) suspects is out to rape his community [1].

Trust is impossible without creative dialogue. And dialogue is very different from "public participation" as defined in 40 CFR 25. Nobody ever choreographed creativity in the Federal Register: What began as a process designed to involve the public in decision-making quickly degenerated into an organized mechanism to notify people about what had already been done and to try to sell the public on decisions that had already been made. The only way to deal with an issue that involves gut feelings as does hazardous waste management is to develop a genuine understanding of mutual needs and overlapping interests. This means real communication in which all parties talk with each other, not at each other.

This is not a pie-in-the-sky fantasy. We are accustomed to extremely hazardous materials in our neighborhoods and even in our homes. We regularly allow such materials to be brought into residential areas by common carrier and discharged into storage tanks by operators that may have no more than an 8th grade education, and using a manifest system that is based 100% on market forces. These storage depots are called gas stations. People were angry and hurt when the steel mills closed down in Youngstown and other cities. Nobody breathed a collective sigh of relief that a number of dirty industrial facilities had stopped producing great masses of hazardous wastes. Many other basic industries are inherently dirty. Oil refineries and many chemical plants are at least as inherently polluting as a well managed hazardous waste treatment facility. The difference is that we understand the steel, oil, and chemicals industries. We do not understand hazardous waste management in the same way, we question its legitimacy.

Even so, responsible environmental groups have begun a serious dialogue with industry, and grass-roots organizations are making clear what they want and need. Representatives of the environmental community have met regularly with spokespersons for industrial generators in Ohio for several years. Last year, this dialogue resulted in a major revision of the legislation that governs hazardous waste management in Ohio that should provide a stable basis for hazardous waste management for several years [2].

Environmentalists in many other states have worked actively with industry and state regulatory agencies to design siting procedures and to serve on siting boards to encourage responsible, well-operated hazardous waste management facilities. Organized grass-roots environmental organizations have spoken in favor of specific projects, as well as against them. One of the most effective of these, the Committee to Oppose Destruction of the Environment (CODE, in Norton, Ohio) has endorsed the concept behind the hazardous waste incinerator proposed by Waste Technologies Industries (WTI) in East Liverpool, Ohio, and for offering the citizens of Ohio an alternative to landfilling of chemical wastes [3].

The Actors in a Dialogue and their Perspectives

In talking about dialogue, it is important to recognize that there are always at least three different "actors," and there may be more than that. These are industry (which may include both the generator and the hazardous waste management facility operator), the community (which may include grass-roots organizations as well as the government), and the State. All have different needs which may complement each other. Industry wants to insure that it can site a facility with a secure technology that will allow it to operate, with an economic base that will assure a profit, with a social climate that minimizes risk of shutdown, and with a control system that minimizes risks of having to pay to clean up after other people's mistakes. The community needs to increase the number and security of jobs, to preserve a satisfactory quality of life, to maintain property values, and to try to keep undesirable facilities out of the community. The State must keep things on track, preferably minimizing hassles.

Each actor sees hazardous waste management differently. Developers see a market for facilities. This is especially true if they are also generators and have wastes to dispose of. They are concerned (especially for treatment plants) that the ground rules for facility operation be fairly constant and not change quixotically. The State is responsible for assisting its industries to deal with the wastes produced within the state as well as safeguarding the health and well-being of its citizens. As a result it is in the schizophrenic position of having to be neutral on every specific site and recognizing that new facilities are essential for the health of the State's industrial base. It is to be expected that the perspectives of the State and the developer should often appear to outsiders as very close to each other.

The community's perspective is more complex, since public opinion is diffuse and does not have clear decision-making nodes. Some people will always respond with the NIMBY attitude. Others will view any new industrial facility as beneficial. Still others may be undecided, willing to listen to a proposal for a hazardous waste management facility, and open minded. We cannot generalize on community attitudes, but it is probably true that the interests of a community are served when it perceives itself as host to a facility rather than as its victims. A community that feels victimized by a project will act accordingly. It is doubtful that any facility can be sited in a community that violently opposes the siting decision. Even if it were, the community would probably find a legal way to shut it down after it had commenced operation; this would be even worse for the developer than losing the siting battle.

Siting Facilities in "My Back Yard"

It is possible for a community to feel host to a hazardous waste management facility. Two communities in Massachusetts have expressed interest [4], as have two in Alberta [5]. There is no magic formula to insure a successful siting process. But for a developer with a plan for a hazardous waste management facility, there are five crucial steps to establish a dialogue with a community and to convince it that it can and will be a host:

1. Go public with the plan: be up-front about what you want to do.
2. Accept the community as a peer: Take its people seriously even if they are different from your usual golfing partners. Remember that a dialogue is impossible if you talk down to them or suggest that they are inferior to you in any way. Recognize that local communities are more diffuse than State or limitations.
3. Make a commitment to negotiate with the community in good faith. You will have to listen to them, understand their needs, and help them understand yours.
4. Work toward a consensus among all parties. This means (among other things) that you will have to involve all of the parties that are in a position to influence a consensus.
5. The consensus position should be the basis of your permit application, which will include mechanisms to insure the continuity of the consensus you have built.

The first step is the most delicate and contains virtually all of the downside risk. If done poorly, it may prejudice a community against even the most appropriate project. But all of the other steps depend on it. It determines first impressions. If a company is too slick, or if it appears to be trying to hide something, or (perhaps worst of all) if a community hears about a project from somebody other than the developer, then the community is likely to assume that the company has something to hide, and it will behave accordingly. Ignoring or talking down to the community in the beginning is likely to be seen as evidence of bad faith. Going public begins a process. A necessary part of that process is for the community to organize itself. Initially, the organizational impetus will be fear and the relationship adversarial. But how long these attitudes last depends on the developer's behavior.

Perhaps the most difficult step is for a developer to accept the community as a peer. How does it do this? A corporation is a centralized agency that is accustomed to going to one place and talking with people who are empowered to make decisions and to deal with it. Communities don't work this way. It isn't enough to win over the mayor. He or she is likely to be most concerned with gauging local public opinion (and the next election); this complicates the decision-making process.

Communities have scores or even hundreds of groups that get involved in proposal for a major facilities. These groups are much more likely to mold public opinion directly than are elected public officials. Public opinion is amorphous enough that it may be efficient for a developer to attend as many meetings of these groups as possible to demonstrate that he understands and respects the way that public opinion is formed in the community. It is difficult for a senior executive in a distant city to relate adequately with local public opinion. This takes somebody in the community -- and preferably from the community -- who can speak for the developer and has the trust and confidence of its highest management.

Accepting a community as a peer means more than being willing to talk with it. Local government has a more vulnerable tax base than any other level of government, and most citizens' groups rely on unpaid volunteers. As a result, communities are limited in ways that corporations and the State are not. They seldom have the money, the organized technical expertise, and the other resources needed to carry on an effective dialogue. Nevertheless, communities need to do their own research and build their own awareness in order to work with a developer.

Reaching A Negotiated Solution

It is strongly in the developer's interest to insure that a community obtains funding to do the necessary research. First reactions to a proposal are likely to be fear and distrust. When a community does no have the opportunity to build a basis for anything other than fear and distrust, those attitudes will remain, and they will determine the community's overall attitude toward the project. Only people who understand the problems of a hazardous waste management facility have the credibility to sign a negotiated consensus. Industry and government would both do well to remember Shakespeare's admonition: "Yond Cassius has a mean and hungry look. He thinks too much; such men are dangerous" [6]. A community that is lean and hungry for information from sources that it trusts will organize well and fight hard.

The goal of negotiation is a consensus between the developer and the community that is acceptable to the State regulatory agencies. Both need to understand the other's fears and needs. Many of the needs are likely to be quite simple, and meeting them may cost little. Communities understand developers' needs, including the need too make a profit. After all, a hazardous waste management company that does not make a profit cannot provide the community with compensation or stable jobs.

Both sides need to understand the other in order to begin a dialogue. This means finding out the other's track record. A large company has plants all over the country, and you can find out their environmental record. One can contact groups like the Sierra Club or the League of Women Voters (or any other active environmental organizations) and find out what's going on. One can check their annual reports and find out where the company is having problems or doing new things. You can contact US EPA to find out what compliance letters have been issued on a particular plant or if any of the company's properties has been suggested as a superfund site [7].

Citizens have track records too. A local organization with a reputation for doing its homework will be taken more seriously by industry and by State regulators than one which does not. One with a reputation as being incapable of agreeing with anything is likely to be ignored by all parties, including other environmental groups that might be viewed as allies in a negotiation.

There are probably many ways to reach a consensus. One especially has had some success. Environmental mediation is a process by which an outsider comes in and facilitates communication among the parties. It begins with a "gelling" process in which the most prominent parties (e.g. the developer, the municipality, and the State) agree to discuss the project, to identify the other key parties, and to make a commitment to participate in a defined mediation process. Funding is arranged in a way that will insure the commitment of each party to mediation (e.g. shared between the key parties), but managed in a way that insures the neutrality of the process (e.g. by oversight through a foundation, University, etc.).

Problems can be addressed with three different kinds of results: The first is mitigation, or reduction of the potential or perceived negative impact of the development. The second is compensation, or community improvements designed to offset costs assumed by the community. These costs may be directly related to the hazardous waste management facility (e.g. road repairs, requirements for water or wastewater treatment, demands on services and infrastructure, etc.) or indirectly related to it (e.g. property values, community image, etc.). Mitigation deals with tangible engineering aspects of the project; compensation addresses more intangible impacts. Finally, mediation may result in institution-building, in which new institutions or mechanisms are built to assure the community of the facility's safe operation (e.g. involvement of local citizens on the facility's board of directors, provision of an Ombudsman, etc.).

From the viewpoints of hazardous waste generators, community groups, recognized environmental organizations, and the like, recognition of self-interest and working toward it does not always mean saying "no." Once reached, a consensus should be written into the permit application and reflected in the actual permit. Furthermore, the State should expedite processing of consensus-based applications as much as possible.

A word must be added about enforcement. A no-win argument on this matter has been running for some time between industry, regulators, and the public. Nobody denies that the regulations governing hazardous waste management are complex, cumbersome, expensive, and not always as good as they might be. Some people would have the US and State EPA's downplay enforcement, perhaps as one way to "get government off the backs of the American People." Neither industry nor government dare forget that hazardous waste scares people. Only a track record of doing it right will legitimate hazardous waste management in most people's eyes.

In the meantime, the goal of the hazardous waste management program must be to reassure people. The most effective way to stop facility siting in the United States would be to convince the public that the US EPA was unwilling to protect them from hazardous wastes. I do not believe that the Reagan administration is trying to cripple U.S. industry. But one of the first results of the Reagan EPA under former Administrator Anne Gorsuch was that an agency loved by few but respected by most became scorned by everybocy. Despite an improvement in its image under Administrator Ruckleshaus, it has not recovered its former credibility. In the State of Ohio, the siting process followed by the State Hazardous Waste Facilities Board appears to have much more public regard and credibility than that of the Federal EPA [8]. Given this breakdown in public confidence in the US EPA, commuinities proposed as sites for hazardous waste facilities are much more likely than they might hve been to insist on safeguards within a negotiated consensus agreement. Fortunately, these kinds of safeguards can be negotiated, and there are many precedents around the country.

It is, perhaps, unfortunate that we come down to funding and enforcement as the last stumbling blocks to negotiated settlement of significant conflicts like hazardous waste management facility siting.

The point I want to leave you with is, that conflict resolution is not only desirable, it is possible. We have not done it well so far, but we know how to do it and we have every reason to believe that it will be done.

REFERENCES

1. This term was pointedly used in this context by the Mayor of Norton, Ohio, in a public meeting called by PPG industries to discuss its proposal to locate a hazardous waste disposal site in its abandoned limestone mine underneath the city.
2. Amended Substitute House Bill 506, as passed by the 115th Ohio General Assembly and signed by Gov. Richard F. Celeste in August, 1984.
3. Richmond, B.R., Testimony on behalf of Citizens Opposed to the Destruction of Our Environment before the Ohio Hazardous Waste Facility Approval Board, regarding the Incinerator proposed by Waste Technologies Industries for East Liverpool, Ohio (January 25, 1983).
4. Sanderson, D., Speech on the Experiences of the Commonwealth of Massachusetts on siting Hazardous Waste Management Facilities; given at the Trialog Conference in Cincinnati (October, 1982).
5. McGlennon, J.A.S. "The Alberta Experience ... Hazardous Wastes? -- Maybe in My Backyard," The Environmental Forum February/1983: 23-25 (1983).
6. Shakespeare, W., "The Tragedy of Julius Caesar" (1600).
7. For more details, see Clapham, W.B., Jr., Searching for Solutions: A Citizen's Guide to Hazardous Waste Management in Ohio. Columbus: Ohio Environmental Council (1983).
8. Compare the transcripts for public hearings on hazardous waste permit applications held under Federal and State aegis.

CHAPTER 3

PUBLIC PARTICIPATION IN OHIO EPA'S SOLID AND HAZARDOUS WASTE PROGRAM

Michael Lewis Greenberg
Ohio Environmental Protection Agency

INTRODUCTION

Proper management of solid and hazardous waste materials, facilities and disposal sites is a major concern for industry, government and the private citizen. The nature and potential impact of solid and hazardous waste on the environment and the numerous complex issues surrounding each Agency decision necessitates the cooperation of elected officials, other government agencies, industry, special interest groups and individual citizens. In order to provide a forum and opportunity for public involvement, Congress and U.S. EPA wrote laws and developed policies for public participation.

The Director of the Ohio Environmental Protection Agency and the Chief of the Division of Solid and Hazardous Waste Management have been supportive of the public participation program which not only meets the legal requirements, but also incorporates additional opportunities for interchange between the Agency and the public. The following sections cover Ohio EPA's program objectives, the legal requirements and opportunities for involvement in the Division's public participation program.

PROGRAM OBJECTIVES

It is the intention of the Division of Solid and Hazardous Waste Management to involve a broad based spectrum of public and private sector citizens in the public participation program. This involvement centers on the Division's five major program objectives for the public participation program:

1. To keep interested persons informed about programs, proposed actions and significant issues.
2. To receive information from citizens regarding programs, proposed actions, and issues of significance to them.
3. To improve the Division of Solid and Hazardous Waste Management's decisions, programs and policies through this

Management of Toxic and Hazardous Wastes, H. Bhatt, R. Sykes, T. Sweeney (Editors)

exchange of information.

4. To enhance acceptance of the Division of Solid and Hazardous Waste Management's programs and policies by providing adequate opportunities for public participation.
5. To develop mutual trust between the Division of Solid and Hazardous Waste Management and the citizens.

It must be recognized that the public participation program is not a transfer of the responsibility for final decision making but an important tool in the decision making process. The actual responsibility for decision making remains firmly with the Agency and ultimately with the elected state officials.

LEGAL REQUIREMENTS

Ohio EPA is required by federal law and policy to conduct a public participation program. The Resource Conservation and Recovery Act of 1976 includes public participation in the development, revision, implementation and enforcement of any regulation, guideline, information or program under RCRA. Since the enactment of RCRA in 1976, additional public participation regulations were enacted on February 16, 1979, and a new policy for public participation programs was developed on January 19, 1981. "The intention of the Policy is to ensure that managers plan in advance needed public involvement in their programs, that they consult with the public on issues where public comment can be truly helpful, that they use methods of consultation that will be effective both for program purposes and for the members of the public who take part, and finally that they are able to apply what they have learned from the public in their final program decisions".

PUBLIC PARTICIPATION OPPORTUNITIES

Proposed New Facilities

Opportunities for public involvement in the permitting process for new hazardous waste management facilities are provided by Ohio House Bill 506, enacted in August, 1984. This bill amended Ohio Senate Bill 269 and made several changes that affect public involvement for proposed new hazardous waste facilities. The bill continues to provide for the Hazardous Waste Facility Board (Approval was dropped from the title) and delineates the times for a public hearing and an adjudication hearing during the permitting process. Additional changes were made to provide public officials and citizens with more information earlier in the permitting process. For example, an applicant for a new facility must "notify by certified mail the legislative authority of each municipal corporation, township and county in which the facility is proposed to be located, of the submission of the application within ten days after the submission. If the application is for a proposed new or modified hazardous waste disposal or thermal treatment facility, the applicant shall also give actual notice of the general design and purpose of the facility to the legislative authority of each municipal corporation, township, and county in which the facility is proposed to be or is located, at least ninety days before the permit application is submitted to the Ohio Environmental Protection Agency." Also, "a representative of the applicant who has knowledge of the location, construction, operation, closure, and post-closure care, if applicable, of the facility shall attend the public hearing in order to respond to comments or questions concerning the facility directed to him by the presiding officer". Finally, the applicant must provide with the copies of the application a short statement of the anticipated environmental impact of the facility, a map of the facility, and where copies of these items will be available for inspection. These additional requirements should provide much needed information early in the permitting process.

The Unregulated/Superfund Sites Program

Unregulated sites is the Division's term for those sites where, previous to the hazardous waste regulations, industrial wastes were treated, stored or disposed. Superfund sites are those unregulated sites which have been submitted by Ohio EPA to U.S. EPA and placed by U.S. EPA on the Naitonal Priorities List. This makes them eligible for monies under the Comprehensive Environmental Response, Compensation and Liability Act (CERCLA) for clean-up.

Naturally, these unregulated sites are of concern to the area citizens. Guidelines developed under CERCLA provide for citizen involvement through a Community Relations Plan specific to each site. This is important because it guarantees that affected citizens will be kept informed and given an opportunity to express their concerns however, it does not meet the needs for early communication between the citizens, the local officials and the Ohio EPA.

Early communication must be established in several ways. The Agency needs to take the initiative by letting citizens, county or city public health officials, the city or county engineers, and other local officials know what information is available regarding the site. This can be done through public meetings, press releases, and working with individuals. Often citizens will organize into a group which can make the Division's job much easier since we can then work with the designated leaders rather than a large, everchanging number of individuals.

The process of cleaning up a waste site is a longer and more complicated process than citizens or local officials realize. It is to the Division's benefit to establish communications and a good working relationship with these local citizes and officials from the outset, assuring them that the Agency is prudently working towards a solution for clean-up.

THE DIVISION OF SOLID AND HAZARDOUS WASTE MANAGEMENT'S ADVISORY GROUP

It would be a monumental task to attempt to work with all the individuals in Ohio who are interested in proper hazardous waste management. Part of the solution, we feel, is the Division's Advisory Group. The Advisory Group consists of approximately 60 representatives from:

- professional organizations such as the Environment Committees of the Ohio State Bar Association, the Central Ohio Chapter of the American Society of Civil Engineers, the Ohio Chemical Council, and the Ohio Municipal League

- business oriented associations, such as the Ohio Chamber of Commerce, Ohio Manufacturers Association, Ohio Chemical Council, Ohio Petroleum Council, and the Ohio Electric Utilities Institute

- governmental agencies such as the Ohio Departments of Health, Natural Resources, Transportation, the Public Utilities Commission of Ohio and the State Fire Marshall

- educational and research groups such as the Ohio Alliance for Environmental Education, the Ohio League of Women Voters, Environmental Resource Center and the Ohio Environmental Council

The representatives from these organizations generally serve on the Advisory Group for a long enough period of time to allow them to understand the functions of the Agency and the programs of the Division. This means the input they can provide from their groups to the Division can be practical and valuable. In addition, many members of the adivsory group have greater expertise in certain areas than do the staff of the Division of Solid and Hazardous Waste Management and it's to the Division's benefit to draw on their knowledge and experience.

The Division works with the Advisory Group through quarterly meetings and through regular mailings which include program updates, recommendations for changes in legislation and task force reports. This dissemination of information encourages advisory group members to present their thoughts on these issues which aids the Division in its program decisions.

Since its inception, the Division's Advisory Group generally has been responsibly for hazardous waste issues. In April, 1984, the Division was reorganized to include the solid waste program. A solid waste sub-advisory group has been formed, and is currently addressing recommendations to strengthen Ohio's solid waste law ORC 3734. A Governor's Task Force was formed recently to address Ohio's solid waste and litter control problems, evaluate recycling practices and resource recovery facilities in Germany and to develop a report to strengthen Ohio's solid waste management and litter control programs. The Task Force will report its findings to the Governor and to the entire Advisory Group with recommendations for statutory change.

The Advisory Group, Sub-Advisory Group and Task Force are valuable to the Agency, Division and the overall public participation program. These groups provide insight and an understanding of the values of the community at large, which might not always be obtained through public hearings and meetings.

SUMMARY

There are many opportunities for public participation in the programs of the Division of Solid and Hazardous Waste Management in Ohio EPA. Some of these are formal, required by federal or states laws and regulations and some, are informal, developed by the Agency and the Division in an effort to work with the many sectors of our statewide community, to make our programs more responsive to varied concerns and to develop mutual trust between the Division of Solid and Hazardous Waste Management and the people of Ohio.

CHAPTER 4

HEALTH AND SAFETY CONSIDERATIONS FOR HAZARDOUS WASTE WORKERS

Lynn P. Wallace, Ph.D., P.E.
Brigham Young University
William F. Martin, P.E.
National Institute for Occupational Safety and Health

INTRODUCTION

In recent years, publicity concerning hazardous substance transportation and hazardous waste disposal activities [1] have increased public awareness, and helped to focus attention on the health and environmental problems associated with the handling of hazardous materials. The public is concerned with how these activities will affect them immediately--should they have to evacuate their homes--or over extended time periods through contaminated water, air, or food. Environmental Engineers are concerned and directly involved with finding solutions to the problems of hazardous waste transportation and disposal. Workers who must handle the hazardous materials have perhaps the greatest concern because of the acute and chronic health problems that could develop from exposure to toxic materials.

Recent disclosure by the Environmental Protection Agency (EPA) of the 512 most hazardous wastes, requiring remedial action [2], and estimates of upwards of 30,000 sites that contain hazardous wastes (1,200 - 2,000 of which present a serious risk to public health [3]) serve to highlight the magnitude of the problem. Not only will workers be involved in the clean-up, but workers are and have been involved in determining the extent and severity of the problem. The estimated number of workers who will be directly involved in these remedial activities varies greatly, but may exceed 40,000. The other workers who may have a high potential for exposure during emergency clean-up operations of hazardous substances spills could include significantly high numbers of emergency response or rescue personnel such as police, fire fighters, and life squads. Regardless of the actual number or type of worker, each individual has a legislated right to be properly protected at the workplace. [4]

Management of Toxic and Hazardous Wastes, H. Bhatt, R. Sykes, T. Sweeney (Editors)

NIOSH TOXIC CHEMICAL/WORKER PROTECTION STUDIES

As a vital part of its studies of occupational environments and worker health, the National Institute for Occupational Safety and Health (NIOSH) has long history of involvement in the study of toxic agents and their effects on workers. NIOSH has established a good record of accomplishments in worker protection and continues to be actively involved in health and safety considerations for those throughout the workforce. Recommendations for standards for good workplace practices have been submitted to government, industry, and labor. These standards address a number of physical and chemical agents in work environments and are presented through technical reports, health and safety guides for employers and employees, and other publications. It is not suprising that the majority of chemicals studied by NIOSH in the past have also been identified by EPA as those of greatest concern at hazardous waste sites or during emergency actions. This experience with toxic materials and work environments was a very important reason why NIOSH was given specific responsibilities under the CERCLA (Superfund) legislation [5] for hazardous waste worker safety and health. These Superfund responsibilities are consistent with the many on-going NIOSH research activities that probe the interactions between worker and work environments.

Toxic substances that enter a worker's body can affect the respiratory, circulatory, and nervous system, including the liver, kidneys, bladder, intestines, etc., depending upon the particular agent to which the worker may have been exposed. These internal effects are not as easily recognizable as a cough, skin burn, rash which can occur at the original site of exposure (lung, skin, or eye). The damage to the body may be acute (immediate) or it may be chronic (long term), and in either case the damage may be permanent. The challenge is to keep the toxic substances from entering the body and the preferred method of worker protection is through engineering controls.

The large volume of air that is breathed and the close interface between this air and the blood in the lungs makes inhalation of toxic materials the major route of entry into the body. Consequently, the primary health and safety consideration for workers involved with hazardous materials is respiratory protection. NIOSH has been very active in determining the detrimental effect of materials in the lungs (silica, coal dust, asbestos, carbon monoxide, etc.) and in testing, evaluating and certifying respiratory protection devices. NIOSH is currently working on an updated respiratory protection guidance document. It is in the peer review process now and will be published as soon as possible.

NIOSH is vitally concerned with the adequacy of currently used respirators under conditions existing at uncontrolled hazardous waste sites and during hazardous materials emergency responses. Improper selection or use of respirators can create situations where the worker assumes protection when in fact adequate protection may not be provided. Respirators should only be used in conjunction with a complete respiratory protection program, which includes proper respirator selection, regular training in that respirator, correct maintenance, frequent inspection and prompt evaluation. NIOSH has an on-going research effort and will continue to be a leader in providing guidance in proper respirator protection for workers who are exposed to breathable hazardous materials.

Another route for hazardous materials to enter the body is percutaneous absorption through the skin or eyes. Splashed and spills of caustic, corrosive, or toxic materials to the skin or eyes can cause effects ranging from minor irritation to blindness or death depending on the material and the extent of the exposure. The problems of toxic entry through the skin or eyes are not limited to splashes and spills. Vapors, smoke, fumes, and mists can also carry toxic materials that cause damage when they contact the skin or eye. The problems can be further exacerbated by acid formation on the skin or eye surface when moisture from sweat, tears, or high humidity is present.

Protective clothing or equipment has been designed to prevent hazardous materials from entering the body through the skin or eyes. Gloves, aprons, boots, face shields, masks, hoods, coveralls, and completely encapsulating suits are few of the specific items of protective clothing and equipment currently available for worker protection. The proper selection of personal protective equipment for workers at hazardous waste sites is critical and must be based on the identity and concentration of materials expected to be encountered.

NIOSH has been concerned with personal protective equipment development and testing for several years. Up-to-date selection criteria for personal protective equipment are currently being developed by the division of Safety Research. Proposals to do permeability, penetration and durability testing of protective clothing and equipment have been submitted for funding. When using protective equipment it is absolutely vital that the actual level of protection be known rather than assumed. Serious damage or loss of life can occur if the worker assumes one level of protection when that protection is not being provided by the equipment used. There is such an array of different chemicals and compounds that can enter the skin or eyes and cause problems or that react differently with the materials used to make the protective equipment, that guidelines must be provided as to what materials should be used against what chemicals and under what donditions. NIOSH will remain active in providing solutions to these problems.

The third route of entry of hazardous materials into the body is through ingestion. Workers eating snacks or lunches that have been contaminated by unwashed hands, clothing, dust, or mist can ingest hazardous materials directly to the digestive tract. Drinks or cigarettes taken on a break can be similarly contaminated.

Generally, the amount of hazardous materials entering the body by these methods is small, but if it is regular and continues sufficiently long, the cumulative effect could be significant. The type of material ingested is also very critical. Small quantities of some hazardous materials can be lethal while large quantities of other material may have no detectable effect. Good personal sanitation practices with adequately protected employee facilities can usually ameliorate the problem.

PERSONAL PROTECTIVE EQUIPMENT PROBLEMS

The very nature of personal protective clothing and equipment causes other problems for workers who must use them. Materials are selected which will provide a barrier of protection for the worker by keeping liquids, vapors, fumes, and smoke from reaching the body. Ironically, these same barriers prevent body generated liquids and vapors from escaping. This can be particularly devastating when workers are subjected to ambient conditions which exacerbate this problem. Simply put, the body is designed to dissipate heat by sweating and sweat evaporation. If the sweat is not allowed to evaporate because of impermeable clothing and hot, humid ambient conditions, and if cooling is not provided, the body cannot get rid of its excess heat. When hard manual work such as lifting, pulling, shoveling, etc., is required, additional stresses are placed on the workers and more heat must be dissipated. As this undissipated heat builds up in the body it causes a variety of strains leading to heat exhaustion and possible heat stroke which can be fatal.

NIOSH has been involved in studies of heat stress and body strain in hot environments and recognizes the magnitude and severity of this problem. When workers are grossly uncomfortable wearing protective equipment while doing hard manual labor in hot humid conditions, their reaction is to remove the protective equipment when it may be unsafe to do so or to refuse to wear it. Unfortunately, these same hot conditions usually increase the volatility of many toxic materials and their ambient concentrations increase making the opportunity for exposure greater.

Working in protective clothing and wearing respirators can place additional stress on the lungs and heart. Workers who have lung or heart conditions should not be allowed to wear this equipment if it could cause damage to their vital organs. Thus, medical screening of workers and medical evaluation to determine if they can handle these stresses is vital to their safety and health. Add to these heat and work stress related problems the psychological stress of being surrounded by unknown toxic materials or the claustrophobic impact of being enclosed in a fully encapsulated suit with increased breathing requirements due to a protective respirator, and the concerns for worker safety and health are significantly increased. Consequently, some people should not be permitted at hazardous waste operations if they cannot wear the special clothing and equipment required for their protection.

The problems of working in protective clothing and equipment have additional ramifications. Excess water vapor inside masks and facepieces can causes fogging and severely reduce vision. The ability to communicate is reduced while wearing a mask or hood. Thick gloves can severely reduce the manual dexterity required to do certain tasks. Special boots or boot covers can increase clumsiness and contribute to slips, trips, and falls. Bulky clothing can easily catch on protruding objects. The bulk and weight of air tanks, hoses and controls carried for air supply can both get in the way and increase fatigue. The ability to freely or quickly exiting during emergencies is often reduced. The problems of heat stress and wearing protective equipment previously discussed are perhaps the most commonly mentioned work-related problems by hazardous waste contractors and their workers. The list of problems with protective clothing and equipment goes on, yet protective equipment is necessary.

PROTECTION MUST ADDRESS THE PROBLEM

The level of protection, decontamination, and control must be based on the identity and concentration of the chemicals or materials encountered. In cases where identification is not possible, or upon first inspection of a site, maximum protection may be required. After a site has been surveyed and samples analyzed, sufficient information may then be available to prescribe the appropriate level of protection commensurate with the problem. High level protective suits and supplied-air respirators should be used only where it has been determined they are necessary since they increase the stress on the worker as well as the effort and expense to complete a task. Isolation of the materials, remote handling, controlled atmospheres and other engineering controls should be the preferred handling methods. Personal protective equipment should be used only when engineering controls are not available or applicable.

To identify and determine the level of contamination, adequate sampling techniques and strategies are necessary. This is complicated at hazardous waste sites because of the mixtures of contaminants that often occur. One current NIOSH project involves the evaluation of direct-reading sampling instruments that were designed for use inside plants where the variety of chemicals is known. In hazardous waste clean-up, the specific chemical or mixture of chemicals may not be known. The application of sample collection techniques, direct-reading instruments, and laboratory analytical methods currently in use in industrial hygiene are being evaluated by NIOSH for utilization in hazardous material situations. The Division of Physical Sciences and Engineering has been working with several manufacturers and consulting firms to develop and evaluate field instruments, such as the portable gas chromatograph, which can detect chemicals in the environment. Experiments are also being done on series and parallel detector tubes for identifying individual substances in mixed environments. Chemical exposure badges for field personnel are being evaluated for efficiency and precision. NIOSH is gathering data for a better understanding of mixture interference problems. With this information, instruments and techniques for sampling and analysis will be improved and new ones developed which will give more accurate and precise analysis of samples containing one or more hazardous components. Better decisions in prescribing the necessary worker protection levels will then be possible because what is actually being dealt with will be known.

MEDICAL SURVEILLANCE

NIOSH is developing medical surveillance protocols that include biological monitoring guidelines for hazardous waste workers. NIOSH is continuing a study of the medical effects on fire fighters who worked on a burning chemical waste dump in Elizabeth, New Jersey in 1980. NIOSH has conducted several other evaluations of hazards for fire fighters who have dealt with PCB transformer fires and chemical warehouse fires. [6] These health hazard evaluations should provide insight to the occupational health hazards associated with these and similar incidents. NIOSH has also undertaken the task of determining the impact on the health of workers responding to chemicl emergencies, cleaning-up abandoned toxic waste dumps, or working at approved waste dump sites.

COMPREHENSIVE GUIDELINES AND TRAINING

NIOSH actively supports the development of a national resource of trained occupational safety and health professionals through training grants and technical support to fifteen Educational Resource Centers (ERC) located at major Universities throughout the United States. The "train-the-trainer" concept which has been successfully used by NIOSH in the past is being used to develop and disseminate the training materials on hazardous waste worker protection that NIOSH has prepared.

Under the Superfund legislation, two Interagency Agreements between the Department of Health and Human Services (DHHS) and the Environmental Protection Agency (EPA) were made. These agreements outline the scope of occupational safety and health activities and provide the funding for their implementation. The two major activities and health and to provide scientific support and training. Since the signing of these agreements, NIOSH has made significant progress in its areas of responsibility.

The comprehensive guidelines are being developed under a Memorandum of Understanding (MOU) signed on December 18, 1980, by EPA, the Occupational Safety and Health Administration (OSHA), the U.S. Coast Guard (USOG), and NIOSH. The four agencies, with NIOSH as lead, agreed to jointly develop a "Worker bulletin" to educate workers at hazardous waste clean-up sites, and a "Comprehensive Guidance Manual" to be used by field supervisors and government authorities.

The "Worker Bulletin", [7] published in December 1982, was written specifically for use by hazardous waste workers. It presents, in layman's terms, guidelines for proper safety procedures during hazardous waste clean-up activities. Ten thousand copies of the Worker Bulletin were distributed to workers in the first two months of 1983. Twenty-two thousand additional copies have been printed and are being distributed.

The purpose of the two-volume "Comprehensive Guidance Manual" is to provide inclusive, consistent, sound, detailed occupational safety and health technical information to all those involved in the management, treatment and disposal of hazardous waste. The Manual is designed to be used by many agencies: Federal, State, and local government; emergency response groups at all levels; clean-up crews; unions and employee groups; and other public interest groups or individuals. The primary users, however, will be those responsible for the management and supervision of workers inivolved in the clean-up and control of hazardous waste. Volume One deals with clean-up operations at remedial sites where hazardous wastes are known to exist and where clean-up will be undertaken. Volume Two deals with worker protection during accidental releases of hazardous substances and hazardous waste emergencies. NIOSH completed final draft of Volume One on September 26, 1984 and submitted it to the other agencies for final review. This manual will be distributed to each state and local control agency as soon as it is completed.

The Memorandum of Understanding also called for the four agencies to develop a strategy for training waste site managers and others who will then train clean-up workers. A training program to implement the transfer of the information in the Manual to expected users is the NIOSH Hazardous Waste Training Program that was presented 17 times in 1984, reaching over 800 people. Pertinent articles are being submitted to appropriate journals to make the information available. A college-level textbook was developed by NIOSH and the University of Michigan. [8] Every effort is being made to get the information in the guidelines out where it can be used to protect workers.

In addition to the "Worker Bulletin", the "comprehensive Guidance Manual" and the training programs, NIOSH is involved in the following hazardous waste related programs that indirectly support Superfund activities.

- The references assembled for the "Comprehensive Guidance Manual" are being compiled into a bibliographic file of pertinent scientific and technical materials. To date, more than 2,200 journal articles and other materials have been obtained. The file will be used as a resource in developing additional publications to guide workers and industry in hazardous waste clean-up efforts. this file will be added to the existing NIOSH biblographic data base "NIOSHTIC" so the information can be retrieved by a large audience.

- NIOSH is cooperating with the National Library of Medicine (NIM) to establish and maintain an automated inventory of literature and data on potential toxic substances. NIOSH has requested that occupational safety and health data fields be added to the NIM toxic substance data base.

The scientific support activity provides for consultation to EPA and state agencies by DHHS physicians, industrial hygienists, engineers and other health professionals for assessing the potential health hazards to individuals working at hazardous waste sites. This assessment process has included reviews by NIOSH scientists of environmental test results and evaluations of safety plans to assure protection of the workers.

TOXIC SUBSTANCE DATA BASES

Recommendations for the addition of toxic substances to existing data bases will be provided by the Subcommittee for Hazardous Waste Information Evaluation of the Department of Health and Human Service's Committee to Coordinate Environmental and Related Programs. This Subcommittee is composed of scientists from most Public Health Service agencies. The DHHS Interagency approach should insure the proper priority setting for determining both the chemicals and the categories of information to be included in the data base. The Subcommittee for Hazardous Waste Information Evaluation will also be responsible for identifying those substances that do not have sufficient health effects information. The National Toxicology Program (NTP) will then be responsible for reviewing these substances and developing protocols for toxicological testing to produce needed health effects information.

SITE EVALUATIONS

NIOSH has gathered a multi-disciplined task force to study the existing hazardous waste program. Over 30 site visits were made to uncontrolled hazardous waste sites to gather information on state-of-the-art practices. [9]

An additional 40 sites are being visited for the purpose of evaluating existing EPA occupational safety and health programs. At eight of the 40 sites, NIOSH is also conducting industrial hygiene sampling to determine the types of hazardous substances present and the levels to which workers could be exposed.

The EPA site occupational safety and health programs at existing hazardous waste sites reviewed to date, indicate some problems, but these can be solved within the resources available. For example:

1. The safety plans are often too general in nature and lack specific tasks that can be checked for compliance.
2. In order to make good industrial hygiene decisions, the capability for on-site sample analysis is needed. Many of the existing plans do not require this.
3. Laboratory practices at the site need greater attention not only for quick, reliable results, but also for the safety and health of the lab personnel.
4. Site safety control is not fully or consistently implemented for all people. Visitors and supervisors are often not protected according to the guidelines of the site safety plan. On the other hand, NIOSH found that some uncontrolled waste site clean-up operations indicate that with good planning, proper equipment, a detailed site safety plan, constant supervision, and adequate occupational safety and health surveillance, the work can be done with the necessary safety for individual workers.

NIOSH has utilized these findings in its field activities and will continue to search for better ways to protect hazardous waste workers.

CONCLUSION

Workers can work safely at a hazardous waste site if they are informed of the hazard involved, receive the necessary training, follow the proper procedures and/or instructions, use the required personal protective equipment, and remain aware of the conditions or situations around them at all times.

The following ten considerations summarize elements of a sample health and safety program for hazardous waste workers.

1. A proper identification and quantification of the materials to be handled.
2. A constant surveillance of the work environment (for example, a knowledge of weather conditions, contaminant levels, and fire/explosion potential).
3. The necessary protective equipment available and properly maintained (that is, both the personal protective equipment and the engineering equipment to provide protection for and/or isolation of the hazard).
4. An appropriate medical surveillance program, including a

record of pre-employment conditions and work-related exposures.
5. A comprehensive program for continual training of workers in all aspects of health and safety commensurate with their work responsibilities.
6. A proper decontamination program (that is, a method of preventing unnecessary worker exposure and eliminating migration of contaminants from the site).
7. A comprehensive site work plan including a fire and spill emergency control plan.
8. A communication/safety program which keeps track of everyone on-site and provides for medical, emergency, and/or community contacts.
9. A site security plan for properly designating and controlling access to and exit from contaminated, decontaminated and safe areas.
10. A proper logistics plan (that is, appropriate arrangements for eating, sleeping, washing and drinking water, compressed air, etc).

REFERENCES

1. Brown, M.; Laying Waste: The Poisoning of America by Toxic Chemicals; Random House; New York, New York, 1979.
2. Proposed rules, Federal Register, Vol. 47, No. 251, Thursday, December 30, 1982.
3. Public Law 96-510, Comprehensive Environmental Response, Compensation, and Liability Act of 1980, 96 Congress, HR 2020, December 11, 1980 (Legislation History, page 6120).
4. Public Law 91-596, Occupational Safety and Health Act of 1970, 91 Congress, S. 2193, December 29, 1970.
5. Public Law 96-510, the Comprehensive Environmental Response, and Liability Act of 1980.
6. HE 80-118, Chemical Dump Fire, Elizabeth, New Jersey, National Institute for Occupational Safety and Health HHE Report.
7. Streng, D.R. et al., Hazardous Waste Sites and Hazardous Substance Emergencies - Worker Bulletin, DHHS (NIOSH) Publication, No. 83-100 (1983).
8. Levine, S.P., and Martin, W.F., Protecting Personnel at Hazardous Waste Sites, Butterworth Publishers (1985).
9. Fiscial Year 1982 Annual Report, Hazardous Waste Project, Division of Standards Development and Technology Transfer, National Institute for Occupational Safety and Health, Cincinnati, Ohio.

CHAPTER 5

HAZARDOUS WASTE MANAGEMENT- AN INDUSTRY PERSPECTIVE

William L. West
Republic Steel Corporation

INTRODUCTION

In late 1976, the Congress of the United States passed the Resource Conservation and Recovery Act of 1976 with the express purpose of regulating the treatment, storage, transportation, and disposal of hazardous wasteswhich have adverse effects on health and the environment. The Act also promoted the demonstration, construction and application of solid waste management, resource recovery, and resource conservation systems which preserved and enhanced the quality of our air, water and land resources. Over eight years have elapsed since Congress passed RCRA and established these noble objectives. I believe it would be helpful to trace the history of RCRA since its enactment to the present date with an eye to evaluating the effectiveness of the implementation of the Act and its regulatory objectives.

RCRA, like its forerunners, the Clean Air Act of 1970 and the Federal Water Pollution Control Act of 1972, established stringent and, in many cases, unachievable time-tables both on the U.S. Environmental Protection Agency, which was responsible for implementing the Act, and the regulated community who had to comply with it. RCRA required EPA to establish, within years of enactment, most of the substantive regulatory processes that would implement the objectives outlined by the Congress. The agency took almost an additional two to four years to adopt these substantive regulations and several of these are still be reevaluated and repromulgated in different forms.

LEGISLATIVE MERRY-GO-ROUND

Now that most of the substantive regulations are in place, American history is suffering a deja vu experience with RCRA similar to those with the Clean Air and Clean Water Acts. Many of the regulatory programs dictated by Congress substantially amended

Management of Toxic and Hazardous Wastes, H. Bhatt, R. Sykes, T. Sweeney (Editors)

then in 1977. Last year, the Congress spent a considerable amount of time debating additional amendments to both the Clean Air and Clean Water Acts, again before the effects of the regulatory programs launched in the 1977 and 1981 amendments were fully implemented. These two Acts appear to be on a legislative merry-go-round wherein we pass amendments and before implementation we adopt additional amendments based on the misconception that the prior amendments have not been totally effective. With substantial Congressional debate on Resource Conservation and Recovery Act amendments in 1982 and 1984, it is apparent that RCRA is also going to join this legislative-regulatory merry-go-round. With the focus of the national press on RCRA's sister legislation, the Comprehensive Environmental Response, Compensation, and Liability Act to 1980, or Superfund, Congress will undoubtedly focus its attention on these two Acts very soon.

In fairness to the U.S. Environmental Protection Agency, a reasonable person must conclude that the time frames establish by Congress under RCRA were totally unrealistic. One would have thought that Congress would apply the U.S. EPA's experience in implementing the Clean Air and Clean Water Acts to establish a reasonable time to accomplish the tasks required by RCRA. In many respects, RCRA was much different from the Clean Air and Clean Water Acts. Both of these Acts essentially establish federal preemption of the regulatory processes that had already been established under state and territorial law. But RCRA, with its focus on hazardous waste could not really preempt existing state programs because the states did not have the regulatory processes required by RCRA in place at the time of its enactment. EPA then had to virtually start from scratch to develop the regulatory processes for the identification, treatment, storage, transportation and disposal of hazardous waste. While it has become quite clear that the Clean Air and Clean Water Acts established unrealistic dates, the Resource Conservation and Recovery Act established impossible dates.

STEEL INDUSTRY WASTE CHARASTERISTICS

Today, the substantive regulations under RCRA have been, in large part, established. I would like to relate to you today, an industrial viewpoint, at least as far as my company is concerned, of the regulatory process as we see it today. As a starting point, I would like to establish how we view our hazardous waste problems. The steel industry generates large quantities of solid wastes which are classified as hazardous by the agency mainly on the basis of the EP-toxicity test. Most of these materials are alkaline materials which do not exhibit the characteristics of leaching as those given by the EP-toxicity test when disposed with similar waste materials. The industry has attempted, through the regulatory processes, to have the agency adopt degree of hazard or monofill regulations which would recognize the uniqueness of disposing large quantities of low toxicity alkaline waste. We have not been very successful in this effort. Essentially, the regulations do not recognize the tremendous difference between our waste and highly toxic materials. This problem becomes a real calamity for the industry when you consider the large quantities of such waste generated by the American steel industry.

DEGREE OF HAZARD ANALYSIS

Our industry believes that the agency should recognize the degree of hazard associated with both the classification and engineering requirements for the disposal of solid wastes. The degree of hazard analysis should focus on the real life conditions that the waste will experience in the actual disposal sites. For our industry, the disposal of alkaline waste can be accomplished in an environmentally safe manner without the extensive engineering controls required for more toxic and leachable hazardous wastes.

WASTE CLASSIFICATION

Another problem encountered by our industry with implementation of RCRA, has been the agency's classification of waste materials. Our industry has for many years recycled much of our solid materials back into the processes to obtain the useful metals and/or fuel value from the waste meterials. The original EPA definition of solid wastes included any material which was or had ever been disposed as a solid waste. This definition would have resulted in much of our basic steel operations being classified as hazardous waste treatment facilities with the accompanying regulatory intrusion into our varied operations. The steel industry can recycle many of the solid and hazardous wastes generated within our plants. However, if the regulatory processes designate our facilities as treatment facilities with the accompanying regulatory requirements, we could be forced to decide between operating our plants for their designated purposes or operating them as disposal facilities. Since the agency's prime responsibility is the treatment of hazardous waste, we would expect that ultimately the regulatory burden of operating such recycling facilities would prove too burdensome and would actually result in the land disposal of more hazardous and solid wastes. Fortunately the agency has recognized this all encompassing definition and has moved to allow the continued recycling of material within our processes regardless of past disposal practices.

PUBLIC AND REGULATORY ATTITUDE

Given the infirmities of the regulatory process, however, we are attempting to establish disposal sites and long-term storage sites for listed hazardous wastes generated by our facilities which cannot be immediately or economically recycled. Our limited experiences concerning the siting of hazardous waste and our examination of the experiences of other industries in establishing sites does not give us much consolation. The present furor at the U.S. EPA in Washington in large part centers around the disposal of hazardous waste. The regulatory climate is such that the American public is convinced that hazardous wastes cannot and will not be handled in a environmentally satisfactory manner. I do not for a second believe that the public perception is an accurate one but it is one that we must recognize. I the Resource Conservation and Recovery Act is to be successful, then the regulatory processes must provide for the construction of treatment, long-term storage and disposal facilities within the United States. We generate tremendous quantities of hazardous waste because of our advanced technology and our standard of living. We should be able to employ that same technology to the safe destruction, storage and/or disposal of hazardous waste. But from our perspective we now see a public and regulatory attitude developing which will, in fact, stymie the objectives of the Resource Conservation and Recovery Act. Let me elaborate.

Firstly, the regulatory process of handling permit applications for treatment and disposal facilities appears to be motionless. It appears from our viewpoint that the regulators recognize the political volatility of issuing hazardous waste permits and have responded with inaction. One must remember that during this period of inactivity on the processing of hazardous waste treatment or disposal permits that the hazardous waste intended for such facilities is being continuously generated. If we are to comply with the objectives of RCRA, we must move quickly to approve and develop hazardous waste facilities.

Secondly, the regulators must recognize that inflammatory statements concerning the treatment and disposal of hazardous wastes are counterproductive. I am not suggesting that the regulators hide or not reveal to the public the consequences of disposing or treatment of hazardous waste. My point is that the regulators must act responsibly when defining the issues for the public. The hazardous waste identification and disposal process is a very complex one which is not easily understood by the general public. While a technically trained person may be able to look at EP-toxicity data or in situ leachate data and determine whether there is an associated environmental risk with such material, the general public is more prone to focus on the mere classification of a material as hazardous or toxic. We must recognize that all materials have hazardous potential. Common table salt can be hazardous to your health if taken in extreme quantities, while no hazard is associated with normal use. I believe the burden of describing in layman's terms the hazard associated with any particular waste lies with both the industry and the regulators. We both must act to responsibly inform the public of the risk associated with any particular waste.

Thirdly, the government must take the lead role in educating the public about hazardous waste disposal and treatment facilities. The general public simply will not accept industrial pronouncements on the hazard issue. In this respect, I believe the government's activities to date have been almost entirely negative. In their haste to focus the public's attention on hazardous waste issues, legislators and regulators have inflamed public opinion against hazardous waste facilities. Time and time again, the public is terrorized into blocking the construction of responsible treatment and/or disposal facilities through either regulatory or judicial processes. One only has to look at a PCB disposal facility in New Jersey which survived the regulatory process but was ultimately blocked from operation in the judicial process after the facility was constructed and ready to operate.

CONCLUSION

We do discern a changing attitude recently in this respect, as regulators have begun the process of rationally informing the public in the positive aspects of responsible hazardous waste management. Included in this new attitude is the agency's recent promotion of the recycle and reuse concepts for hazardous wastes.

We as a nation cannot close our eyes to the hazardous waste issue. We must be prepared to correctly identify hazardous wastes that present health or environmental risks. Such waste should be treated, contained, or disposed in permitted facilities as expeditiously as the permits can be issued. We should not delude ourselves that additional legislation and/or regulations will somehow improve our regulatory processes. The program must be initiated now. We cannot afford to go through another ride on the legislative-regulatory merry-go-round. The time to move forward with the regulatory process is now. We can evaluate that regulatory process as these programs unfold and make whatever adjustments may be necessary when such adjustments are demonstrated by the facts rather than the present speculation.

CHAPTER 6

THE PARTNERSHIP APPROACH TO HAZARDOUS WASTE FACILITY SITING

Stephen H. Sedam
Ohio Environmental Council

OHIO ENVIRONMENTAL COUNCIL

For those of you who aren't familiar with the Ohio Environmental Council, we are a federation of 60 diverse environmental interest organizations. We were founded in 1969 and serve as the coordinating body for Ohio's environmental community. Our efforts to promote environmentally sound state hazardous waste policies and to develop broader public awareness of hazardous waste management has extended over a five year period. Perhaps many of you have seen our publication. Searching for Solutions: A Citizen's Guide to Hazardous Waste Management in Ohio.

INDUSTRIAL WASTE MANAGEMENT

Since the advent of the Resource Conservation and Recovery Act and the regulations and state laws that followed, business and industry have been forced to comply with a new code of hazardous waste management. Government has forced industry to reexamine its waste management practices. Certain businesses have found an opportunity for expansion, while others, after the permitting process was underway, have found that they can no longer legally accept the hazardous waste they once handled. Business and industry have begun to look around for places to store, treat, and dispose of their waste legally.

A great deal of this waste is handled on site. At least 33 percent is handled on site in Ohio alone. And efforts are afoot to encourage generators to reuse and recycle the waste, make volume reduction changes and utilize waste exchanges. You have probably heard of the Ohio EPA task force that is recommending policy changes to encourage these practices.

However, the need and the disposal industry's desire to increase off-site treatment, storage, and disposal opportunities is real. Just as real is what many consider to be the ultimate bugaboo of hazardous waste management - the issue of faciltity siting.

Management of Toxic and Hazardous Wastes, H. Bhatt, R. Sykes, T. Sweeney (Editors)

FACILITY SITING PROCESS

Before a facility is built or expanded, it must first run the course of a facility siting process. The process is supposed to operate just as it appears on an agency's facility siting flow chart, the one with the neat boxes and smooth, flowing lines. Bus as we all know, there are many social, political, and other factors that impinge upon that siting process and create kinks in the flow of the system designed to give us additional disposal capacity.

A facility siting process is largely determined at the state level by laws, regulations, and guidance documents which vary from state to state. There are nearly as many approaches to citizen involvement in these processes as there are siting procedures. In examining the siting processes around the country it is obvious that no one has the answer yet. There are varying degrees of success. Some siting programs are on the books but are as yet untried. Others have failed, and from the developer's point of view, some of these established siting procedures have failed miserably. One key element of most siting formulas is the authority of the state to override any local zoning that prohibits the development of a hazardous waste facility.

COMMUNITY INVOLVEMENT - KEY TO FACILITY SITING

Regardless of what the siting procedure is, the hard knocks of siting hazardous waste facility lies in community involvement. The community, local citizens, and the facility developer need to see themselves as partners in this venture. This is a too-important and potentially volatile issue to be treated in any other way. Many a time, public relations firms advise an industry of various strategies of siting facilities. They speak of the community and local citizens in terms of teh "opposition." This type of approach creates an adversarial relationship that fosters in the minds of the community's citizens the notion that they will become victims of a facility rather than its hosts.

Most businessmen don't consider themselves to be bullies and most concerned citizens don't consider themselves to be obstructionists. But this is how townspeople and developers consider each other when there is little or no effort made to consider each other as peers in the siting process.

The public has developed a sizeable distrust of government and of the hazardous waste industry due to regular media attention to improper hazardous waste practices. Before an application for a permit for a new facility is even filed, the community is organized and asking themselves: "Is this really necessary?" and, "Isn't there a better way?"

Ohio, the nation's third largest hazardous waste producing state, proves to be a fertile territory for discovering the pitfalls and pluses that can be used in attempting to site hazardous waste facilities.

CASE HISTORIES

A couple of years ago, a Fortune 500 company attempted to site a large hazardous waste facility in northeast Ohio. They call the facility a storage operation even though the waste would be placed underground. There was virtually no discussion with the city fathers and the townspeople on how the facility would be managed before the plans were finalized. On top of this, the company exhibited a poor public attitude toward the community's concerns. The company's suspect environmental compliance history further compounded the community's fears. Overnight, dentists, housewives, mechanics, and high school students became crative community organizers. Town meetings of over 200 people became media events. Speakers were brought in to remind the town of the horrors of past hazard waste mishaps.

Meanwhile, discussions with the company were negligible. The company had come into town with their plans announcing the facility's grat benefits to the ocmmunity - more employment and an expanded tax base. The company even offered tours of the proposed facility. But there were no meaningful attempts to embrace the community's concerns and alter the plans accordingly. The company's performance indicated to the townspeople that they wanted no part of this facility. The public reaction to the proposed development was so intense that the townspeople voted to raise their own taxes to fund opposition to the facility. It never became operational. The company has even shelved plans to apply for a permit.

In the rolling hills of rural central Ohio, a similar scenario was unfolding in 1983. Community residents woke up one morning to a small announcement in the local paper about an industrial park that would be receiving hazardous waste for disposal. Later announcements revealed plans for a land disposal facility.

Considering the technical merits of the facility as a separate item, one must ask if this company was truly serious about siting a facility using this approach to the community. Where was the public discussion of 1) the need for the facility, 2) the design of the facility, 3) spill prevention and response capabilities, 4) designated transportation corridors for waste hauling, 5) the company's history in hazardous waste management, 6) mitigation of noise, and 7) the prevention or response to ground water contamination.

Rather than advancing an interest in the community's concern about the facility, the company had given the community the notion that they meant to proceed with their development, regardless. As you might expect, the community responded with overwhelming rejection of the site. County commissioners hired consultants, a city council donated funds to a citizens' group, the citizens' group hired an attorney, and regular meetings attracted hundreds of people in opposition to the facility. Ultimately, the facility could not be sited as it was ruled technically non-feasible by Ohio E.P.A.

There are also some pluses in Ohio's young hazardous waste facility siting history. Although this facility has yet to be sited, the companies seeking to develop the regional waste incinerator in East Liverpool, Ohio have found their host community to be an important part of their facility development process.

Although it has taken considerable sums of money and a couple of years, the effort on the part of the developer to receive the community's concerns in a constructive fashion has been applauded by many. The Ohio Environmental Council Board of Directors has even written the company to commend them on their efforts to educate the public and respond to their comments and concerns. Public participation and community involvement have been an integral part of this facility's evolution.

This company had a number of things going for it, the most obvious of which was the philosophy toward the local community. The community was considered the host of the facility. Plans were altered to allay community concerns. Transportation corridors were identified. A 24 hour emergency response team was included in their plans. The company shared documents on the operation of the proposed facility with city fathers and concerned community leaders They established an office in the host community and conducted numerous meetings with local residents, the media and others. They have apparently realized that providing employment and an expanded tax base is not enough. In the case of hazardous waste facilities, the protection of public health and the environment are integral parts of a facility's operation.

SUMMARY

In summary, the peer or partnership approach to facility siting is paramount to successful siting. A company cannot assume it knows all of the community's concerns and at the same time the need for additional facilities cannot be denied.

Business and industry need to have expanded hazardous waste treatment, storage, and disposal opportunities available. They also need to take a more open view towards the siting process than has been the norm to date. Developing a partnership with the host community makes just plain good business sense.

CHAPTER 7

SOLID WASTE FACILITY SITING - COMMUNITY ASPECTS AND INCENTIVES*

Harry E. Smail, AICP
Battelle Columbus Laboratories

INTRODUCTION

With few exceptions, the siting of new facilities is the single most difficult waste management problem facing local government and industry officials today. In addition to technical engineering criteria, financial requirements and compliance with environmental regulations, social and political concern have evolved as equally important considerations in landfill siting programs. Typically, such social and political concerns are expressed by the statement: "Nobody wants a landfill in their backyard or wants to pay for waste disposal facilities in their community."

This chapter provides a generic overview of nontechnical socioeconomic/institutional aspects of landfill siting requirements, issues and processes from the following perspectives:

- historical and contemporary aspects of waste disposal sites
- relevant waste management laws, regulations and standards
- social aspects of the Waste Management Facility Siting Process including community planning programs, characteristics of siting processes, identification of community issues, and public involvement programs

---

* Text of this chapter is from a 1981 U.S. EPA Region II Technical Assistance Panels Program research report conducted by Battelle-Columbus for the U.S. Environmental Protection Agency under contract 68-01-6002, Work Assignment No. 22, entitled "Solid Waste Disposal Feasibility Study for Erie County, New York". The content of this chapter was effective as of March 1983 when this chapter was presented at the Third Ohio Environmental Engineering Conference, Columbus, Ohio. The contents of this chapter are based on information presented in the several reference publications listed in the bibliography.

Management of Toxic and Hazardous Wastes, H. Bhatt, R. Sykes, T. Sweeney (Editors)

- common social/economic/political issues and opposition tactics; and
- various management and institutional approaches, incentives and mitigative measures for increasing the public acceptance of a proposed waste management site such as facility redesign measures, operational management measures, and compensation measures.

Selected events from previous specific local community waste management siting processes are referenced to exemplify the generic discussions presented in the text. Although the information presented in this chapter focuses on the siting of solid waste facilities, the information and concepts discussed are also generally applicable to hazardous waste siting programs. However, hazardous waste regulations and siting requirements are more complex and negative public sentiment towards hazardous waste facilities is much stronger than towards solid waste facilities.

HISTORICAL AND CONTEMPORARY ASPECTS OF WASTE DISPOSAL SITES

From a social siting perspective, landfills are simultaneously an objectionable, undesirable land use and a public necessity. The social characteristics of landfills are similar to other types of locally unwanted land uses such as power and sewage treatment plants, prisons and airports. Few people desire to live near such land uses which characteristically are noisy, dangerous, unaesthetic, polluting, etc. Since such locally unwanted land uses also create other secondary impacts such as lowering adjacent residential property values, neighboring citizens resist the siting of facilities next to them. Likewise, from a community perspective, local citizens usually must bear most of the economic and environmental costs of constructing and operating such facilities. The cost of disposing of wastes is expensive both in terms of direct costs for equipment, manpower, land, etc., and indirect costs on society.

Landfills are, without question, a necessary public land use in most areas, given the quantities of wastes that are generated today that must be properly disposed of in order to protect the health and safety of the public and to prevent degradation of the environment through uncontrolled disposal. Generally, the vast majority of a community's residential, commercial and to a lesser degree industrial waste stream is in a solid and nonhazardous form and is typically comprised of garbage, paper, metals, molded plastics, wood, glass, rubber and inert materials such as concrete and ash. A portion of these waste streams are frequently biodegradable or recyclable and can be safely and economically disposed of in properly designed and operated sanitary landfills without any additional threat to human health or the environment.

During the last few decades, because of national recognition of the need to conserve natural resources and because of economic incentives, certain waste streams increasingly have been reduced by reuse and energy and/or resource recovery options. Indeed, waste reduction and recycling are, when economically and technically feasible, preferred options to landfilling. However, recycling facilities typically can have or are perceived as having similar characteristics to waste disposal facilities (e.g., increased traffic, noise, dust, odor) which also cause local siting opposition. While these waste management options can to some extent significantly reduce the needed annual disposal capacity and extend the economic life of an existing landfill, landfills still remain as a necessary component of a community's waste management program. Waste reduction and recycling processes themselves produce residual wastes and cannot completely accomodate all types of wastes. In the foreseeable technological future, the requirement for new landfills will not disappear regardless of the extent to which new energy and resource recovery programs are instituted.

Waste disposal facilities were once generally accepted with benign resignation by community leaders and residents who recognized the community needs and benefits associated with such facilities. However, times and conditions have changed for sponsors and developers of waste storage, recycling and disposal facilities. Today, as communities (both urban and rural) value their own character and the quality of their environment and public services, citizens and officials almost unanimously fear the consequences of and object to the siting of any new waste management facilities within their jurisdictional boundaries.

In the recent past, blanket opposition to and total rejection of locally unwanted land uses by enactment of prohibitive zoning regulations has been a common response to such new proposed community facilities. This typical non-responsive action by local governments did not address the fact that waste disposal is a community responsibility and simply dumping in some adjacent community is not always a practical or possible solution to their problem. This approach also made the community vulnerable to potentially high dumping fees, long travel distances, and limited, non-guaranteed disposal capacities. Yet at the same time, it must be recognized that some municipalities may not be able to solve their own disposal problems within their own jurisdictional boundaries because of the lack of available undeveloped sites which meet the various state and Federal site suitability and design standards and regulations enacted to protect public health, safety and the environment.

A nationwide trend in recent waste management facility siting legislation is that of requiring local communities to respond to new facility proposals rather than swiftly terminating their involvement by simply rejecting a proposal or over-ruling deficiencies in local ordinances by political fiat. The use of eminent domain powers and creation of statewide siting boards with powers to issue permits over-ruling local objections to the construction and operation of technically sound waste management facilities has created a new decision-making climate. Moreover, most State and Federal siting procedures for needed public facilities have provisions which ensure that community concerns will be addressed and that risks and impacts will be minimized, and frequently require negotiations between communities and developers during the siting and licensing permit process. Some siting procedures also now have provisions enabling the benefits associated with such facilities to be increased by compensation options so as to conceivably make a community better off by accepting rather than rejecting these facilities. This trend is likely to continue during the 1980's and beyond.

The advent of stringent regulatory controls on waste disposal siting and oeprations during the past decades has pushed physical site suitability into the forefront of land use siting decisions. Regulations prohibiting the establishment of new solid waste disposal sites in areas not physically suitable for waste disposal facilities has created a situation in many communities in which new disposal sites must compete for more expensive undeveloped land (such as agricultural land) and land that traditionally has been reserved for more "socially acceptable" land uses. Before the establishment of these more stringent siting regulations, unwanted or remote parcels of land were traditionally chosen as waste disposal sites. Since waste disposal sites have historically been an extremely undesirable land use option in comparison to residential, ocmmercial, industrial or recreational developments, areas were selected as community dump or landfill sites because "they weren't good for anything else." Today, then, land cost and competition for more desirable land have also become a focus for local community concern and opposition to landfill siting proposals.

RELEVANT WASTE MANAGEMENT LAWS, REGULATIONS AND STANDARDS

The handling and disposal of solid wastes is subject to a wide array of State and Federal laws and regulations which have a significant impact on waste management options available to municipalities. Briefly cited in the following paragraphs are Federal and State laws relevant to operations at existing waste management facilities and the siting of new facilities. However, specific provisions of these laws are not discussed given the generic, non-technical objectives of this paper and because the promulgation and enforcement of regulations and standards to enact provisions of these laws has been slow and many standards are being changed for technical, economic and environmental reasons. Without question, however, a detailed assessment of current and proposed

Federal, State and local regulations, standards and procedures and the identification of jurisdiction of authorities are a necessary prerequisite at the outset of any siting effort to ensure compliance within the regulatory framework.

At the Federal level, the Solid Waste Disposal Act of 1965, as amended, and the Resource Conservation and Recovery Act of 1976 (RCRA), as amended, are the major laws pertaining to solid waste management practices throughout the country. Regulations and standards promulgated to implement provisions of these laws establish landfill inventory requirements, environmental performance for sanitary landfill criteria and require states to develop compliance schedules to bring unsatisfactory sites into compliance, or close them down. Additionally, at the Federal level, the U.S. Environmental Protection Agency has published regulations in the Code of Federal Regulations (40 CFR 25) establishing procedures for meaningful public involvement programs by providing opportunities for public interest groups, private citizens, and elected officials to become extensively involved in the siting process at the planning, design and decision-making stages. States must observe these federal minimum requirements in order to qualify for grant funds under Subtitle D of RCRA.

Within state governments, rules and regulations of Departments of Health, Natural Resources and Environmental Protection frequently govern solid waste management facilities. Generally, comprehensive and often times stringent rules established by such state agencies specify technical siting and operational requirements with which a solid waste management facility must comply from its construction to final closure, as well as providing for governmental level, most communities further enact guidelines through local zoning ordinances which restrict the location and operational aspects of solid waste facilities.

SOCIAL ASPECTS OF THE FACILITY SITING PROCESS

Integration of site location, facility design and the expected waste stream volume and composition are fundamental components of a successful solid waste management program. A well-matched site and facility offer the opportunity for a safe, economical and publicly acceptable way to dispose of community wastes with minimal, if any, adverse health or environmental effects. Government regulations may specify particular site requirements which exclude some areas from consideration as landfill sites, and sound technical engineering designs can go a long way towards insuring the proper disposal of solid wastes at new facilities. However, proper location (i.e., a site with good natural conditions) is critical even for a well designed facility. Reliance on technology and design factors at deficient or inappropriate sites will likely lead potential opponents and public officials to conclude that the proposed facility/site package represents an unreasonable and unacceptable risk. Given this perspective, the following paragraphs explore some major social aspects of community planning programs, the siting process, the identification of significant community issues and public involvement programs.

Community Planning Programs

A good, long-range community planning program in combination with adequate zoning regulations and enforcement may greatly aid the siting of undesired but necessary public facilities. Accurate long-range population and community composition forecasts upon which to project waste volumes, genuine awareness of public concerns, and plans for adequate utility and transportation infrastructures are examples of ways in which a good community planning program can facilitate the siting process. Proposed sites that are publicly owned typically do not encounter as strong public opposition as privately owned sites; however, public acquisition of a site before subsequent announcement of intentions to use the site as a landfill to avoid escalated land costs and opposition merely creates more intense opposition and distrust of public officials and agencies.

The siting and land acquisition process that has transpired during the past three years for the proposed new Franklin County, Ohio, 218-acre land fill near State route 665 and I-71 in Jackson Township presents a classic example of land acquisition complexities for developing new waste disposal sites in urban areas. Specifically, the Franklin county commissioners, recognizing the need for new landfill capacity, public opposition to landfills in any local comunity, and escalating land prices if the intended use of the site became public knowledge, quietly purchased the site from a retired farmer through a third-party holding company for $1.75 million in December, 1980. Immediately upon disclosure that the land was purchased for a landfill, public outcries of anger and opposition were voiced by the former property owner and local citizens. Plans for the landfill have subsequently been in litigation since that time, including a Governor's veto of State legislation which would have made the project and other similar projects subject to a township referendum. Only recently, given the critical need for the timely construction of the site, has a modified Ohio EPA permit been issued which allows the County to complete final plans and obtain construction bids while the litigation continues.

From a theoretical land use planning perspective, the location of a community's locally unwanted land uses have traditionally been either concentrated into one or a few specific areas (such as industrial parks) or dispersed throughout the community. The concentration of locally unwanted land uses is least socially objectionable if the concentration area is not readily visible by the general public, has natural or man-made features which buffer other land use patterns, or the facilities have characteristics which blend into the existing environment. However, the concentration strategy often times tends to force locally unwanted land uses into areas least able to resist them such as poor or minority neighborhoods, politically under-represented communities, and unincorporated and thinly populated areas. The dispersion of unwanted local land uses, on the other hand, often tends to ignore valid environmental, economic and social impacts to specific areas.

Siting Process

Four generic characteristics of local waste management facility siting processes are:

- multiple parties - with different concerns, values, goals; and with different authority, power, status, etc.
- multiple issues - created by perceived impacts of proposed actions on multiple systems in the environment; and scientific and technical issues vs. value-laden assessments
- uncertainty and risk - associated with attempts to predict or forecast specific impacts of proposed actions, especially those with public health and environmental implications
- emotional overtones - public anxiety arising from divergent perceptions that some impacts may have health effects or irreversible environmental impacts.

A solid waste management facility siting process should, therefore, be designed to achieve consensus among parties-at-interest on an optimal balance among technical, environmental, economic, social, political and other critical factors. An effective siting process is one in which affected parties, from both the host community and adjacent communities, are able to voice their concerns to each other share information and through an open forum, discuss appropriate questions and issues. As such, the siting process is not a win/lose proposition, but a win/win process in which views of impacted parties are presented, concerns are addressed and a logical, defensible process is followed to reach a decision. Therefore, "stakeholders" must understand the siting process and feel that they have some influence on the final decision. With all interested parties participating in the solid waste management siting process from the beginning, upon the completion of the study a decision can be confidently and appropriately made to support or reject a proposal, to make facility design changes, to institute measures, to mitigate potential impacts, and/or to agree to compensation payments. At the end of this siting process, the local community may still oppose a particular facility plan and vote to reject it, the proposal may be revised or withdrawn by teh facility sponsor or the state may decide to exercise its preemptive regulatory authority (if available). Others, proponents or opponents, may resort to the courts. If an effective siting process has been followed, these decisions will be made with a full understanding of the facility design and an understanding of its consequences at that particular site, and they should be able to stand the tests of state administrative and judicial review. As such, and effective siting process conserves resources and does not waste the participants' time or money.

No proposed site is likely to be "perfect" from every perspective, and therefore, the community must cope with the strengths and weaknesses of a given site. Occasionally, the element of time may become an important factor in the decision to accept a well-designed facility rather than looking for another°° "slightly better, less flawed" site when the probability and extent of possible impacts from continued improper or inadequate disposal or indiscriminate dumping are considered.

Community Issue Identification

Effective "Scoping" of community issues to accurately identify and define the significant public concerns of relevance to the proposed facility is very important from the outset of the siting process. Effective scoping will pin-point those issues--impacts, risks, benefits, and costs--that are of primary concern to the affected parties, thus potentially reducing costs and delays by streamlining the factors to be intensively studied, by building a consensus of the range of alternatives to be considered, and perhaps by reducing the propensity for environmental conflict and litigation through resolution of disputes before they become too complex.

Although the risks and costs of solid waste management facilities are often well perceived by the public, the regional and community benefits from such facilities are not usually so perceived. Such benefits can include:

1. Comparatively lower waste disposal costs and better operational efficiencies since waste are disposed locally rather than being shipped elsewhere.
2. Enhancement of the environment and public health and safety since waste are not being disposed of by other less satisfactory alternatives (e.g., illegal open dumps or inadequate, poorly sited, poorly run landfills).
3. Creation of tax revenues and local jobs.
4. Promotion of the community image and enhancement of the growth attractiveness of the region for new development.
5. Protection of a community's viability and control over the availability and cost of adequate disposal facilities.
6. Potential for compensation (e.g., payments and other benefits) when negotiating with facility sponsor/operator.

Inclusion of applicable social factors in the siting methodology criteria to distinguish between impacts of alternative proposed sites and disposal plans will also likely facilitate the siting process. In addition to physical and engineering characteristics of the site and facility, detailed site evaluation criteria usually include some socially oriented considerations such as the population density of areas neighboring the facility site and delivery routes, zoning and land use patterns, health and safety risks and operational cost estimates. Not usually examined at all or in as much detail, however, are other social factors whose consideration could provide a clear distinction between proposed alternative sites and technological processes, and the public acceptance of such proposals. These other local social factors may include secondary and tertiary fiscal impacts on the community, management/operational procedures, short and long-term land reuse potentials and other issues of local concern.

Public Involvement

A public involvement program, now required by law for most siting programs, often serves two primary and very important functions in the siting process:

- informing the general public about what the problem is, how the siting study is being conducted, what the alternatives are for solving the problem, and what the impacts, risks, cost and benefits are of the alternatives are for solving the problem, and what the impacts, risks, cost and benefits are of the alternative solutions; and,
- providing mechanisms for parties-of-interest to actively participate in the siting process.

The purpose of a public involvement program is not to provide information to overcome public opposition but rather to establish parity of knowledge among concerned parties and to solicit and utilize their inputs in a timely and meaningrul manner. Disagreement about the acceptability of a site may be inevitable, but resolution of subjective issues is difficult to achieve if relevant objective facts have not been shared and understood.

No single public involvement program will be applicable to every siting study given local community situations. In some communities, the public involvement program effort might be on a small scale; elsewhere, it may require a large commitment of time and resources. Nevertheless, an open public involvement program from the outset will reduce the likelihood of total rejection or facing needless delays and skyrocketing site costs. Public involvement delayed until the traditional hearing process is widely regarded as an exercese in futility, since it typically breeds frustration, distrust and anger on the part of concerned citizens who recognize that very important decisions have already been made and that their involvement comes too late to have a meaningful influence on the project, except to perhaps stop it altogether by court action. Alternatively, public involvement programs may include public workshops, periodic newsletters, hotline telephone numbers, establishment of a local advisory board and other appropriate activities from the outset of the siting program.

An effective, open public involvement program can ameliorate the problems of misinformation and confusion that so often have eroded public confidence of siting studies in the past. Public uncertainty about specific details of the alternatives and the process by which decisions are being made, leads to conflicts and distrust among affected interest groups. Specifically, local officials and citizens are often uncomfortable with letting a private developer or sponsor judge the strengths and weaknesses of various sites, and tend to become bewildered and distrust plans when a myriad of complex technical terms are used. The general public also tends to believe the press over developers and politicians if they first hear about the proposed facility from other unreliable (or at least ill-informed) sources. Additionally, since individual citizens must eventually pay for the solid waste management collection and disposal program, it is essential that they understand and approve plans that they must financially support. A public involvement program usually also brings out other public issues (such as adequacy of refuse collection and potholes in the streets) which have little or no direct relevance to the siting study.

COMMON SOCIAL ISSUES AND OPPOSITION TACTICS

Generally, a necessary community locally unwanted land use, such as a sanitary landfill, is eventually sited, but because of public opposition, the siting process tends to become longer and more expensive than anticipated, bureaucratic, acrimonious, and prone to stalemate. Any proposed waste management facility--landfill, transfer site, resource recovery plant--can be subject to fierce citizen opposition. The degree to which the local public has become involved in a waste management siting proposal has varied from community to community with bitter conflicts arising in some cases and seemingly complete indifference in others.

To a minor degree, the technical complexity of the proposed plan, site characteristics, and the threat to public health, safety and the environment determine when and where general public concern and involvement will arise. More importantly, the social and political history of solid waste management programs within the community may be more of an indicator of likely general public opposition. For example, if the municipality and/or facility operator have a good neighbor image and a reputation for trustworthiness and responsiveness, then the majority of citizens may see no need to dramatize their concerns and become involved. However, general public acceptance of waste management facilities lessens and opposition intensifies with each discovered incident of improper disposal or inadequate operation at existing facilities.

In general, the people who are most likely to become involved in local waste management siting programs are those who will likely be directly impacted (e.g., forced to relocate or, more commonly, who will be located adjacent to the proposed site) and those citizens who believe that their health or pocketbooks will be affected. For example, citizen concerns frequently focus on increased traffic near the site, fears that property values will decline, and general mistrust of government and industry assurances that the facility will be safe. However, not all of the people who may be affected by a proposed facility become actively involved in the siting process. Many people do not become actively involved in day-to-day community affairs even though they are genuinely concerned about the community, but these same people will "come out of the woodwork" if their assistance and support is critically needed.

Thirty-five socially, economically and politically oriented substantive issues, grouped into nine generic categories, which are frequently used by groups to support their opposition to the siting of waste management facilities are listed in Table I. Additionally, the two basic factors which stand out as the primary causes of public conflict in recent waste management facility siting experiences in the United States are inadequacies of the management of the siting process (e.g., unresolved substantive issues and inability to accommodate unplanned factors) and secondly, the lack of understandable, credible and easily accessible information to address issues and concerns. Appropriate measures that can be creatively employed to mitigate these types of public opposition issues are discussed in the following subsection of this paper.

Four tactics frequently used by opponents to a proposed waste management facility plan include:

- publicity - to make their position known to solicit support via multimedia news releases, bumperstickers, and testimony at public meetings and hearings
- legal action - initiation or threat of lawsuits to delay, modify or stop the siting effort
- hiring of outside experts - to refute sponsoring data or findings
- political action - to persuade local officials to use zoning controls to preclude siting of such facilities and/or to deny necessary construction and operation permits.

In essence, it is essential to consider seriously all public opposition to any proposed facility as these concerns are, for the most part, sincere and not imagined, and methods of citizen recourse in the siting process are frequently effective.

Table I. Substantial Issues Frequently Voiced by Public Groups in Opposition to the Siting of Waste Management Facilities

---

Site and Facility Characteristics

- Soil permeability
- Problems of site operations: odors, fires
- Lack of contingency plans
- Possibility of pollution, especially ground water contamination
- Alternative uses of the site
- Lack of long-term maintenance and decommissioning plan
- Insufficient data on important or unknown parameters
- Costs (e.g., construction costs, operating cost/disposal fees, etc.)

Siting Process

- Failure to identify other superior sites
- Use of subjective siting criteria especially overly strong reliance on technical and economic selection criteria to the virtual exclusion of socio-political/human criteria
- Lack of substantive public input
- Inadequate public access to information
- Failure to notify relevant parties public of siting process and their role in it
- Public not informed or misunderstanding the complete siting procedure

Sponsor and Regulatory Agency Credibility
(Failure to build trust and respect)

- Arrogance and overconfident attitudes by sponsors
- Previous performance record in the community
- Previous record at other sites
- Failure to be open and honest concerning unknowns

Table I. (Continued)

Transportation Problems and Risks

- Access routes to facilities
- Frequency or number of trips per day

Types of Wastes to be Disposed on Site

- Wastes accepted at facility and from whom
- Community image as "dumping ground"
- Resistance to accept waste from other areas
- Choice of disposal technology

Direct and Indirect Impacts on Surrounding Areas

- Area too populated
- Property values will decline
- Aesthetics and quality of life decline

Economic Issues

- Loss of Property taxes for publicly-owned facilities
- Few jobs for local community
- Loss of local agriculture
- Lack of compensation for nuisances and risks

Local Control of Land-Use

- Zoning ordinances
- Site-specific issues (impact on neighboring industry)

External Factors

- Elections
- Publicized improper disposal

Adopted from: New England Regional Commission, A Decision Guide For Siting Acceptable Hazardous Waste Facilities in New England.

GENERIC MEASURES FOR INCREASING THE PUBLIC ACCEPTANCE OF SOLID WASTE MANAGEMENT SITES

The typical siting process may reveal several likely or potential impacts and viable community concerns that are associated with a proposed solid waste management facility at a specific location. Public acceptance of the proposed facility will likely depend upon efforts to reduce the risk (probability of occurence) and upon efforts to reduce the magnitude (severity) of the impacts. Many of the impacts and concerns about the proposed facility amy be reduced or even eliminated through creative facility redisgn and facility operational/management procedures. However, to obtain public and community acceptability of the proposed facility, it may be necessary for the facility sponsor to "compensate" the community and impacted individuals for accepting the residual risks and impacts still associated with the proposed facility after redesign and mitigation agreements have been reached. Environmental conflict resolution experiences over the last few years have demonstrated that avoiding disputes by replacing litigation with opportunities for negotiation will often times save money and time, and result in better and more accommodating decisions. All such redesign, mitigative and compensation measures must be reached within the general bounds of state (and, where applicable, federal) standards and regulations so as to ensure public health, safety and welfare and protection of the environment. The following paragraphs describe a few of the endless number of possible redesign, mitigative and compensation measures which can be negotiated between developers and communities in order to reduce or eliminate public impacts and risks and improve community benefits, thereby improving facility performance and public acceptability of a proposed solid waste management facility.

Facility Redesign Measures

After initial siting reviews have been completed, it may be appropriate to reconsider the incorporation of energy and resource recovery systems as part of the total solid waste management system. Over the years, U.S. manufacturing techniques have largely been based on the use of virgin materials and a very small percent of the municipal waste volume has been salvaged for reuse or recycling. Today, however, more companies are reusing wastes as the enforcement of air, water and waste disposal regulations are limiting sites for and increasing the costs of conventional solid waste disposal methods. Likewise, with escalating energy prices, many companies and communities are exploring municipal trash burning power plants as a viable alternative for reducing solid waste volumes and management costs. Numerous methods and technologies for recovering materials and energy from solid wastes in ecologically acceptable and (sometimes) economically profitable ways have been devised. However, initial capital costs for front-end handling equipment, such as shredders, balers, conveyors, magnetic separators, air classifiers, etc., are high; markets and market prices are variable; waste stream compositions are variable; labor costs can be high; and, compliance with stringent health, safety, and environmental regulations must be achieved. Similar qualifications apply to most energy recovery systems. Nonetheless, resource and energy recovery systems can provide a long-term source of municipal revenue to help reduce solid waste management costs and greatly extend the life of a landfill by reducing annual disposal volumes.

In some communities, the possibility of collecting and using gases produced during decomposition in the landfill cells may also warrant investigation. In landfills, particularly ones whose disposed waste streams contain large amounts of biodegradable materials in deep cells, potentially malodorous and hazardous gases such as carbon monoxide, hydrogen sulfide and methane are generated. Regulations may require the monitoring, collection, and venting of gas pressure build-ups from landfill sites for public safety reasons. If such gases can feasibly be collected from landfill cells, some community revenues may be realized. Some companies (notably in California) have removed $CO2$, $H2S$ and other impurities from landfill gases and have sold the remaining combustible hydrocarbon gases (namely methane) to industrial and residential customers. For example, the Central Contra Costa Sanitary District in Contra Costa County, California uses landfill methane gas (LFG) as a major source of energy for operating its wastewater treatment plant.

Examples of other (and less substantial) facility modification measures which can be used to reduce impacts and improve facility performance include:

- creation of buffer zones to reduce impacts to adjacent parcels
- reduction of facility size to protect sensitive adjacent sites
- reorientation and layout of site plans to achieve compatibility with the existing land uses
- incorporation of earthen berms, vegetation fences, and other landscaping measures to improve the aesthetic attractiveness of the facility during its operational life
- construction of alternative access routes for public safety and minimal traffic impact on the community.

Operational/Management Measures

Through its police powers (e.g., zoning and pollution control laws) local governments usually have the power to approve or reject a proposed solid waste management facility and/or to place operating conditions on the facility. Compliance with standard construction and operationl regulations and procedures established primarily by State agencies must be achieved. Such regulations pertain to cover materials, cell size, rodent and bird control, fire and emergency equipment, gas and leachate monitoring requirements, closure and sometimes post closure maintenance. Local governments also may impose more stringent environmental standards and additional operational and management procedures designed to reduce adverse impacts and provide secondary level back-up systems to prevent damage from unanticipated malfunctions or accidents.

Governmental agencies may assume total responsibility for the siting, construction, operation and maintenance of the facility. Alternatively, because of philosophical, operational, economic or political reasons, local governments may encourage the private sector to build, maintain and operate the components of the community's solid waste system. While there are several management and economic pros and cons of private versus public sector construction and operation of disposal facilities, public trust in and the reliability of the facility operator must also be given major consideration. In some communities, independent citizen boards have been estalished to oversee and/or to review the community's solid waste management program, or the oepration of a specific facility.

Examples of other types of operational/management measures which can be applied (or are in fact required) at solid waste facilities to reduce impacts and improve facility performance and public acceptance include:

- limitation of operating hours to reduce noise and traffic impacts
- noise attenuating devices and other truck safety features (e.g., covers) to limit noise and other impacts and to improve public safety
- revegetation of disturbed areas as quickly as possible and use of sprinkling systems to control dust and to improve site aesthetics
- security systems to prevent unauthorized dumping and insure public safety
- groundwater monitoring systems to detect leachates and to protect underground aquifers
- surface water runoff collection systems to protect water supplies
- periodic inspection programs by independent consultants (in addition to required governmental inspections) to measure facility compliance
- special training programs for municipal fire departments and other community service organizations.

Compensation Measures

Compensation is the payment made to a community or individuals for direct risks and costs borne by them for the benefit of others for indirect or intangible impacts on a community's character or amenities by a disposal site which redesign or mitigation measures cannot offset. Conversely, in some cases, a community may have to offer private firms "incentives" and "guarantees" to establish and operate a waste disposal facility within their jurisdiction. Compensation may be a one-time, lump sum payment or periodic payments in many forms: e.g., monetary, new public facilities, land exchange or dedication, or bond/insurance guarantees. Compensation measures are established through negotiations involving all parties having a stake in the outcome of a particular issue but who do not initially agree on how the issue should be resolved.

Landfills represent potential long-term risks, and public officials today generally require some method of compensation as insurance to cover the expensive costs of remedial actions which may be required in the event of an accident, either during the operation of the facility, or after closure of the facility when the site may possibly be in different ownership. Such compensation may take the form of surety bonds, liability and environmental impairment insurance packages, performance bonds, or a combination of these and other mechanisms.

Similarly, a major negotiable item is the use and ownership of the completed landfill site. In the past, landfill operators have reclaimed waste lands for more "productive" uses, most commonly park and recreational areas, but also industrial parks, transportation and utility corridors and even (rarely) residential areas or agricultural fields. The short-term and longer-term envisioned use of a completed landfill site may necessitate special engineering measures during the operational life of the landfill. Examples of such engineering measures which can be included as part of the negotiated package include gas ventilation systems, depth of cover materials, and placement of cells to form a hill in relatively flat areas for aesthetic as well as recreational purposes (i.e., Mt. Trashmore near Virginia Beach, Virginia).

Other examples of compensation measures commonly negotiated include:

- agreement to stop operations if specified performance limitations are exceeded or standards are not achieved
- agreement to clean up and rehabilitate existing abandoned dump sites in the locality
- Purchase of local bonds at interest rates more favorable to the locality than prevailing market rates
- payment to neighboring land owners to compensate for diminished property values or land value guarantees (conditional payments)

- payments to the local government's general fund as gross receipt taxes on facility revenues, in lieu of tax payments or "tipping fees"
- payment for community costs of services in support of the facility such as road maintenance or sponsor improvement of these public services
- provision of an alternate water supply system to avoid groundwater withdrawals in the area
- "goodwill" dedication of land for public facilities such as recreational areas, emergency facilities, etc.
- free or reduced disposal costs for community residents
- provision for planning grants or other grants for independent consultant validation assessments.

Again, such compensation measures are designed to provide some level of equity between those who benefit and those who suffer by locating a solid waste management facility in a certain location, and therefore, are unique to each individual site proposal. None of the above devices are directly related to the public risks posed by the facility, but represent creative ways of addressing real, if intangible, problems which arise when controversial facilities are placed into an existing community setting.

SUMMARY

As local community officials and citizens begin to undertake and evaluate a proposed solid waste management and facility siting program, the need quickly arises for information to understand the complexities of solid waste management programs (e.g., siting methodologies, regulations, technologies, etc.) so that informed public decisions can be made regarding specific questions about procedures, risks, impacts, costs and other aspects of the proposed program. Accordingly, the generic overview of nontechnical socioeconomic and political/regulatory aspects of solid waste management programs presented in this chapter can assist public officials and citizens at the outside in formulating an effective and workable approach for evaluating and solving their waste management concerns.

REFERENCES

1. Burch, Charles G., "There's Big Business in All That Garbage" Fortune, April 7, 1980.
2. Columbus Dispatch (various articles on proposed Jackson Township Landfill, Dec. 1980-present).
3. Erie County Solid Waste Action Plan Steering Committee, Erie County Solid Waste Management Implementation Plan, Final Report, Buffalo, New York, Dec. 1980.

4. Hagerty, D. Joseph, Joseph L. Pavoni, and John E. Heer, Jr. Solid Waste Management, Van Nostrand Reinhold Company, New York, 1973.
5. Illinois Environmental Protection Agency Sanitary Landfill Management, 1979.
6. Janis, James R., "The Public and Superfund", EPA Journal, June 1981.
7. New England Regional Commission (Handbook Series on Siting Acceptable Hazardous Waste Facilities in New England), Boston, Massachusetts, November 1980.
Negotiating to Protect Your Interests
An Introduction to Facilities for Hazardous Waste Management
Criteria for Evaluating Sites for Hazardous Waste Management
A Decision Guide for Siting Acceptable Hazardous Waste Management Facilities
8. New York State Department of Environmental Conservation, 6NYCRR Port 360: Solid Waste Management Facilities, Albany, New York, Jan. 1981.
9. New York State Department of Environmental Conservation, Solid Waste Management Facility Content Guidelines for Plans and Specifications, Albany, New York, August 1977.
10. Ohio Environmental Council A Citizens Guide to Hazardous Waste Management in Ohio Columbus, Ohio, May 1982.
11. Popper, Frank J., "Siting LULUS (Locally Unwanted Land Uses)", Planning, April 1981.
12. "Resource Recovery", Environmental Comment, March 1981.
13. McDonald H. Steve and Clark L. Weddle. "Landfill Methane Gas Utilization at a Wastewater Treatment Plant--A Case History". Proceedings of the 55th Annual Conference of the Water Pollution Control Federation, Washington, D.C., October 1982.
14. U.S. Environmental Protection Agency, Washington, D.C.
Socioeconomic Analysis of Hazardous Waste Management Alternatives: Methodology and Demonstration (EPA-600/5-81-001) July 1981.
Siting of Hazardous Waste Management Facilities and Public Opposition (SW-809), Nov. 1979
Solid Waste Guide: Environmental Information Sources For State and Local Elected Officials, Feb. 1977.
Financial Methods for Solid Waste Facilities (SW-766), 1974.
Assessment of the Impact of Resource Recovery on the Environment (EPA-700/8-79-001), Aug. 1979.
Decision-Maker's Guide in Solid Waste Management SW-127, 1974.
15. Williams, Edward A. and Alison K. Massa. Siting of Major Facilities: A Practical Approach. New York, 1983

SECTION II

IMPACT ON GROUNDWATER

CHAPTER 8

GROUNDWATER CONTAMINATION CONTROL AND TREATMENT, ROCKY MOUNTAIN ARSENAL COLORADO

Paul MacRoberts,
C.B. Hagar, and
Harry L. Callahan
Black & Veatch, Engineers-Architects

PROJECT DESCRIPTION

General

Contaminated groundwater at Rocky Mountain Arsenal, Figures 1 and 2, will be contained, removed from two aquifers, treated, and returned to an aluvial aquifer by this project. The Black & Veatch Industrial/Special Projects and Civil-Environmental Divisions, with geotechnical consultant services from Earth Sciences Associates (ESA), Palo Alto, California, and Ft. Collins, Colorado, designed this innovative system for the U.S. Army Corps of Engineers, Omaha District. Major components of the system are: (1) 54 dewatering wells valved and manifolded to selectively intercept and permit separate treatment of three identified zones of contamination; (2) a 6,740-foot length of ground water barrier keyed into bedrock; (3) granular activated carbon filters for organic contaminant removal (designed by others); (4) activated alumina columns for fluoride removal; (5) 38 ground water recharge wells downgradient of the barrier to reinject treated water into an alluvial aquifer; and (6) an arrangement of monitoring wells, located on Arsenal property, designed to provide water quality and ground water level data to permit optimization of system effectiveness.

INTRODUCTION

Ground water contamination has apparently been occurring at the Rocky Mountain Arsenal (RMA) since its inception as a facility for producing chemical and incendiary munitions in 1942. In 1946, a major segment of the centrally located manufacturing facility at RMA (Figure 2) was leased to industry. Liquid wastes generated by the industries were discharged into several unlined holding lagoons. In 1954, several farmers north (downgradient) of the

Management of Toxic and Hazardous Wastes, H. Bhatt, R. Sykes, T. Sweeney (Editors)

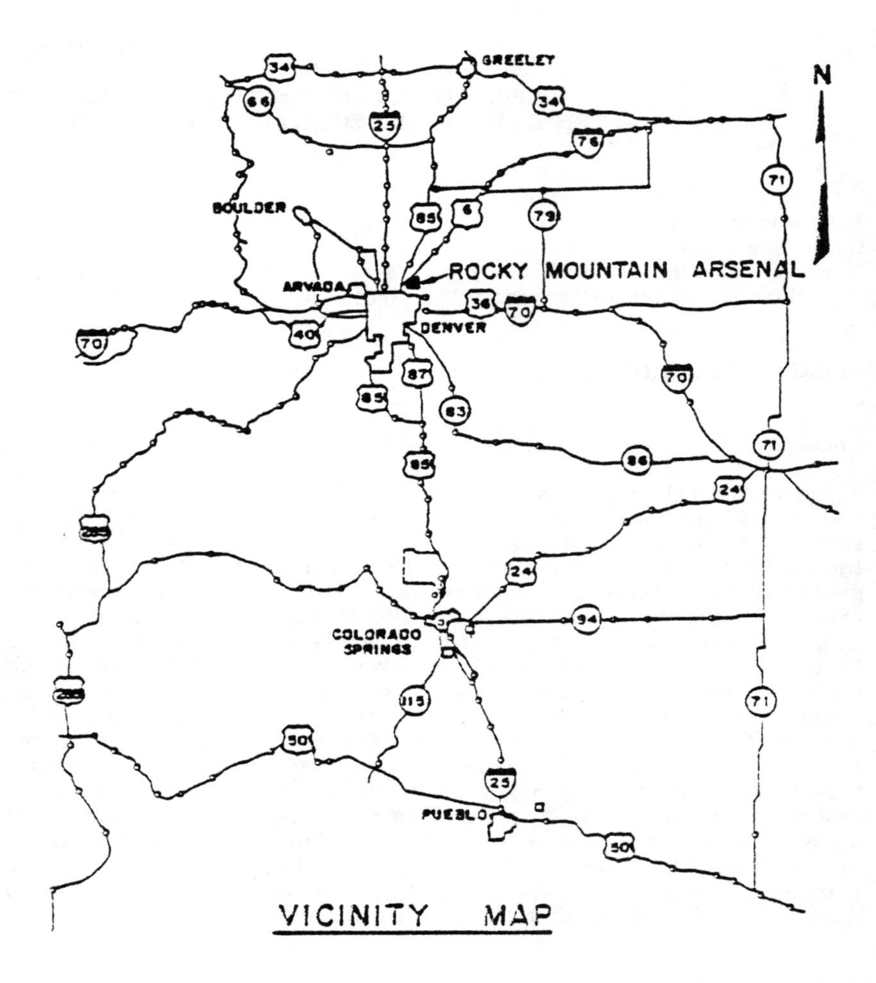
GREELEY
N
BOULDER
ROCKY MOUNTAIN ARSENAL
ARVADA
DENVER
COLORADO
SPRINGS
PUEBLO
VICINITY MAP

FIGURE NO. 1

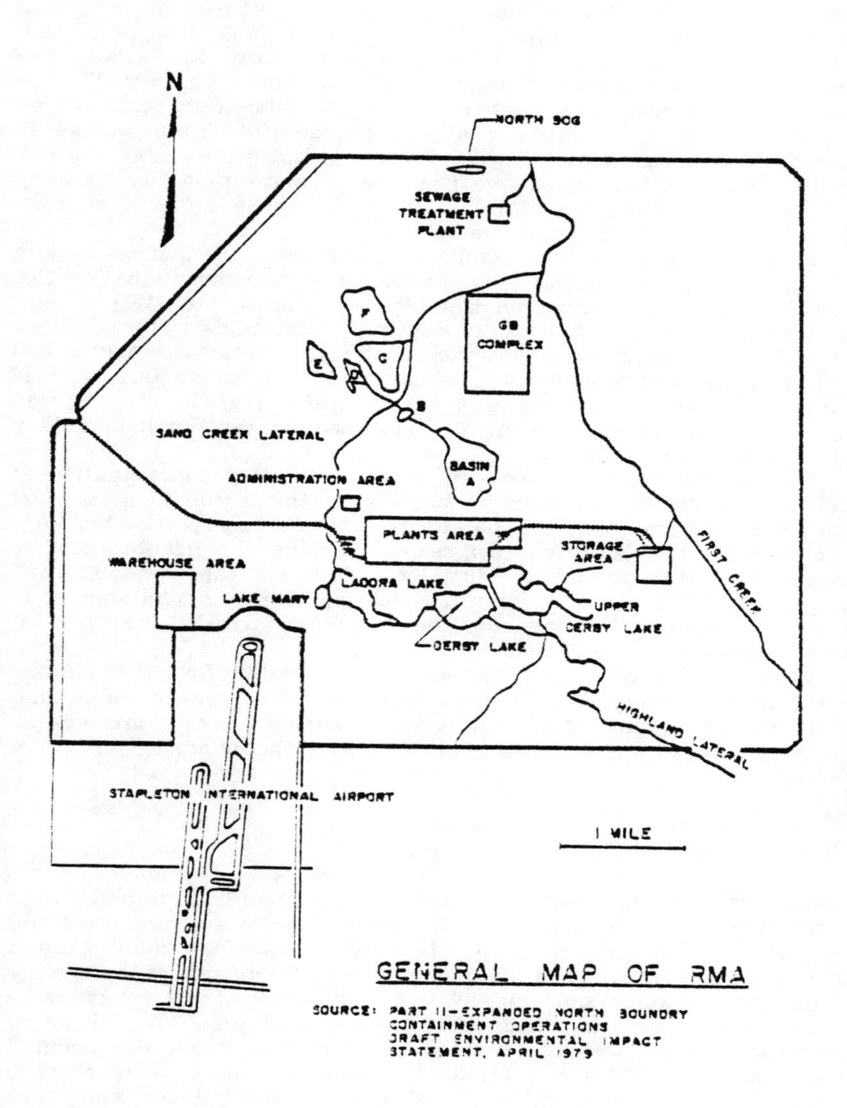
N
NORTH BOG
SEWAGE TREATMENT PLANT
F
E
C
D
GB COMPLEX
B
SAND CREEK LATERAL
BASIN A
ADMINISTRATION AREA
PLANTS AREA
STORAGE AREA
FIRST CREEK
WAREHOUSE AREA
LADORA LAKE
LAKE MARY
UPPER DERBY LAKE
DERBY LAKE
HIGHLAND LATERAL
STAPLETON INTERNATIONAL AIRPORT
1 MILE
GENERAL MAP OF RMA
SOURCE: PART II—EXPANDED NORTH BOUNDRY
CONTAINMENT OPERATIONS
DRAFT ENVIRONMENTAL IMPACT
STATEMENT, APRIL 1979

FIGURE NO. 2

Arsenal, using shallow alluvial aquifers as a source of irrigation water, complained about the loss of crops. It was thought at that time that seepage from unlined waste holding ponds was the principal source of ground water contamination; so in 1957, the government designed and constructed an asphaltic-lined waste lagoon (Basin "F") to hold liquid wastes. Storage of liquid wastes in unlined lagoons was discontinued with the completion of the Basin F facility, and ponding areas previously used were breached following the transfer of their contents to Basin F. These areas were then allowed to revert to natural conditions.

In 1974, water quality samples were taken at a man-made bog area located in the proximity of RMA's north boundary; the samples proved that diisopropyl-methylphosphonate (DIMP) and dicyclopentadiene (DCPD) were present in the bog's water. These particular chemicals are traceable to the production of a chemical warfare agent and insecticides, respectively. Because the bog was known to respond to changes in ground water levels, it was concluded that these contaminants had reached the north boundary via a ground water route.

In response to the discovery of ground water contamination at Rocky Mountain Arsenal's north boundary, the Arsenal, in concert with the U.S. Army Toxic and Hazardous Materials Agency (USATHAMA), developed and implemented programs to define the problem and to develop corrective measures. In 1978, Black & Veatch was engaged by the Omaha District of the U.S. Army Corps of Engineers to provide engineering services for the design of remedial measures at three locations on the Arsenal.

This paper discusses Black & Veatch's design for an expanded containment barrier, additional dewatering and recharge wells, and a treatment plant for removing fluoride from intercepted contaminated ground water migrating across the north boundary of the Arsenal.

DESIGN DESCRIPTION

For the concept design, five alternative control methods were investigated. These consisted of four hydraulic systems involving the extension of a previously installed 1,500-foot pilot ground water barrier system by a hydraulic barrier comprised of dewatering and recharge wells and one system comprised of an impermeable barrier with dewatering and recharge wells (Figure 3). Although hydraulic systems were determined to be functional, it was decided to use a bentonite-soil barrier because it would offer a more positive cutoff in the unlikely event of power failure; and more importantly, the bentonite-soil barrier would essentially eliminate repumping of already treated ground water.

Barrier Design

Prior to design, Black Veatch and ESA conducted a ground water hydraulic analysis, utilizing a finite difference model developed by ESA. Flow estimates made by ESA were compared to estimates made by various other investigators and found to be in general agreement. The areal distribution of the contaminated ground water flow migrating off the Arsenal's north boundary was determined by the model and further assessed by mass flux analysis of avilable water quality data. Through these analyses, the lateral extent and general alinement of the barrier were determined. An extensive geotechnical program was also conducted during this project to obtain information for designing the barrier depth.

This program was developed to comply with engineering instructions provided by the Corps of Engineers. THe implemented program procuced a complete geophysical analysis of 30 boreholes along the centerline of the proposed ground water barrier and included two electric logs (spontaneous potential and resistivity) and three radiation logs (natural gamma, neutron and gamma-gamma, and a caliper log). Continuous core samples were recovered and preserved for use by the successful bidder; standard penetration tests and material gradation analysis were also performed. Shear strengths were obtained from selective unconfined compressive strength tests. Boring logs and ground water level observations collected at other locations near the proposed system were also obtained and used during the design of the containment barrier system.

The geohydraulic and geotechnical analyses indicate that a 30-inch barrier width of select, low permeability (10^{-7} cm/sec or less) backfill is adequate to impede ground water flow inder the anticipated gradients imposed by the operation of dewatering and recharge wells.

The lateral limits of the containment barrier were defined from data obtained from both field and modeling studies. The depth of the barrier was ascertained from geotechnical data and pump test results. For the most part, the barrier extends through the alluvial aquifer and identified segments of the weathered claystone bedrock units, Denver Sands, when they were determined to be in contact with the alluvium aquifer. Isolated sand units identified as being contaminated are intercepted by deep wells.

The actual constructed width of the installed barrier was dependent on the construction method (equipment used) but was specified to be a minimum of 30 inches (Figure 4). Based on the design width, flows through the barrier will be less than 2 percent of the total estimated contamination flow crossing the Arsenal's north boundary even if subjected to an hydraulic gradient of unity (1.0 ft/ft). Therefore, under an arbitrarily selected severe flow condition, we would expect that less than 10gpm of treated water will be recirculated if a reverse gradient is maintained.

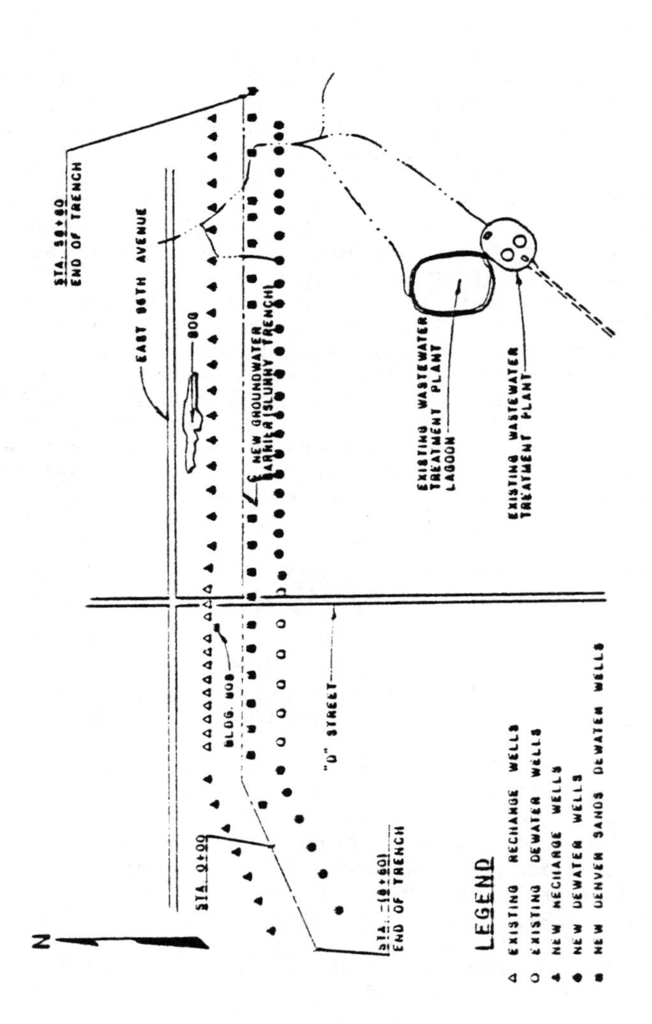
N
END OF TRENCH
EAST 96TH AVENUE
BOG
STA. 0+00
"D" STREET
END OF TRENCH
NEW GROUNDWATER BARRIER (SLURRY TRENCH)
EXISTING WASTEWATER TREATMENT PLANT LAGOON
EXISTING WASTEWATER TREATMENT PLANT
LEGEND
EXISTING RECHARGE WELLS
EXISTING DEWATER WELLS
NEW RECHARGE WELLS
NEW DEWATER WELLS
NEW DENVER SANDS DEWATER WELLS
NORTH BOUNDARY EXPANSION LAYOUT

FIGURE NO. 3

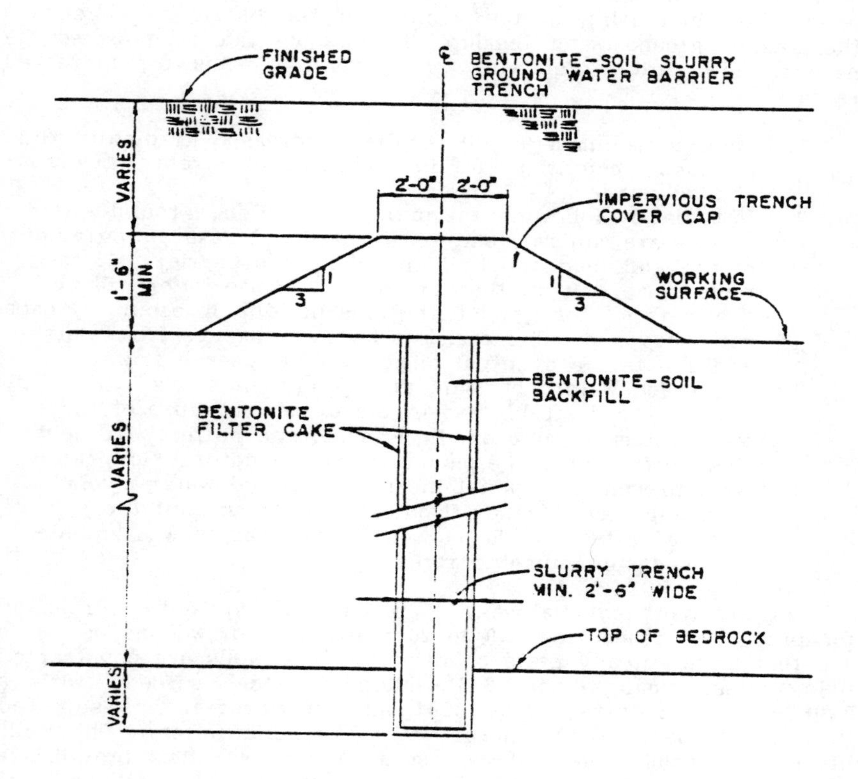

TYPICAL BENTONITE-SOIL SLURRY GROUND WATER BARRIER TRENCH SECTION

FIGURE NO. 4

Well Design

A computer analysis of the proposed system determined that ground water flow through the alluvial aquifer at the north boundary is about 440 gpm. Once the barrier is in place, this flow must be pumped from the aquifer in order to keep ground water levels from increasing in the vicinity of the barrier. To ensure there is no ground water buildup, the pumping rate must exceed the natural flow rate. Following are several reasons for increased rates:

1. Long-term pumpage will result in increased gradients, thus producing conditions for increased ground water velocities and volumes.
2. Immediately following the start of pumping, ground water effects are minimal, and some water will pass downgradient of the well field and build up along the barrier. If the wells pump only at the natural rate, some water will be drawn from storage, allowing the buildup to occur. Pumpage rates have been increased to handle pumpage from storage in addition to the eventual pumpage of bypassed flows.
3. Decreased water levels on the upgradient side of the barrier are desirable, so in the event of failure of the well system, a zone exists as a storage buffer. To achieve this buffer, pumpage must exceed the natural flow rate.
4. As a precaution against increased ground water levels resulting from surface flooding, and to account for localized effects, a 50 percent safety margin was included in the design pumping rates.

Based on these analyses, 35 wells are required. Projected pumping rates range from 1.0 to 26.2 gpm with drawdowns of 1.4 to 4.8 feet under steady state conditions. The wells are constructed using steel casings and 316L stainless steel screens with a 0.060-inch slot size; the 316L screen material was selected because of the corrosive nature of the ground water. The wells have been installed in a 16-inch minimum borehole, thus providing a minimum 4-1/2 inch gravel pack around the 6-inch diameter well. Screen lengths and well depths were designed from conditions observed during the geotechnical field work; however, screen placement and well depths for all wells were based on conditions observed at each well site during actual construction.

Wells designed for the deeper sand units were located based on geotechnical and water quality considerations. A total of 19 wells were designed to pump a total of 31 gpm, which is approximately 10 times the estimated natural flow through these routes. The design was based on two pump tests and on drawdown calculations. The relatively large number of wells used to pump the small volume of water is necessary in order to develop a deep sink for intercepting contaminants during short pumping periods. Because of severe boundaries, interference and drawdown calculations are approximate, and as such, allowances were made for estimating errors.

Since these wells were designed to dewater at low pumping rates, slotted 4-inch diameter PVC casings were selected. Milled slots of 0.064 inch were designed for the screened portion of the well. The wells were installed in 9-inch boreholes providing a 2.25-inch annular space for gravel packing. A 10-inch PVC conductor casing placed in a 16-inch well bore and grouted in place with cement was used to seal the well from the overlying alluvial aquifer.

The sands dewatering system was designed to operate intermittently on an as-needed basis and to allow monitoring of contaminant levels.

Thirty-eight recharge wells were required to ensure adequate recharge capacity without causing significant alterations in the natural ground water patterns. Using the computer model analysis, flows at the treatment plant were apportioned among these wells with the objective of restoring ground water to a natural condition without causing surface flooding. Recharge rates range from 0.4 to 37.0 gpm with an increase in head from 0.5 to 2.3 feet.

All recharge wells have been constructed in the alluvial aquifer and consist of 12-inch steel casing and 304 stainless steel screens with 0.060-inch slots. Wells are placed in 24-inch boreholes, thus producing a large effective radius.

Manifold System

Water quality flux analyses have indicated rather definitive plumes of contaminants in the vicinity of the barrier. Under the instructions of the user agency, and with approval of the Corps of Engineers, Black & Veatch designed a three-manifold collection system. The physical extent of each collection line, i.e. manifold, is determined by a series of 18 valves positioned along the main collection line. These valves are used to either increase or decrease the length of any particular manifold, thus effectively providing the user with the flexibility to intercept and treat ground water of similar quality. Denver Sand wells are connected to a single collection pipe and fed into one manifold. Each manifold is connected to a specific influent wetwell, which in turn is connected to a specific treatment process stream (Figure 5).

Ground Water Treatment

During the design phase of this project, ground water treatment at the north boundary was being accomplished by an activated carbon system leased from Calgon, Inc. Hydrocarbons found in the ground water include DIMP (diisopropylmethylphosphonate) and DCPD (dicyclopentadiene). Since the inception of the ongoing abatement

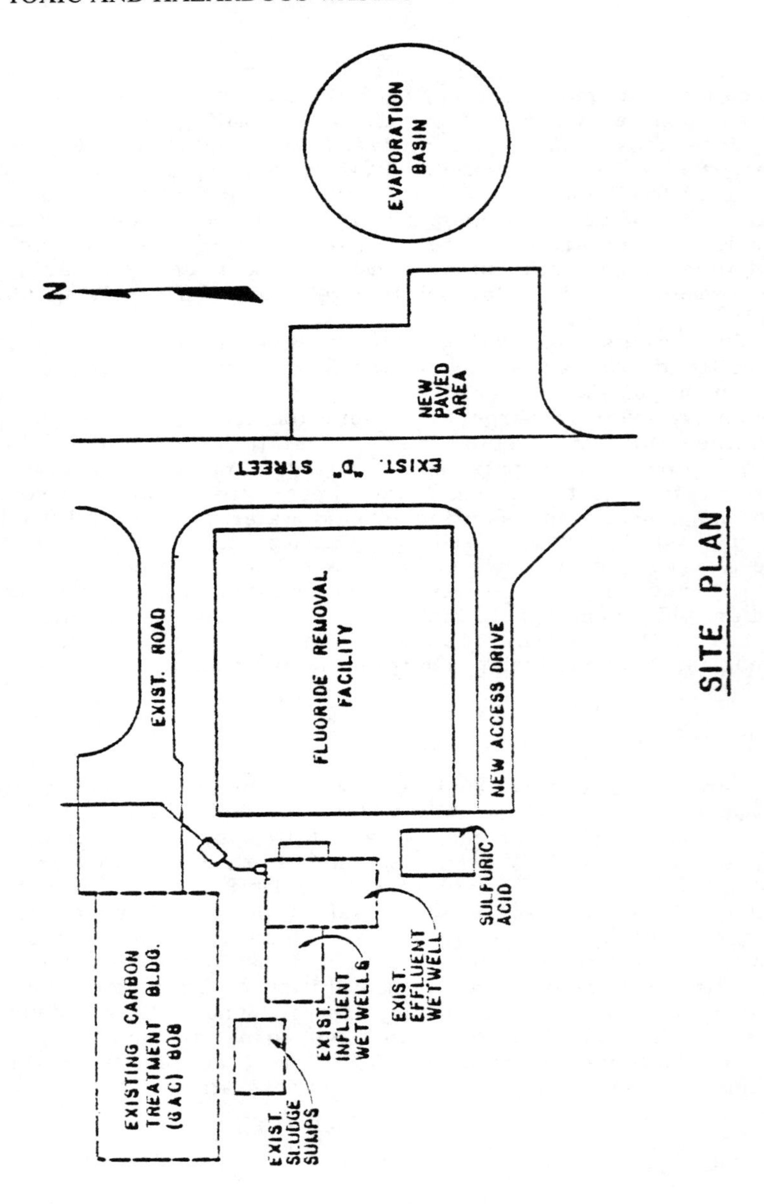
EVAPORATION BASIN
N
NEW PAVED AREA
EXIST. "D" STREET
EXIST. ROAD
FLUORIDE REMOVAL FACILITY
NEW ACCESS DRIVE
SITE PLAN
SULFURIC ACID
EXIST. EFFLUENT WETWELL
EXIST. INFLUENT WETWELLS
EXISTING CARBON TREATMENT BLDG. (GAC) 808
EXIST. SLUDGE SUMPS

FIGURE NO. 5

programs at RMA, other contaminants have been detected in the ground water in the vicinity of the north boundary. These organic contaminants are effectively being removed from the ground water by the activated carbon pilot plant system.

Because the State of Colorado requires that reinjected water be subject to drinking water standards, including a fluoride criteria of 2.4 mg/l, fluoride became a design requirement for the north boundary project. The following discusses the fluoride removal system design.

Several alternatives were evaluated for fluoride removal, including ion exchange, lime softening, bone char adsorption, and activated alumina adsorption. While ion exchange of fluoride is possible, fluoride is last on the exchange selectivity series for monovalent ions. Thus, the removal of fluoride with ion exchange is not feasible because of low media selectivity for fluoride. Excess lime softening coprecipitates calcium fluoride with magnesium hydroxide. However, the combination of process equipment requirements, space requirements, chemical use, and waste sludge handling problems eliminated softening as a practical solution for fluoride removal. Bone char will adsorb fluoride, but repeated regeneration will cause the bone char to permanently lose its fluoride adsorption capacity. Therefore, bone char was also eliminated from consideration. Activated alumina adsorbs fluoride and retains its adsorption capacity through repeated regeneration. On the basis of fluoride removal efficiency and process reliability the activated alumina adsorption process was selected.

The following criteria were used in the design of the activated alumina columns:

Hydraulic loading	
Normal	5 gpm/ft^2
Maximum	7 gpm/ft^2
Superficial velocity	
Normal	0.67 ft/min
Maximum	0.95 ft/min
Minimum empty bed contact time	5.3 min
Adsorption capacity	104gm F/cu ft media

The flouride level in the manifolded stream to be treated is 3.7 mg/l. The one manifold stream that will be treated to remove fluoride is expected to average about 110 gpm; however, the flow rate of this stream could go as high as 150 gpm. To provide adequate treatment facilities in case of high fluoride water in more than one manifold stream, provisions were included in the treatment plant to treat a maximum of 450 gpm. As will be discussed later, these provisions included both physical facilities and changes in operating procedures.

The regeneration waste treatment system recirculates several effluent streams to the activated alumina column system. These waste streams increase the influent fluoride level to 4.8 mg/l when the waste regenerant is being processed. These return streams total 10gpm. Both the increased fluoride level and the flow were taken into account in the process design.

Based on the column design criteria, the adsorption capacity of the activated alumina, the fluoride levels, and flow, the columns were designed as follows:

Number of columns	3
Diameter	6.5 ft
Bed depth	5 ft
Sidewall depth	10ft

The sidewall depth is provided to allow expansion of the bed during backwashing. The tanks are constructed of steel to meet the ASME design code for 125 psig. (Normal operating pressure is 25 to 30 psig.) The tanks are rubber lined. All internal piping is PVC, including the influent distributor and effluent collection piping.

Flow through the fluoride removal facility is relatively simple. Effluent from the granular activated carbon (GAC) plant flows to an influent sump in the fluoride treatment building and, depending on the flow rate, is pumped to one or more of the activated alumina columns. Prior to reaching the columns the flow is automatically adjusted to an optimum adsorption capacity pH level of 5.5 with sulfuric acid. After passing downflow through the alumina, the water flows to an effluent sump at the GAC building to be recombined with other treated manifolded streams prior to reinjection. Provisions have been made to adjust the pH of the combined recharge stream upward using sodium hydroxide if it becomes necessary.

The fluoride treatment system utilizes two parallel columns operating in a "staggered-exhaustion" mode to treat the entire flow up to a maximum of 160 gpm (150 gpm influent plus 10 gpm recycle). A portion of the GAC effluent is slipstreamed and mixed with the effluent of two columns operating in parallel when the flow is between 160 and 460 gpm. The third column is required so that it can be placed in service while one of the previously used columns is being regenerated. The treated effluent fluoride level varies from 0 to 1.6 mg/l when 150 gpm or less is being treated; for flows above 150 gpm the effluent fluoride level will be fixed at 2.2 mg/l using the variable slipstream to maintain a constant effluent fluoride concentration. The value of 2.2 mg/l was selected to provide a cushion below the limit of 2.4 mg/l.

Column regeneration is an eight-step process. The first cycle is backwashing the bed: the accumulated suspended solids are removed from the columns and the media is expanded, rearranging the orientation of the alumina particles to minimize channeling. The remaining seven cycles involve removing the fluoride from the bed and preparing the bed for further service.

Regeneration is accomplished in Cycle 2 by treating in an upflow direction the exhausted bed with a 1 percent solution of sodium hydroxide. The bed is then rinsed with water to flush out the fluoride in Cycle 3. It should be noted that all water used at high pH, either for diluting sodium hydroxide or rinsing it from the bed, is softened water. This is done to prevent fouling the media through precipitation of calcium and magnesium. Cycle 4 consists of simply draining the column to reduce the dilution of the next sodium hydroxide treatment and Cycles 5,6, and 7 are identical to Cycles 2,3, and 4, respectively, except the flow is down rather than up. Cycle 8 is used to adjust the pH of the bed prior to returning it to service. During the rinse process, the pH of the rinse water is adjusted to 2.5 with sulfuric acid and maintained at this level until the column effluent pH has been reduced to 6.5. The rinse water is then adjusted to a pH of 5.5 for the remainder of the cycle.

The fluoride removal process waste comes from two sources: regeneration of the alumina columns and regeneration of the water softener. Wastes from the alumina columns include sodium hydroxide and sulfuric acid used in the regeneration of the alumina media. These wastes have a high fluoride content (100 to 125 mg/l). The softener regeneration waste contains high levels of sodium and calcium chloride.

Several alternatives were considered for disposal of wastes from the fluoride removal process. Water losses generally associated with waste disposal practices were unacceptable because the using agency is required to return essentially all of the water removed from the aquifer. Thus, the alternative selected includes chemical precipitation of the fluoride using a mechanical sludge dewatering system. This system and its operation are described below and are shown on Figure 6.

A lined concrete precipitation of the regeneration waste. The basin has treatment capacity for two complete regeneration waste volumes. A 30-foot diameter basin with 14.5-foot sidewater depth and a 2-foot freeboard allowance provides the necessary volume. The basin is lined with fiberglass reinforced epoxy and is baffled to facilitate complete mixing and to prevent vortexing. basin equipment includes an influent distribution weir, two 75 horsepower bridge-mounted rapid mixers, and a rotating sludge collection system for removal of the precipitated solids. A decant mechanism consisting of a flotation supported flexible hose is included to provide an adjustable drawoff point for recycling reactor supernatant. Chemical addition points are located above each rapid mixer to facilitate complete dispersion of sulfuric acid and calcium chloride throughout the basin. Sludge formed during the

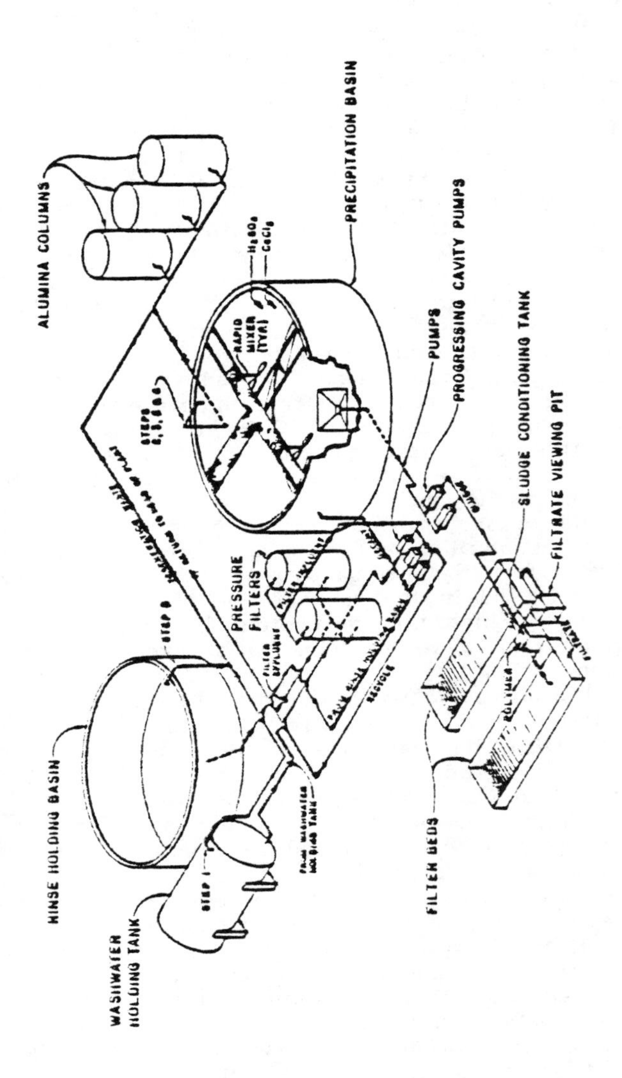
FLUORIDE WASTE TREATMENT SYSTEM
ALUMINA COLUMNS
PRECIPITATION BASIN
RAPID MIXER (TYP.)
PUMPS
PROGRESSING CAVITY PUMPS
SLUDGE CONDITIONING TANK
FILTRATE VIEWING PIT
PRESSURE FILTERS
STEP 8
STEP 1
WASHWATER HOLDING TANK
FILTER BEDS

FIGURE NO. 6

precipitation reaction is collected by the rotating sludge collection system and directed to a central sludge pit for pumping to the gravity dewatering filter bed.

The selected chemical treatment system utilizes calcium chloride addition and pH adjustment to reduce fluoride levels in the regeneration waste. The calcium fluoride precipitate formed during the reaction of calcium chloride with fluoride is a very stable, insoluble compound. Once formed, calcium fluoride does not ionize in water and therefore can be easily and safely disposed. To facilitate maximum fluoride removal, the regeneration waste is adjusted with sulfuric acid to a pH of approximately 6.5 immediately following calcium chloride addition. This step permits additional fluoride removal through aluminum hydroxide precipitation/adsorption.

The waste from the final neutralization of the alumina beds is held in a lined concrete basin prior to recycle to the plant influent wet well. The basin is unequipped except for the recycle pump suction.

Two 5-foot diameter pressure filters are provided to remove suspended solids from (1) alumina bed backwash, (2) dewatered sludge filtrate, and (3) precipitation basin decant flows prior to recycle. One filter is used to filter the recycle flows, while the second filter functions as a standby. The filters are sized for a maximum solids loading of 1 pound suspended solids per square foot of media area. Mixed media (coal/sand/garnet) is specified because of its high surface area per unit volume, resistance to breakthrough, and high solids storage capability. Total media depth is 36 inches, and a surface wash system is utilized to obtain maximum solids removal during backwash.

To permit solids removal from the system, a 70-foot diameter evaporation basin is provided outside the treatment building. Filter backwash could be discharged to the basin once per month. The basin is sized for a net annual evaporation rate of 26 inches and has storage capability for five consecutive years of low net evaporation. Provisions for discharge of filter backwash to the sludge sump at the head of the GAC plant or to the filter backwash evaporation basin are included for operational flexibility.

Gravity sludge dewatering is utilized to reduce the sludge volume prior to disposal. Two 8-foot by 25-foot wedgewire filter beds are employed. Sludge formed during chemical treatment of two complete regeneration waste volumes is applied to one of the filter beds; the second bed serves as a standby. A conditioning tank is included for mixing polymer with the sludge pumped from the precipitation basin. The conditioned sludge then flows by gravity to the filter bed. The polymer system is designed to use either liquid or dry polymer. Provisions are made for discharging filtrate to the pressure filter influent, the sludge sump at the head of the GAC plant, the precipitation basin, the filter backwash evaporation basin, or directly to the fluoride removal system influent wetwell.

When the sludge has reached a consistency that permits it to be handled easily (typically 1 to 2 days), removal of the cake is initiated. Cake removal consists of hand-cleaning and placing into containers. The filled containers are sent to the disposal site, and the screen is hosed down to remove any accumulated solids.

Total fluoride removal system water losses (i.e., losses attributable to sludge removal and filter backwash evaporation) represent approximately 0.1 percent of total system related water production (at 150 gpm plant flow).

PROJECT STATUS

The north boundary expansion project, although contracted as a single engineeriong design, was submitted to the Corps of Engineers in two packages. The first submittal consisted of bid documents covering the dewatering, containment, and recharge facilities including appurtenant structures such as roads, wetwells, retrofitting of existing wells, and expansion of the GAC building to house additional piping.

The fluoride treatment portion of the project has been designed and submitted to the Corps of Engineers. The State of Colorado, recognizing that fluoride removal may not be necessary once the total barrier system becomes operable, has allowed for a delay in construction of this facility. The State has stipulated that once the barrier system becomes operable, close monitoring for fluoride will be performed. In the event fluoride levels are found to be near the drinking water standards of 2.4 mg/l, the decision may be made not to require fluoride removal. This assessment is presently ongoing.

FUTURE WORK

The installation restoration programs at Rocky Mountain Arsenal will continue to require major efforts by the various concerned agencies/commands. The cleanup of contamination at Rocky Mountain Arsenal may take many years to accomplish. However, the right course is being taken by the government in rectifying a problem generated in the past.

CHAPTER 9

STATISTICAL EVALUATION OF HYDRAULIC CONDUCTIVITY DATA FOR WASTE DISPOSAL SITES

Wayne R. Bergstrom and
George R. Kunkle
Neyer, Tiseo & Hindo, Ltd

INTRODUCTION

Hydraulic conductivity, or permeability, is a performance-related characteristic of natural soils that is widely used to evaluate site suitability for the containment of waste materials. Many governmental agencies, ranging from local to state and federal units, require the determination of hydraulic conductivity for evaluating waste disposal sites in natural soils. For this reason, substantial hydraulic conductivity data was obtained during the investigation of eight existing or proposed waste disposal sites in Michigan. As many as 28 conductivity tests were performed on soil samples from a single site. Due to the volume of hydraulic conductivity test data, a statistical approach was judged appropriate for analysis of the results.

As expected, the hydraulic conductivity data from the various sites exhibits a wide variation, which greatly affects the interpretation and use of the findings. Because most regulatory agencies set a maximum value for hydraulic conductivity, there may be a tendency to reject a site based on one or two values that exceed the maximum. However, many variables such as sample disturbance can radically affect a single test result. Thus, rejecting a site on the basis of a few high values may not be warranted.

Another evaluation technique might be to average the data and use the arithmetic mean to select or reject a site. This approach assumes that the arithmetic mean of the test results adequately models the behavior of the site materials while at the same time ignores the variability of the test results.

Neither of the above approaches offers a rational scientific basis for evaluation of the data. The purpose of this paper is to present a statistical approach for evaluation of hydraulic conductivity data and to demonstrate its use. This approach can ultimately be used to:

Management of Toxic and Hazardous Wastes, H. Bhatt, R. Sykes, T. Sweeney (Editors)

1. Reject erratic test data.
2. Provide a rational method to evaluate the test data.
3. Formulate decisions from the test data based on a chosen level of confidence.
4. Provide parameters for use in an expanded risk analysis.
5. Formulate new testing programs for other sites based on a chosen level of confidence.

GEOLOGIC ENVIRONMENT

The hydraulic conductivity data analyzed herein are derived from soil investigations conducted at eight existing or proposed solid waste disposal facilities in southeastern Michigan. The area in which these sites are located is shown in Figure 1. Each of the counties of Oakland, Wayne, Macomb and Monroe contain at least one of the sites.

Southeastern Michigan may be characterized as a glaciated area consisting of glacial moraines (both ground and terminal) and a glacial lake plain. The major areas of moraine and lake plain were defined by Martin [1] and are presented in Figure 1.

In southeastern Michigan both the moraine and lake plain areas are underlain by thick deposits of glacial till. Clayey tills were deposited by the Erie Ice Lobe that advanced into the area from the Lake Erie Basin. To the north, the glacial tills become more sandy, a characteristic that is common to tills deposited by the Saginaw Ice Lobe. This ice lobe advanced into southeastern Michigan from the Saginaw Bay area.

When the glacial ice retreated from southeastern Michigan, those areas below Elevation 800 feet were inundated by meltwater. The resulting proglacial lakes persisted for many thousands of years although the lake level underwent a progressive lowering. During this geologic period, erosion of the till surface created a lake plain characterized by little topographic relief. Lake deposits such as lacustrine clays and beach sands were deposited over the glacial till. Of these deposits, the granular sands are found in greater abundance. Few deposits of lacustrine clay appear to have survived the progressive lowering of lake levels.

DESCRIPTION OF THE TILL DEPOSITS

The glacial deposits investigated for potential waste containment are all identified as overconsolidated glacial tills. These deposits vary from low plasticity clays possessing only a minor fraction of sand and gravel to relatively coarse tills

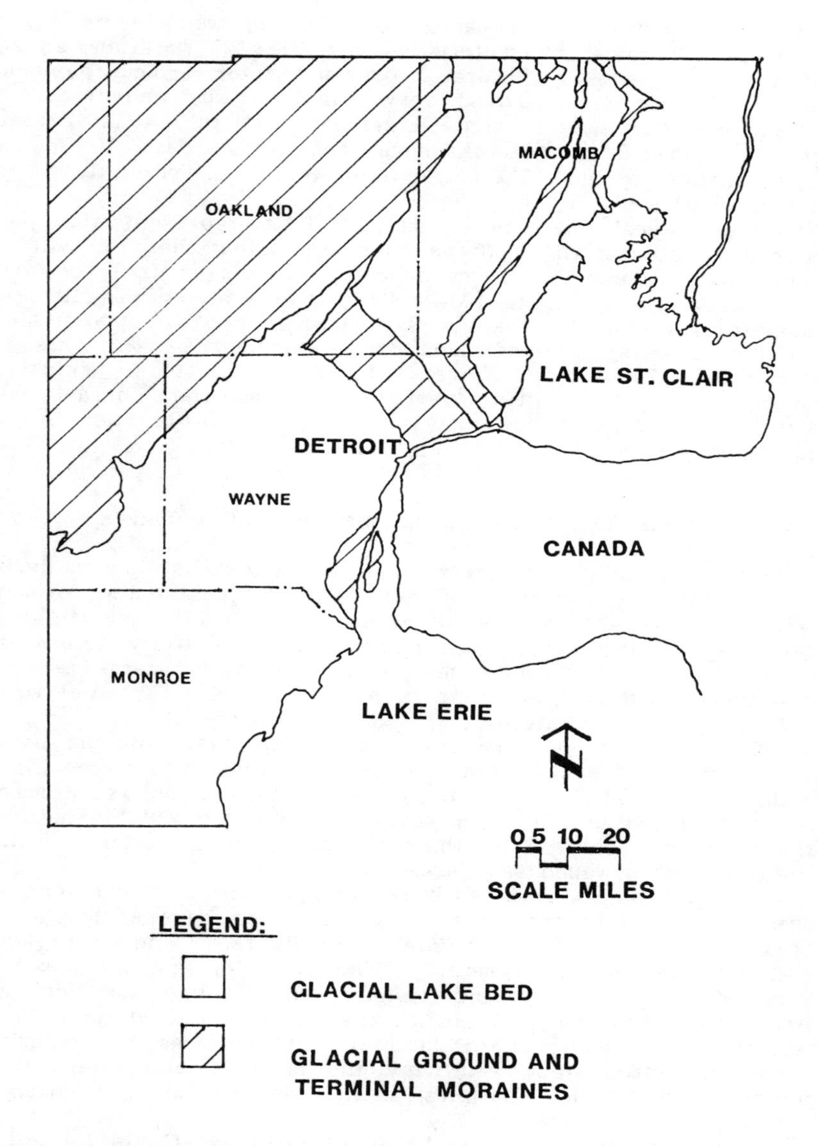

FIGURE 1. Region of study in southeastern Michigan.

possessing primarily fine sand and silt-sized constituents. Table I presents typical characteristics for the till materials at each of the eight sites. Figure 2 presents their typical grain-size distribution curves. These curves indicate the well-graded but poorly sorted nature of the till deposits. Note that the material ranges from CL and CL-ML to SM according to the Unified Soil Classification system [2]. The deposits are almost entirely gray and unweathered.

Mineralogical analyses of samples from representative glacial tills in southeastern Michigan have indicated that the silt and clay-sized component of these deposits consists primarily of quartz and carbonate minerals [3]. From these analyses, illite and kaolinite were found to be the clay minerals in greatest abundance.

The deposits discussed herein range in thickness from less than 15 feet to 70 feet or more. They appear to be relatively homogeneous, with little variation in material characteristics either vertically or laterally. Isolated granular pockets are occasionally encountered.

SOIL SAMPLES AND HYDRAULIC CONDUCTIVITY TESTS

The hydraulic conductivity data presented herein are derived from laboratory tests on intact soil samples using primarily falling head test methods. Details of this test method are presented in [4]. A limited number of conductivity values were back-calculated from one-dimensional consolidation tests [5]. As discussed later, these results have not been included in the analysis presented herein.

The soil samples were, with few exceptions, obtained by one of two methods. A limited number of soil samples were obtained with the use of 3-inch diameter thin-walled samplers, commonly called Shelby tubes [6]. However, the overconsolidated nature of these deposits as well as their gravel inclusions often precluded the use of thin-walled samplers.

Most of the soil samples were obtained with a 2-inch diameter, thick-walled drive sampler. This sampler, possessing a split barrel, was modified to contain brass liners, 3 inches in length and 1.37 inches in diameter. The resulting soil samples were retained in the liners and tested under falling-head conditions prior to extrusion. While this sampling methodology is not considered to result in an "undisturbed" soil sample, experience with overconsolidated, relatively insensitive, low plasticity tills indicates that such samples can yield useful and somewhat conservative test data.

The relative merits of these sampling methods have been extensively studied [6]. Practical considerations, most particularly the difficulty in penetrating these overconsolidated till deposits with thin-walled samplers, have resulted in common and widespread use of thick-walled drive samplers with liners.

TABLE I
Typical Till Characteristics for Eight Sites

Site	Dry Density lbs/ft^3	Moisture Content Percent	Liquid Limit	Plasticity Index	Percentage Finer Than #200 Sieve	Unified Soil Classification
1	113	18	27	11	85	CL
2	126	12	18	5	64	CL-ML
3	136	9	13	3	44	SM
4	128	11	22	9	73	CL
5	128	11	18	6	59	CL-ML
6	127	12	24	8	68	CL
7	115	17	30	14	82	CL
8	130	11	23	8	71	CL

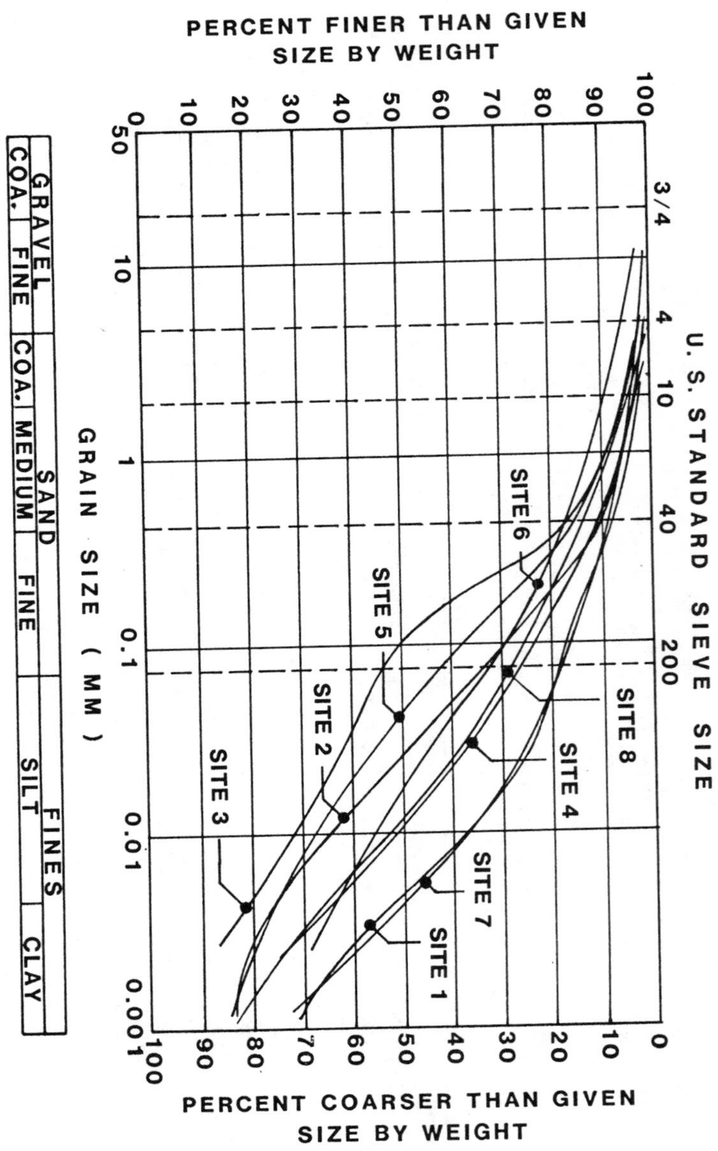

Figure 2. Typical grain size distribution curves.

Other methods of sampling these deposits, which may prove to be more advantageous, are becoming more widely practiced and it is hoped that future use of these methods will result in a body of data which can be compared to the data presented herein.

STATISTICAL CONCEPTS

Prior to presentation of the test data and the resulting statistical analysis, a brief description of several statistical concepts is warranted.

For any number of tests, the mean value resulting from those tests is defined as the sample mean. If all possible soil specimens could be obtained and tested, a population mean could be determined. Naturally, this last value is the objective of a test program but cannot truly be determined in most cases for obvious reasons. Hence, the best we can hope to obtain is a sample mean from which the population mean can be estimated.

Likewise, there is a true population variance (the square of the standard deviation) which cannot generally be obtained. The sample variance, obtained from a limited number of tests, can be used as an estimate of the population variance.

One additional concept concerns the distribution of data about the sample mean. Test data can be plotted as a frequency histogram. The enveloping curve of this histogram can ultimately be modelled by a mathematically defined probability distribution function. Examples will be presented later. The shape of the resulting distribution is useful in data evaluation and analysis. The most commonly reported probability distribution function is the normal distribution which results in a "bellshaped" curve, symmetrical about the mean. Many other probability distribution functions can be fitted to test data with differing degrees of success. The distribution presented and used herein is the lognormal distribution. The data in this case, when converted to their natural logarithms, follows a normal distribution.

REJECTION OF OUTLINERS

The conductivity data, described herein and summarized in Table II, were first inspected for test values which were sufficiently outside the expected range of values to warrant rejection. Such extreme values, called outliers, may have resulted from severe sample disturbance, testing error, or misclassification of the soil type.

A simple statistical tool was used to aid in the evaluation of outliers. The rejection of anomalous data was based on Chauvenet's criterion [7]. Under this criterion, a data point is rejected if it deviates from the sample mean by an amount greater than a

TABLE II
Summary of Hydraulic Conductivity Data for Eight Sites

Site	Total Number of Tests	Outliers Rejected	Falling-Head Data: Number of Tests	Falling-Head Data: Sample Mean Conductivity K_f (cm/sec)	Falling-Head Data: Coef. of Var.	Consolidation Data: Number of Tests	Consolidation Data: Sample Mean Conductivity K_c (cm/sec)	Consolidation Data: Coef. of Var.	K_f/K_c
1	28	1	17	3.8×10^{-8}	47%	10	1.0×10^{-8}	51%	3.8
2	16	0	13	3.7×10^{-8}	38%	3	2.1×10^{-8}	37%	1.8
3	18	1	15	6.9×10^{-8}	67%	2	4.1×10^{-8}	34%	1.7
4	14	0	12	4.6×10^{-8}	62%	2	2.3×10^{-8}	81%	2.0
5	14	0	14	5.4×10^{-8}	106%	0	-	-	-
6	18	0	13	9.2×10^{-8}	116%	5	3.2×10^{-8}	75%	2.9
7	26	0	21	3.1×10^{-8}	126%	5	2.6×10^{-8}	13%	1.2
8	11	2	9	2.0×10^{-8}	39%	0	-	-	-

specified limit. Use of the criterion is predicated upon the assumption of a normal probability distribution function.

As presented later, the hydraulic conductivity test data described herein were not found to follow a normal distribution. However, conversion of the data to natural logarithms yielded a normal distribution. Hence, the hydraulic conductivity data for each of the eight sites were first converted to natural logarithms, and then Chauvenet's criterion was applied. Outliers which significantly violated this criterion were rejected. Others which were very close to the limits defined by this criterion were not rejected unless other corroborating information was available, such as evidence of severe sample disturbance.

It was interesting to note that systematic evaluation of the test data according to this approach sometimes contradicted, and sometimes confirmed, subjective decisions made by individuals who had previously reviewed the data. Clearly, purely subjective evaluation of suspicious data points can be inconsistent. A rational, systematic approach can help to evaluate those spurious test results in a more consistent manner.

COMPARISON OF TEST METHODS

After evaluation of the test data from each site for outliers, the distribution of the results was examined. Figure 3 presents an example of the data frequency distribution for Site 1. All test results for this site are included in the distribution. It illustrates that the conductivity results from the falling-head tests are similar but slightly greater than the results from the consolidation tests. In other words, the conductivity data from the consolidation tests may represent a slightly different population than the data from the falling head tests.

Hence, the data from each site where both test types were performed was evaluated to determine whether the test types represented a different population of conductivity values. It was noted that the mean conductivity values obtained from the falling-head tests were consistently greater than the mean values from the consolidation tests. Table II presents the ratio of these two mean values for each site. Olson and Daniel [8] report such ratios ranging from approximately 2 to 1,000 for highly overconsolidated clays. The mean conductivity values from each of the two test types were statistically compared at each site according to Student's "t" test [7]. The variances obtained from each test type were compared by using the "F" test [7].

These comparison tests indicated that the two test types yielded conductivity values which probably represent slightly different populations on three of the sites, Sites 1,4 and 5. Hence, the falling-head and consolidation test results were not combined for evaluation. All conductivity data indirectly determined from the consolidation tests have been excluded from the

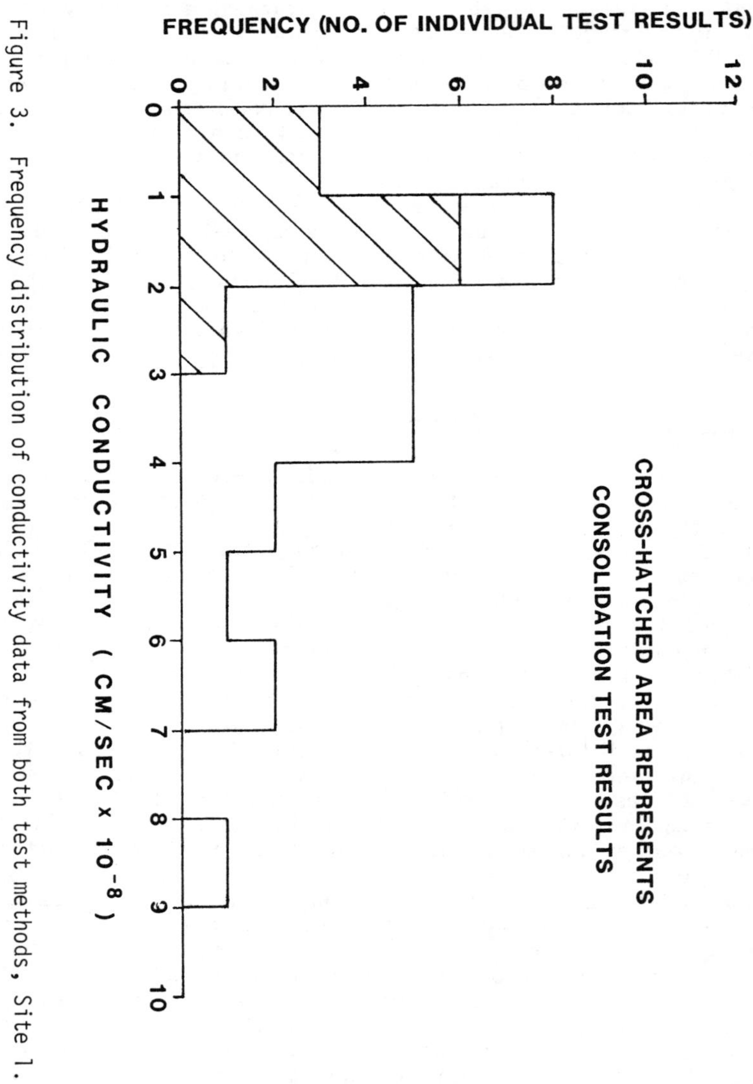

Figure 3. Frequency distribution of conductivity data from both test methods, Site 1.

remainder of the analysis presented herein. Figure 4 presents the data for Site 1 with the consolidation test results removed.

TEST DATA DISTRIBUTIONS

After removing the consolidation test results, the distribution of data from each site was evaluated. The resulting distributions were all skewed positive or right, meaning that the long tail of the data distribution was on the side of greater permeability. This is one attribute of the lognormal probability distribution function. Additionally, the lognormal distribution allows for only positive values of a parameter. Conductivity cannot be negative.

Hence, the lognormal probability distribution function was chosen as a reasonable model for the test data from each site. As examples, the data distribution for three of the eight sites are presented in Figures 5 through 7 with the lognormal probability distribution function plotted as well. The actual mathematical form [9] of this function is beyond the scope of this paper. Note that the curve in each case appears to fit the data rather closely.

A measure of the accuracy with which the lognormal distribution fits the data is provided by the "W" test [10]. This test was chosen because it is used to verify normalcy (after conversion of the conductivity values to their natural logarithms), and it is useful when only a relatively small number of data points are available (less than 30). For all eight sites, the lognormal distribution was shown to be a reasonable model for the conductivity data.

COEFFICIENT OF VARIATION

In attempting to characterize these till deposits, it was found that little could be inferred from the test results concerning the tills as a whole. An attempt to combine the test results from the various sites using the "t" test and "F" test previously mentioned and a more powerful technique known as Bartlett's test [7] was made. These tests indicated that a combination of sites into one or two overall conductivity populations was probably not reasonable.

However, one statistical parameter has been shown to be somewhat applicable when evaluating different geological deposits which are similar in nature and origin to each other. This parameter is the coefficient of variation [9]. It is defined as the ratio of the standard deviation to the mean. Table II presents the coefficients of variation for the conductivity data at each site. While the range in these results may initially appear to be excessive, this parameter is nevertheless useful as a starting point in future investigations. This will be demonstrated in the next section of the paper.

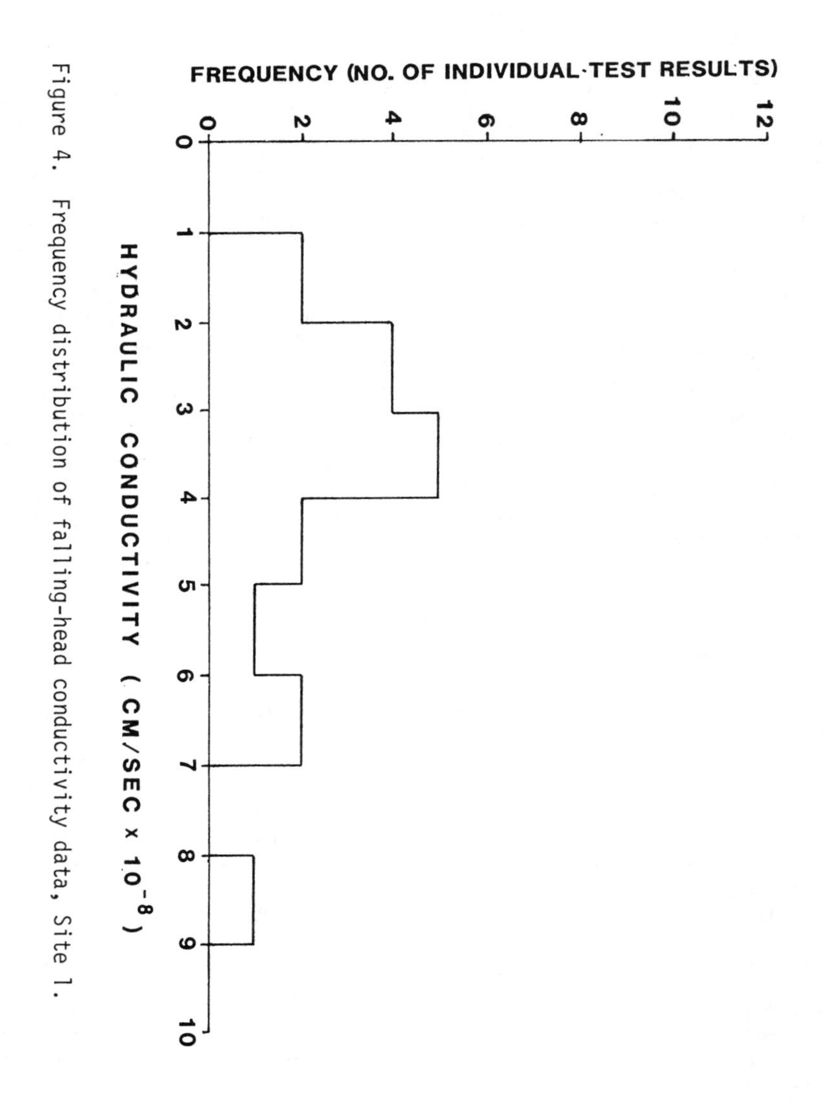

Figure 4. Frequency distribution of falling-head conductivity data, Site 1.

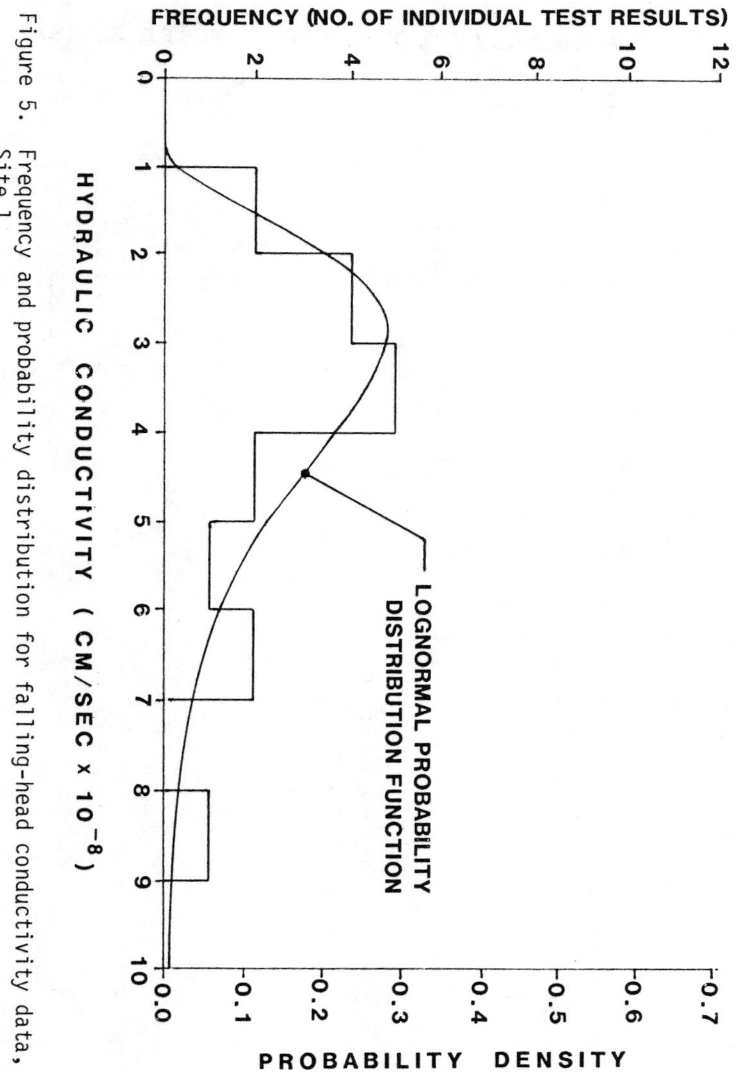

Figure 5. Frequency and probability distribution for falling-head conductivity data, Site 1.

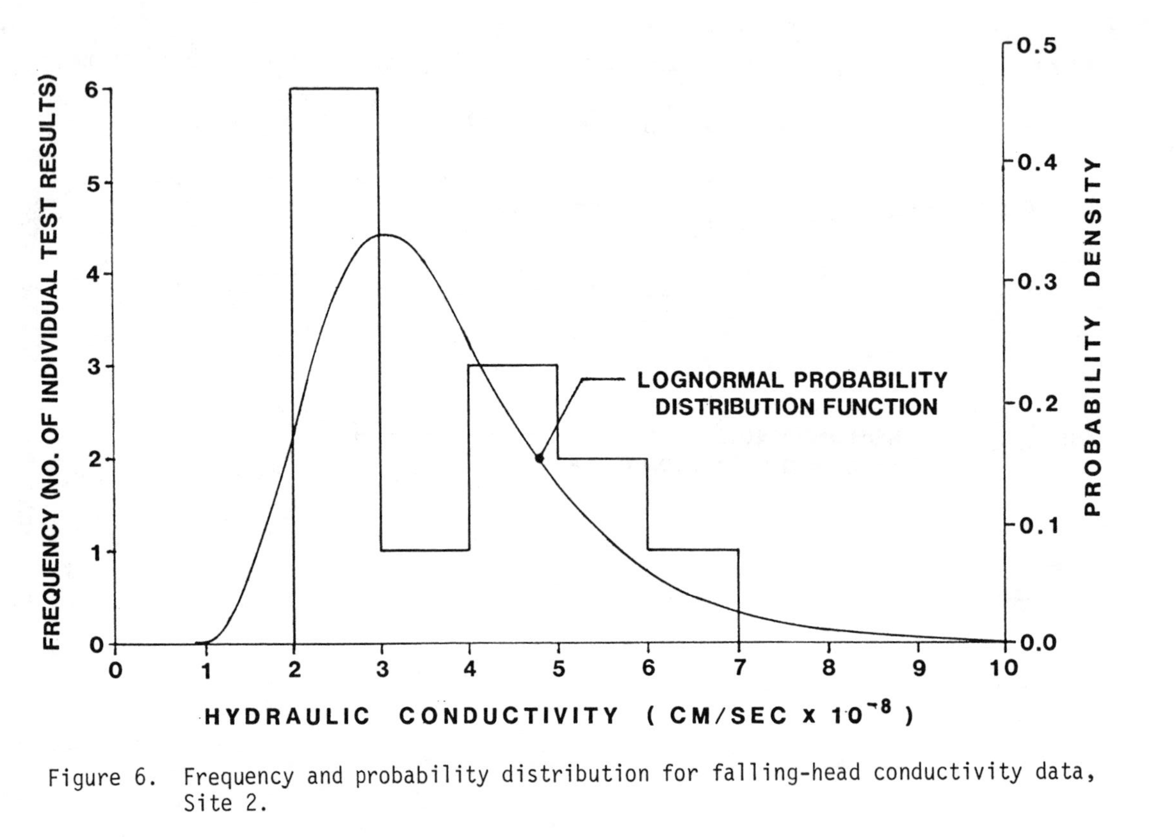

Figure 6. Frequency and probability distribution for falling-head conductivity data, Site 2.

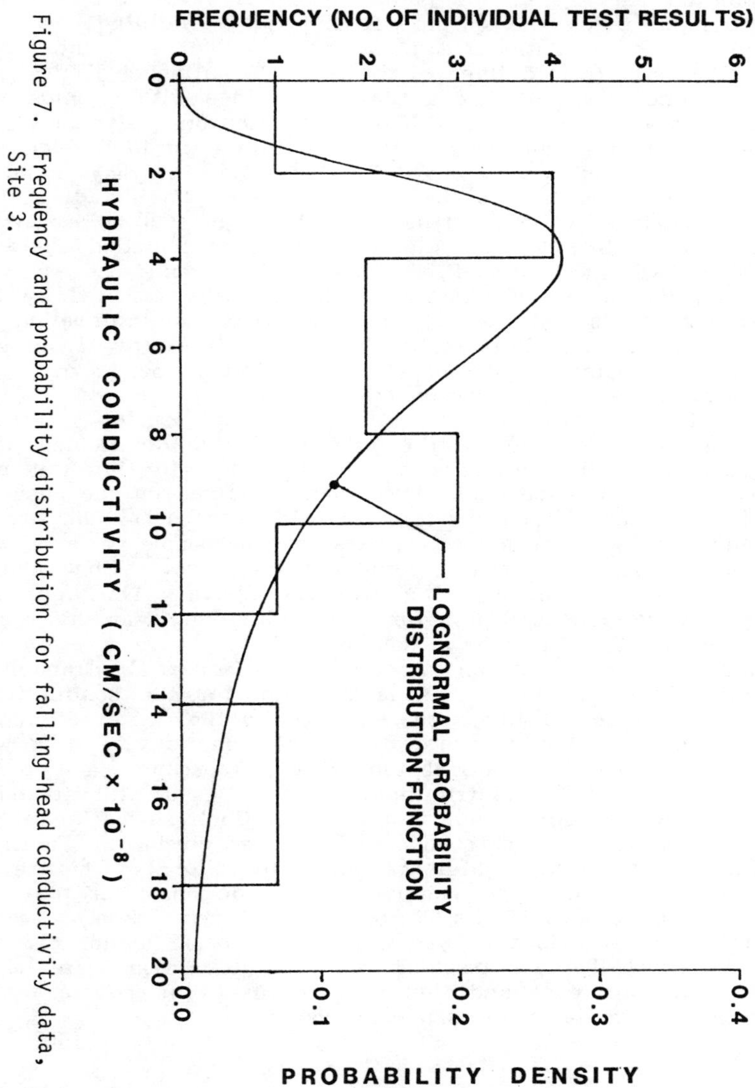

Figure 7. Frequency and probability distribution for falling-head conductivity data, Site 3.

PRACTICAL APPLICATION

Estimate of the Population Mean

One of the goals of the testing programs described herein was to obtain an estimate of the population mean value for hydraulic conductivity. Using the sample statistics just presented, such an estimate of the population mean can be made for each individual site. This estimate is defined in the form of confidence limits.

Confidence limits define a range of values within which the population mean should occur. They are based on a given level of confidence. Hence, the mean may be estimated within a relatively narrow range of values with a confidence of 50%. The range must be much larger in order to be 90% confident of including the population mean value. The range must be larger still to achieve a 99% level of confidence that the mean value is included therein. Obviously, these limits depend upon the level of confidence we wish to achieve, the degree of variation in the data (as measured by the variance or standard deviation), and the amount of information we know about the distribution of the data. However, it is not necessary to establish the shape of the distribution in order to define confidence limits as demonstrated below.

Site 1 was used to demonstrate this application. First, confidence limits for the mean conductivity were determined with the assumption that the form of the data distribution was not known. The confidence limits were then redefined on the basis of the known lognormal data distribution. In both cases, the sample mean and sample variance were known and the sample variance was taken as an estimate of the population variance. Chebyshev's inequality [9] was applied to the first case above. This approach resulted in confidence limits for the population mean hydraulic conductivity at Site 1 as shown on Figure 8.

Since it has been shown that the lognormal probability distribution function is a reasonable model for the conductivity data, the confidence limits were redefined by the use of Student's "t" distribution. For this approach, the conductivity data was transformed to its natural logarithm, forming a normal distribution. The "t" distribution was then used with standard tables [9] in a manner similar to the use of Chebyshev's inequality above. A narrower set of confidence limits was obtained. Figure 9 presents the resulting confidence limits for Site 1. Figure 10 compares the results from each method and demonstrates that the mean hydraulic conductivity could be estimated more closely because the probability distribution was determined to be lognormal. It should be noted that the resultinig estimated mean values are based on levels of confidence and therefore are useful in risk analyses performed with contaminant transport models.

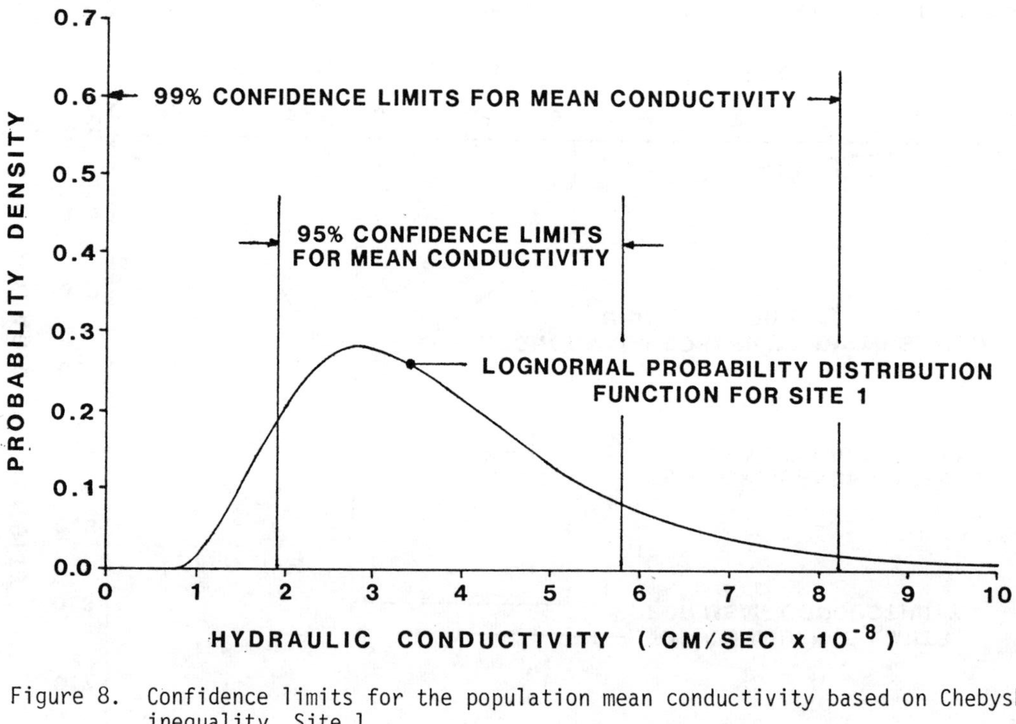

Figure 8. Confidence limits for the population mean conductivity based on Chebyshev's inequality, Site 1.

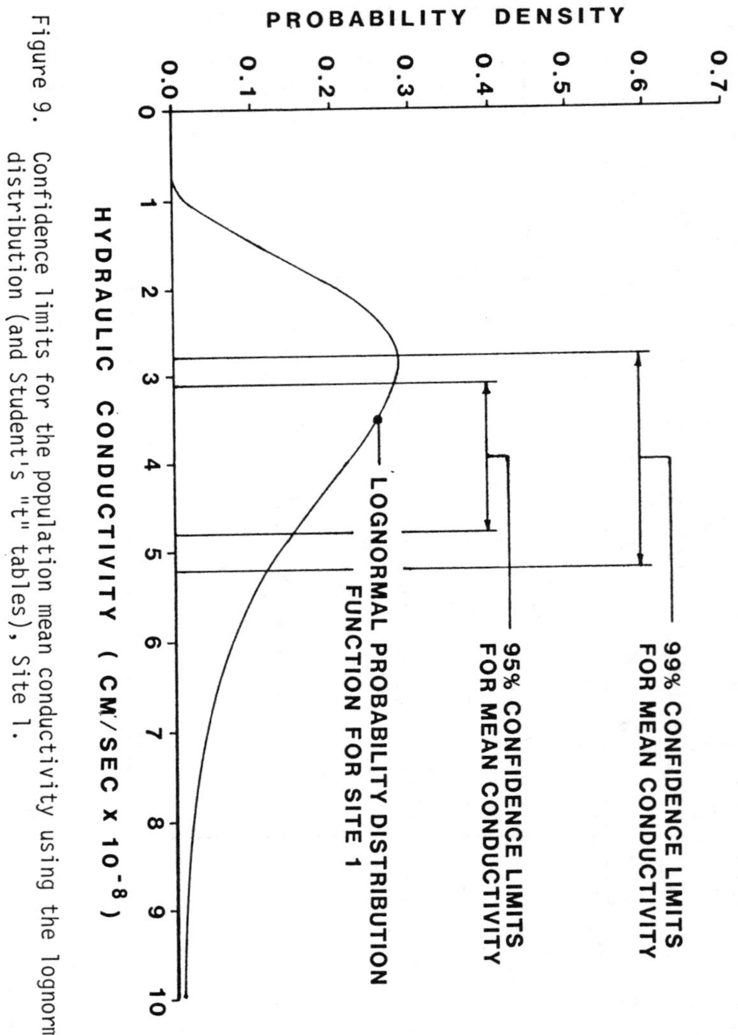

Figure 9. Confidence limits for the population mean conductivity using the lognormal distribution (and Student's "t" tables), Site 1.

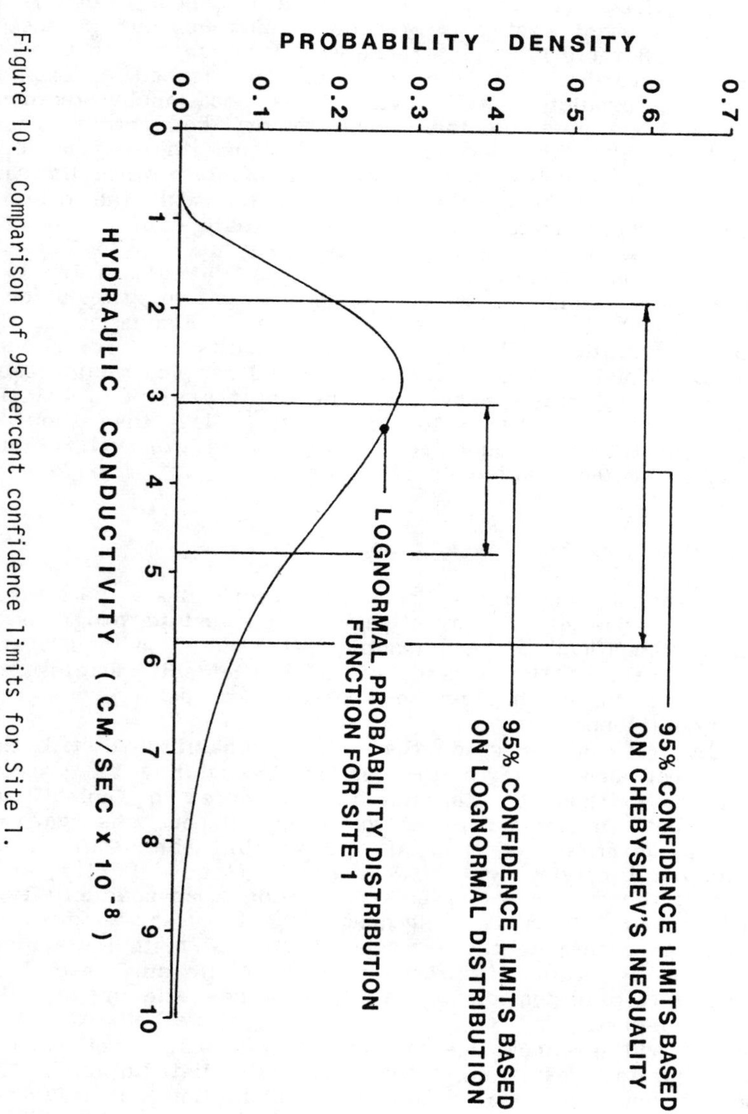

Figure 10. Comparison of 95 percent confidence limits for Site 1.

Verification Testing

Quality assurance procedures during facility construction generally require verification that the natural deposit which is to form the containment system is of proper thickness and possesses the proper characteristics. Usually, classification testing is sufficient to verify that the proper soil deposit exists as anticipated or predicted. However, it is occasionally required that a sample be obtained for a confirmatory conductivity test. Care must be taken in these situations because there is a finite probability that a single test result may violate some limiting value, and still not prove to be inconsistent with the original testing program performed during the site investigation.

This will be demonstrated by once again using Site 1 as an example. The mean conductivity from the test data was 3.4×10^{-8} centimeters per second (cm/sec). Assume a limiting criteria for a single test result is set at 5×10^{-8} cm/sec. Evaluation of the lognormal distribution [9] for this site, results in a conclusion that approximately 1 out of every 5 individual samples would result in a conductivity value exceeding the arbitrary limit. If the limit was set at a maximum value of 1×10^{-7}, less than 1 out of 100 test results would be expected to exceed the limit. This concept is presented graphically on Figure 11.

Development of New Site Investigations

The statistical techniques discussed herein can also be used to aid in the development of investigations and testing programs at new sites. Specifically, these techniques provide the investigator with an estimate of the number of individual soil samples and conductivity tests which may be necessary to achieve a target degree of confidence.

Initially, it is assumed that the probability distribution function is not known. From the data obtained during the previous investigations in these till deposits and presented in Table II, we anticipate that the coefficient of variation will be conservatively near 100%. Further, we roughly guess that the sample mean hydraulic conductivity will be 5×10^{-8} cm/sec. Lastly, it is decided that the maximum allowable population mean conductivity is 1×10^{-7} cm/sec. Chebyshev's inequality can be used, as discussed previously, to determine the required number of tests for a given confidence level. Table III presents the requisite number of tests to achieve different confidence levels for a new site investigation in a similar deposit.

Now we add the knowledge that the program will, if performed as described herein, result in a lognormal data distribution. This knowledge allows the use of the "t" distribution as previously described. As can be seen from Table III, the number of tests required to estimate that the population mean hydraulic

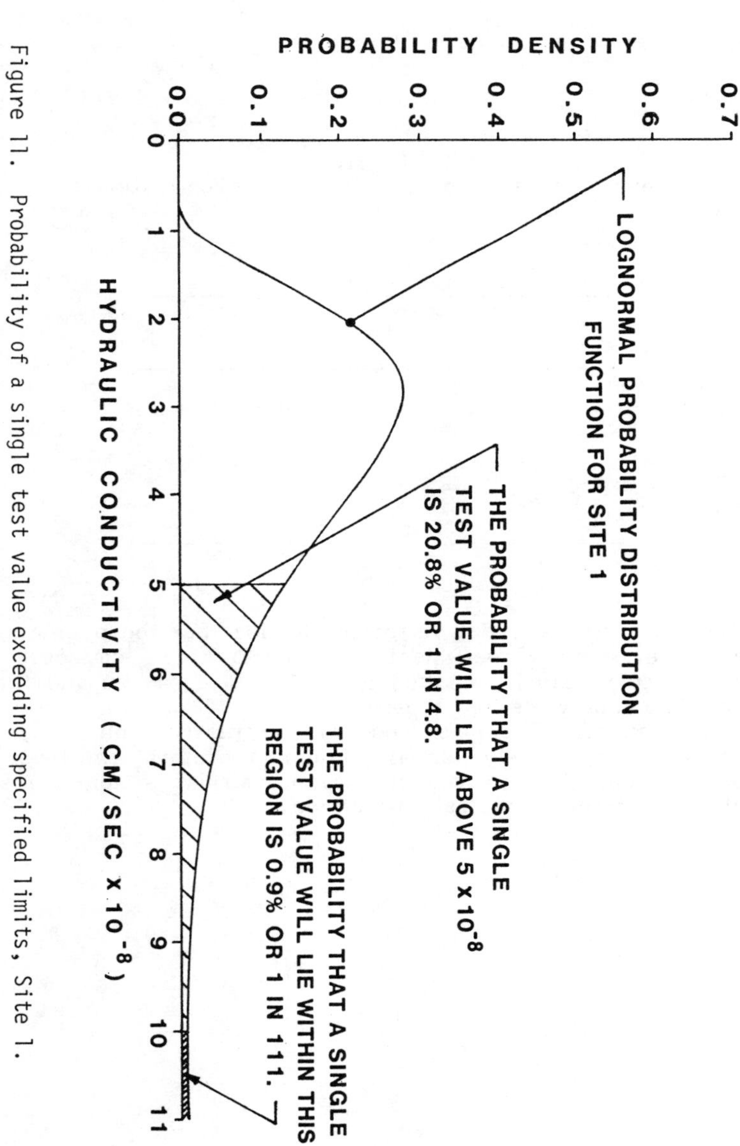

Figure 11. Probability of a single test value exceeding specified limits, Site 1.

TABLE III
Number of Tests Required for New Investigations*

Method	Level of Confidence 95%	99%
Chebyshev's Inequality	20	100**
Lognormal Distribution	6	11

* Based on the desired level of confidence that the population mean permeability coefficient lies below 1×10^{-7} cm/sec. The sample mean is assumed to be 5×10^{-8} cm/sec and the coefficient of variation is assumed to be 100%.

** If the number of tests exceeds 30, the distribution of means will be approximately normal [9] and, therefore, the level of confidence is improved in a manner similar to that allowed by the lognormal assumption.

conductivity lies below 1 x 10^{-7} cm/sec with any given level of confidence is reduced. Hence, definition of the lognormal probability distribution function, shown to be applicable to these till deposits on eight different sites, can reduce the requisite number of tests in similar investigations at sites with similar deposits.

CONCLUSIONS

A statistical evaluation of the hydraulic conductivity date described herein resulted in several definitive conclusions. First, it was shown that the conductivity values obtained from falling-head tests and those values backcalculated from one-dimensional consolidation tests probably did not represent exactly the same population of hydraulic conductivity. The difference can probably be attributed to testing method as well as sampling method. The falling-head results indicated a slightly greater mean conductivity than the mean value represented by the consolidation test results. A similar result has been reported by others [8]. It should be noted, however, that the use of the greater falling-head conductivity values is more conservative in the deposits discussed herein.

Secondly, a range of values for the estimated coefficient of variation for hydraulic conductivity was established. This range of coefficients can be applied, for preliminary estimating purposes, to investigations undertaken with similar techniques in soil deposits similar in nature and origin.

Third, the test data at each site was found to result in a distribution which reasonably approximates a lognormal probability distribution function. Knowledge that the lognormal probability distribution function is a reasonable model for these test programs resulted in the ability to more closely estimate the mean hydraulic conductivity for each site. An estimate of the coefficient of variation was also combined with the lognormal distribution to demonstrate how new similar investigations can be planned to achieve an approximate chosen level of confidence.

It must be remembered, however, that the hydraulic conductivity data presented herein are not unique values. They are affected by a number of factors, including sampling and testing methods. Hence, the population mean conductivity discussed above is defined as the mean which would be expected if all possible soil specimens from a deposit were obtained and tested in a manner similar to the methods described herein. The actual relationship of the laboratory-determined values of hydraulic conductivity to an in-situ mean conductivity value is beyond the scope of this publication. Similarly, the apparent applicability of the lognormal probability distribution to the test data does not necessarily imply that the actual in-situ hydraulic conductivity follows a similar distribution. Indeed, definition of the "true" in-situ conductivity and its probability distribution may be beyond the realm of possibility given the current state of testing methodology.

It should also be remembered that hydraulic conductivity is only one property of the investigated deposits. Many other factors must be examined to determine the suitability of a deposit for waste containment. Specific waste types may result in different hydraulic conductivity properties in these deposits. Indeed, they may even be entirely incompatible with these deposits. A statistical evaluation of the laboratory-determined conductivity will certainly not overcome these and other factors. However, the approach will allow rational evaluation of one site characteristic which is currently deemed important in site suitability investigations.

REFERENCES

1. Martin, H.M. "Map of Surface Formations of the Southern Penninsula of Michigan", Geological Survey Division, Department of Conservation, Michigan, Publication 49 (1955).
2. "Standard Test Method for Classification of Soils for Engineering Purposes", ASTM D 2487, Philadelphia, Pennsylvania (1969).
3. Eschman, D.F. Unpublished data, Department of Geological Sciences, University of Michigan, Ann Arbor, Michigan (1983).
4. Lambe, T.W. Soil Testing for Engineers, John Wiley & Sons, New York (1951).
5. Taylor, D.W. Fundamentals of Soil Mechanics, John Wiley & Sons, New York (1948).
6. Hvorslev, M.J. Subsurface Exploration and Sampling of Soils for Civil Engineering Purposes, ASCE and Waterways Experiment Station, Vicksburg, Mississippi (1949).
7. Neville, A.M., and J.B. Kennedy Basic Statistical Methods for Engineers and Scientists, International Textbook Company, New York (1964).
8. Olson, R.E., and D.E. Daniel "Measurement of the Hydraulic Conductivity of Fine-Grained Soils", Permeability and Groundwater Contaminant Transport, ASTM STP 746, pp. 18-64 (1979).
9. Harr, M.E. Mechanics of Particulate Media, McGraw-Hill, New York (1977).
10. Shapiro, S.S., and M.B. Wilk "An Analysis of Variance Test for Normality (Complete Samples)" Biometrika, V. 52, pp. 591-611 (1965).

CHAPTER 10

GROUNDWATER MONITORING SYSTEMS - ONLY AS GOOD AS THE WEAKEST LINK

David E. Johe, C.P.G.
ERM-Midwest, Inc.

INTRODUCTION

Chemical analytical techniques are so sophisticated today, and the detection limits for numerous given parameters are so low, that we are finding many of them virtually everywhere we look. Almost every day you can pick up the newspaper or news periodical and find a news release such as "Exotic Chemical Solvent Found in Area Wells(s)" or "Pesticides Found in New Municipal Well". These headlines alarm most of the local residents and often lead to statements such as "local groundwater unsafe" or "local aquifer contaminated, not fit for use."

Quite often the conclusions are not valid but nobody bothers to find out why the sample contained the identified substance. Everybody, including the media, pays a great deal of attention to the chemical analysis of water samples but rarely do they bother to find out where a sample came from, how it was collected, what it was placed in for delivery to the laboratory, how it was preserved, and how long it was held prior to shipment. In following up some alleged contamination incidents, it was discovered that in some instances, the contaminant was from the solvent cement used to glue casing joints of PVC pipe, or that the pesticide in a sample resulted from a driller using contaminated surface water for makeup water during drilling of a well. Upon completion, the well was not developed long enough to purge the contaminated water from the system and the contaminants showed up in the analysis.

The opposite condition also occurs. I have seen examples of wells contaminated with volatile organic compounds but the contaminants did not show up because the sample was air lifted from the well or placed in the wrong container with an air space. By the time the sample got to the lab, the volatiles were "gone."

The point of this paper is not to downplay the importance of the chemical analysis but rather to place more emphasis on those elements of the program that are often overlooked. If the results of chemical analyses are to be meaningful, we as professionals must be responsible for seeing that the entire monitoring/sampling program is as sophisticated as the chemical analysis.

Management of Toxic and Hazardous Wastes, H. Bhatt, R. Sykes, T. Sweeney (Editors)

There are several elements of a monitoring/sampling program that I would like to discuss in this paper:

- the well
- sampling equipment
- sample containers
- collection and preservation of samples
- custody and shipping
- chemical analysis
- interpretation of results
- quality assurance

A lot of good material is available in the literature for detailed discussion of each of the elements listed and I do not intend to dwell on any point in particular. My intent here is to list the elements, expand on them and to stress the importance of the actions of the professional in ensuring that all of the elements are addressed each time we undertake a monitoring or sampling program.

The Well

Quite often reports of investigation list analytical results of water samples collected from a well or wells and all that is presented, or known, is the sample source. The results are meaningless unless we know exactly where the sample came from. Table I lists some of the minimal well information that we must

Table I. Critical Well Information

Owner Nema:
Location: (description and map)
Date Sampled:
Elevation: (top of casing or ground surface)
Depth to Water: (corrected to elevation)
Amount of Casing: (cased to what depth)
Diameter of Casing: (inside diameter)
Open Hole (consolidated formation) or Screened Interval:
Length of Screen and Depth (top and bottom) of Screen:
Formation Containing the Water Being Sampled:
Casing/Screen Material:

know in order to make sense of the analytical results.

In addition, it is very helpful to know who installed the well(s), when, and all of the construction details such as:

- drilling technique
- drill water source
- grout interval(s)
- backfill interval(s)
- surface cement interval
- method of development
- volume of fluid removed during development

The information listed in Table I is essential if we are to determine the direction and gradient of groundwater flow and where the sample point is in relation to suspected sources of contamination. Table I data coupled with the other construction information help us to evaluate whether a sample is representative of the water from the aquifer of concern (or zone within an aquifer).

Sampling Equipment

Selection of appropriate equipment depends upon several factors including diameter of the well, depth to water, volume of water to be displaced prior to collecting the sample, what parameters you will be analyzing, and available budget. As a general rule, the depth to water and volume of water that is to be removed can be used to determine whether you want to use a pump or bailer: If the depth to water is deep, and the volume is more than a few bailers will hold, it is much less strenuous and much faster to set a submersible pump than to raise and lower a bailer several hundred times. On the other hand, if the well is amall diameter and the volume of water is small, a simple bailer is probably much more expedient. The new small diameter submersible pumps are very useful; however, they pump at low rates and are generally quite expensive. If you do a lot of sampling of small diameter wells, they are worth the expense.

After yu have determined how you are going to collect the sample(s), you must decide what the bailer is to be constructed of or the pump is to be lined with. The most accepted materials are stainless steel (for samples to be analyzed for parameters other than metals), PVC, and teflon. The tubing from the pump to the surface should be an "inert" material such as teflon, if possible.

An area that is often overlooked is the rope to which the bailer is attached. As the bailer is used, the rope normally becomes wet; as the bailer is withdrawn from the well, the rope is often allowed to lie on the ground. When the bailer is lowered the next time, the rope, covered with adhering "dirt" easily introduces contaminants into the well. Extreme caution should be exercised to ensure that the rope does not come into contact with the ground.

Direct air lift methodologies* should never be used to obtain groundwater samples. This method results in erroneous laboratory analyses because the parameters you are analyzing for may be oxidized or driven off (volatiles).

Table II summarizes and compares advantages and disadvantages of the conventional methods of collecting groundwater samples. In general, sampling equipment that is easy to clean should be utilized (to avoid cross contamination of wells); arrange to sample the least contaminated well(s) first (if possible to determine ahead of time), or, dedicate a sampling device to each well.

Sample Containers

The importance of using the proper sample container cannot be overemphasized. Many analytical results are invalidated because the sampler used the wrong containers, failed to fill them properly, or failed to follow correct bottle cleaning procedures.

All samples for organic chemical analyses should be placed in clean, amber glass bottles with teflon lined lids. Samples for volatile organics should be collected in properly prepared 40 ml glass bottles with teflon lined septum caps (<u>no</u> air bubbles permitted in bottle). Samples for inorganic chemical analyses should be placed in clean polyethylene bottles.

Table III lists the correct type of container, sample volume required, preservatives and maximum holding times for specific analytes. <u>Standard Methods</u> describes the correct bottle cleaning technique as follows:

> "Sample bottles must be cleaned carefully before each use. Rinse glass bottles, except those to be used for chromium or manganese analyses, with a cleaning mixture made by adding 1 1 of conc H_2SO_4 slowly, with stirring, to 35ml saturated sodium dichromate solution, or with an alkaline permanganate solution followed by an oxalic acid solution. Rinsing with other concentrated acids may be used to remove organic matter. Detergents are excellent cleansers for many purposes; use either detergents or concentrated HCL for cleaning hard-rubber and plastic bottles. After the bottles have been cleaned, rinse them thoroughly with tap water and then with distilled water." (Ref. Standard Methods For the Examination of Water and Wastewater, 15th edition, APHA-AWWA-WOCF, 1980.)

*Note: Direct air lift should not be confused with the air driven bladder type small diameter pumps which do not expose the water to air. These pumps are acceptable and in general use although they, like the small diameter conventional submersible pump, pump at a very low rate.

Table II. Methods of Water Removal and Sampling (Scalf, 1981)

Description	Advantages	Disadvantages
Bailer. A weighted bottle or capped length of pipe on a rope is lowered and raised by hand.	Can be constructed from a wide variety of noncontaminating materials compatible with the parameter of interest. It is economical and convenient enough that a separate bailer may be dedicated to each well to minimize cross contamination. No external power source is required.	Sometimes impractical to evacuate stagnant water in a well bore. Transfer of water sample from bailer to sample bottle can result in aeration. Cross-contamination can be a problem if equipment is not adequately cleaned after each use.
Suction Lift Pumps. Centrifugal pumps are highly portable, may pump 5 to 40 gpm and usually require a foot valve on end of suction pipe. Peristaltic pumps are low-volume, suitable for shallow, small diameter wells and may require electricity. Hand operated diaphragm pumps are portable.	Generally, suction lift pumps are readily available, relatively portable, and inexpensive.	Sampling is limited to ground water depths less than about 20 feet.
Submersible Pumps. Electric pumps require 4-inch or larger diameter casing. Pneumatic pumps fit 2-inch diameter casing.	May be used at depths greater than 20 feet. Pumps are portable if a portable generator is available. Small diameter pumps allow sampling from 2-inch diameter wells.	Not suitable for sampling for organics with conventional pump construction materials. Small diameter pumps, although available, are very expensive ($2000-5000).

Table III. Sample Volumes Required, Containers, Preservatives and Holding Times for Specific Analyses[a]

Analyte	Volume (ml)	Container[b]	Preservative	Maximum Holding Time
Fluoride	300	P,G	None	7 days
Nitrate (as N)	100	P,G	Cool, 4°C 2 ml conc. H_2SO_4/1	24 hrs >24 hrs
Sulfate	50	P,G	Cool, 4°C	7 days
Chloride	50	P,G	None	7 days
Dissolved Metals	200	P,G	Filter on site HNO_3 to pH <2	6 months
Dissolved Mercury	100	P,G	Filter on site HNO_3 to pH <2	P - 13 days G - 38 days
Phenols	500	G	Cool, 4°C H_3PO_4 to pH <4 1 g $CuSO_4$/1	24 hrs
Pesticides[c,d]	1000	G	Cool, 4°C 0.008% $Na_2S_2O_3$(f)	7 days (before extraction) 30 days (after extraction)
TOC	25	P,G	Cool, 4°C H_2SO_4 or HCl to pH <2	24 hrs
TOX[c,e]	250	G	Cool, 4°C 0.008% $Na_2S_2O_3$(f)	7 days (before extraction) 30 days (after extraction)
Radioactive Constituents[c,d]	1000	P,G	HNO_3 to pH <2	6 months
Coliform Bacteria[c,d]	500	P,G	Cool, 4°C 0.008% $Na_2S_2O_3$(f)	6 hrs
pH	25	P,G	Measure on site	6 hrs
Specific Conductance	100	P,G	Cool, 4°C	24 hrs
Temperature	25	P,G	None	Measure immediately

(a) "Methods for Chemical Analysis of Water and Wastes", EPA-600/4-79-020, except as noted.

(b) Plastic (P) or glass (G). For metals, a polyethylene bottle with an unlined polypropylene cap is preferred. Caps for glass bottles have teflon liners.

(c) Federal Register, December 3, 1979, p. 69574.

(d) Standard Methods For the Examination of Water and Wastewater, 15th Ed., APHA-AWWA-WPCF, 1980.

(e) Total Organic Halide, EPA Interim Method 450.1, November 1980.

(f) 0.008% $Na_2S_2O_3$ should only be added in the presence of residual chlorine.

Collection of Sample

The literature is full of procedural descriptions for the collection of representative groundwater samples. I will therefore not dwell on the subject, but, rather, will summarize the components and procedures for collecting representative samples.

Immediately before sampling, determine the total depth of the well and the depth to water. This can be accomplished with a clean chalked tape or an electric sounder. Calculate the volume of water standing in the well using the formula Volume $(ft^3) = r^2h$ where r=1/2 inside diameter of well and h= height of water standing in the well.

Take pH, specific conductance and temperature measurements on the first groundwater removed using the procedure shown in Table IV. The equipment used for these measurements must be calibrated prior to each day's field operation.

Remove five times (five equivalent volumes) the calculated volume of water standing in the well using the procedure selected from the section Sampling Equipment. In some instances the well may be pumped or bailed dry before five equivalent volumes are removed. If this occurs, allow the well to recover and collect the samples. Before collecting the samples (or laboratory chemical analysis), take another set of pH, specific conductance and temperature measurements.

Place the samples for laboratory analysis in the containers described in the previous section, Sample Containers, as appropriate. Sample bottles and caps should be rinsed twice with the water to be sampled before the laboratory sample is collected. Never touch the inside of the sample container or lid liner.

Sample volumes required, type of container, recommended preservatives and holding times for specific analyses are listed in Table III.

It is very important that the person or persons collecting the samples be familiar with, and adhere to, the correct procedures. Pay particular attention to maximum holding times and ship the smaples in a manner that will allow the laboratory to complete the analysis within the time limits. In some instances it may be necessary to perform some of the analyses in the field, or to make arrangements with a local approved laboratory due to very limited time constraints.

A standardized label or tag containing the following information should be placed on each sample container:

- sample number (unique)
- date and time
- sample location (well number)
- sampling technique (bailing, pumping)
- preservatives used
- analyses required
- name of collector(s) and witness(es)

Table IV. Temperature, pH and Specific Conductance (EPA 1979)

Measurement	Method/Instrument	Calibration
Temperature	Any good grade mercury-filled or dial type centigrade thermometer, or a thermistor. Report to nearest °C. See Method 170.1 of EPA (1979).	Device should be routinely checked against a precision thermometer. See Standard Methods for the Examination of Water and Wastewater, 15th Edition, p. 124, Method 212 (1980).
pH	Determined electrometrically using either a glass electrode in combination with a reference potential or a combination electrode. Field model pH meter (a variety of commercial models available). After calibration, place at least 25 ml of water in a clean glass beaker (or enough to cover sensing elements). Report to nearest 0.1 unit measure. See Method 150.1 of EPA (1979).	Follow manufacturer's instructions for calibration at a minimum of two points that bracket the expected pH of samples, at least 3 pH units apart. If instrument design involves the use of a "balance" or "standardize" dial and/or a slope adjustment, see EPA (1979) for calibration procedures.
Specific Conductance	Measured with a self-contained conductivity meter or wheatstone bridge-type. Temperature correction must be made for samples not analyzed at 25°C. See Method 120.1 of EPA (1979).	Instrument must be standardized with KCl solution before daily use.

Each sample container should be placed in a sealed plastic bag prior to placement on ice in an insulated shipping container (to prevent the label from soaking loose or becoming illegible during transport). The containers should be packed in such a manner that breakage during transport can be prevented.

At the time of sampling, bound sample control data notebooks should also be logged with the same information recorded on the sample bottle label or tag. Additionally, pertinent field data (pH, specific conductance, temperature and water levels) and reason for sampling (initial, annual, semi-annual sampling or special problem sampling in conjunction with contaminant discovered in nearby domestic well) should be logged along with any subjective field observations, such as meteorological conditions, unusual sample colors/odors, etc.

Immediately after you have collected, preserved, and packed the samples, clean and flush the sampling bailer or pump with copious amounts of distilled water to prevent cross contamination between sampling wells. Materials incidental to sampling such as bailer ropes and tubing must also be flushed with distilled water. Protect sampling equipment from ground surface contamination. If there are any detectable traces of contaminants on the sampling equipment after rinsing (oily residue, odor), do not use in another well. Use another sampling device.

Custody and Shipping

The field sampler is responsible for the care and custody of the samples collected until properly dispatched to the receiving laboratory or turned over to an assigned custodian. He must assure that each container is in his physical possession or in his view at all times or stored in a locked place where no one can tamper with it.

As few people as possible should handle the samples from collection through analysis. Chain of custody records should be established, attached to the outside of each shipping container(s), and should contain the following information:

- number of samples (containers)
- date samples collected
- name of person collecting the sample(s)

Each form should be signed and dated by the person who collected, preserved and packed the samples. Each container, with attached chain of custody form should then be sealed so that the container cannot be opened without breaking the seal. When transferring the sample shipping containers, the transferee should sign and record the date and time that he accepted custody on the chain of custody record. Custody transfers, if made to a sample custodian in the field, should be recorded for each individual shipping container sample. To prevent undue proliferation of custody records, the number of custodians in the chain of possession should be minimized. If samples are delivered to the laboratory when appropriate personnel are not there to receive them, the samples should be locked in a designated area within the laboratory so that no one can tamper with them.

Each laboratory should have a sample custodian to maintain a permanent, bound log book in which he records for each sample the person delivering the sample, the person receiving the sample, date and time received, source of sample, sample number, how transmitted to the lab, and a number assigned to each sample by the laboratory. Distribution of samples to laboratory personnel who are to perform analysis shuld be made only by the custodian. The custodian should enter into the log the laboratory sample number, time date, and the signature of the person to whom the samples were given.

As was mentioned in the section on Collection and Preservation of Samples, the field sampler(s) are responsible for seeing that all samples are shipped in a timely manner that allows adequate time for transportation, receiving (by the lab custodian) and laboratory analysis within the maximum time constraints specified in Table III.

Chemical Analysis

This paper is not going to dwell on the methods of analyzing for specific parameters except to say that all mehtods used should be EPA approved or recommended. Table V provides a summary of analytical methods.

All samples should be scheduled for analysis as soon as possible upon receipt at the laboratory, and the analyses must be completed within the maximum holding times listed in Table III. During analysis, good laboratory practice dictates that the quality of data be monitored. Each laboratory should have on file a written quality assurance plan, detailing instrument maintenance frequency, calibration procedures analysis of field replicates, split samples, spiked samples, method blanks and blind standards. A recommended quality control program should consist of 15 to 20 percent of the total sample load. Field blank samples should be collected with and without preservatives, so that the laboratory analysis can be performed to show that neither the containers nor sampling techniques caused sample contamination. More detailed guidance for acceptable quality assurance programs for chemical analysis may be found in "Methods for Chemical Analysis of Water and Wastes,"EPA-600/4-79-020 and "Handbook for Analytical Quality Control in Water and Wastewater Laboratories," available through the Office of Technology Transfer, Washington, D.C. 20460.

Table V. Summary of Analytical Methods

Analyte	Reference[a]	Technique
Fluoride	340.1	Colorimetric, SPADNS with Bellack Distillation
	340.2	Potentiometric, Ion Selective Electrode
	340.3	Colorimetric, Automated Complexone
Nitrate (as N)	352.1	Colorimetric, Brucine
Sulfate	375.1	Colorimetric, Automated Chloranilate
	375.2	Colorimetric, Automated Methylthymol Blue AAII
	375.3	Gravimetric
	375.4	Turbidimetric
Chloride	325.1	Colorimetric, Automated Ferricyanide AAI
	325.2	Colorimetric, Automated Ferricyanide AAII
	325.3	Titrimetric, Mercuric Nitrate
Arsenic	206.2	Furnace Atomic Absorption Spectroscopy (AAS)
	206.3	Hydride Generation AAS
	206.4	Spectrophotometric - SDDC
	206.5	Digestion for 206.3 and 206.4
	200.7 (Interim)[b]	Inductively Coupled Argon Plasma Spectroscopy (ICAP)
Barium	208.1	Direct Aspiration AAS
	208.2	Furnace AAS
	200.7 (Interim)[b]	ICAP
Cadmium	213.1	Direct Aspiration AAS
	213.2	Furance AAS
	200.7 (Interim)[b]	ICAP
Chromium	218.1	Direct Aspiration AAS
	218.2	Furnace AAS
	218.3	AA, chelation-extraction
	200.7 (Interim)[b]	ICAP
Iron	236.1	Direct Aspiration AAS
	236.2	Furnace AAS
	200.7 (Interim)[b]	ICAP
Lead	239.1	Direct Aspiration AAS
	239.2	Furnace AAS
	200.7 (Interim)[b]	ICAP
Manganese	243.1	Direct Aspiration AAS
	243.2	Furnace AAS
	200.7 (Interim)[b]	ICAP

Table V. (Continued)

Analyte	Reference[a]	Technique
Mercury	245.1 245.2	Manual Cold Vapor AAS Automated Cold Vapor AAS
Selenium	270.2 270.3 200.7 (Interim)[b]	Furnace AAS Hydride Generation AAS ICAP
Silver	272.1 272.2 200.7 (Interim)[b]	Direct Aspiration AAS Furnace AAS ICAP
Sodium	273.1 200.7 (Interim)[b]	Direct Aspiration AAS ICAP
Phenols	420.1	Spectrophotometric, Manual 4-AAP with Distillation
	420.2	Colorimetric, Automated 4-AAP with Distillation
	420.3	Spectrophotometric, MBTH with Distillation
	604[b], 510D[c]	Gas chromatography
Pesticides (Organochlorine)	608[b], 509A[c]	Gas chromatography
TOC	415.1	Combustion or Oxidation
TOX	450.1 (Interim)[d]	Microcoulometric, titration
Gross Alpha	703[c], 900.0[e]	Internal Proportional Counting
Gross Beta	703[c], 900.0[e]	Internal Proportional Counting
Radium	705[c] 706[c], 707[c]	Precipitation Liquid Scintillation
Coliform Bacteria	980[c]	Multiple Tube Fermentation

(a) "Methods For Chemical Analysis of Water and Wastes," EPA-600/4-79-020, except as noted.
(b) Federal Register, December 3, 1979, "Guidelines Establishing Test Procedures for the Analysis of Pollutants."
(c) Standard Methods For the Examination of Water and Wastewater, 15th Ed., APHA-AWWA-WPCF, 1980.
(d) "Total Organic Halide, Interim Method 450.1," U.S. EPA, Office of Research and Development, Environmental Monitoring and Support Laboratory, Cincinnati, Ohio, November 1980.
(e) "Procedures for Measurement of Radioactivity in Drinking Water," EAP-600/4-80-032.

Interpretation of Results

After the results of the chemical analysis of the samples are available, it is time for a qualified individual to interpret the results. That person should be knowledgeable of groundwater transport mechanisms, the site where the samples were collected, have a good understanding of the local hydrogeological conditions, and know the history of the site or area being monitored (e.g., suspected spill of chemicals A,B,C, suspect gasoline storage tank leak, or suspect leachate from a land disposal site). All too often a person totally unfamiliar with the situation will simply look at an analysis form and jump to some erroneous conclusion.

The person interpreting the results must be willing to look at all available information before drawing conclusions. Groundwater gradient and direction, source and type of suspected contaminants and background water quality (ambient conditions) must be established. Conclusions should never be based on the results of a single sample or parameter. The intepreter should compare the results of down gradient samples with those of estabished background (up gradient) samples.

Establish a form on which you can plot the results of the chemical analysis of each well on a separate sheet. List the chemical parameters in columns and the date sampled in rows. Look for changes in the concentration of individual parameters and overall water quality with time; establish whether there are any trends emerging from the data. Compare the results of the down gradient samples with those of established up gradient (background) samples and then, form your conclusions.

Quality Assurance

The objective of a quality assurance (QA) program is to provide data for which the bounds of uncertainty are known and from which confident conclusions may be drawn. This objective encompasses all aspects of the project and requires data which are complete, representative, precise, and accurate (within specified limits). Documentation provides records of traceability and adherence to the prescribed protocols.

Quality assurance throughout the entire monitoring program is critical to the success of the program and validity of the conclusions. QA procedures should be estabished for each element of the program and a qualified individual should be assigned to see that the procedures are followed.

Summary

The intent of this paper has been to make the point that a groundwater monitoring program involves a lot more than just collecting a sample and getting it analyzed. A good monitoring program, one that we can have confidence in, and know that the conclusions are valid involves numerous elements which have been described herein. Emphasis has been placed on those elements that are most often overlooked or slighted.

The "bottom line" is that you, the professional, need to know and understand all of the elements of a good monitoring program and should never try to shortcut the program. If any element is ignored or downplayed in the interest of saving time or money, it could result in questionable or dubious results. In the long run, this could prove to be very expensive to you in terms of credibility and cash flow. It is better to do the job right the first time than to have to redo it at your own expense.

REFERENCES

- Scalf, M.R., et al. 1981. Manual of Ground-Water Quaity Sampling Procedures. Robert S. Kerr Environmental Research Laboratory Office of Research and Development. U.S. EPA, Ada, Oklahoma, NWWA/EPA Series.
- Standard Methods for the Examination of Water and Waste Water, 15th ed. APHA-AWWA-WPCF, 1980.
- U.S. EPA (Environmental Protection Agency). 1979. Methods for Chemical Analysis of Water and Wastes. EPA-600/4-79-020, Cincinnati, Ohio.
- "Guidelines Establishing Test Procedures for the Analysis of Pollutants," Federal Register, December 3, 1979, p. 69574.
- U.S. EPA (Environmental Protection Agency), Office of Research and Development, Environmental Monitoring and Support Laboratory, "Total Organic Halide, Interim Method 450.1," November 1980, Cincinnati, Ohio.
- U.S. EPA (Environmental Protection Agency), "Procedures for Measurement of Radioactivity in Drinking Water," EPA-600/4-80-032.

CHAPTER 11

PROBLEMS IN ASSESSING ORGANICS CONTAMINATION IN GROUNDWATER

Robert A. Saar
Geraghty & Miller, Inc.

INTRODUCTION

Organic contamination of the groundwater near hazardous waste facilities and of municipal or industrial supply wells is widespread and undoubtedly has been for decades. However, a general awareness of this problem has come only in recent years. This awareness results from the widening availability of analytical instruments with low detection limits and from an increased understanding of the way contaminants move in the ground.

During investigations of groundwater contamination, water samples are collected and analyzed by one or more laboratories. The results are frequently baffling. Organic compounds that are expected in water samples may be absent, and unexpected compounds may be present. Even if the expected compounds do appear, their relative concentrations may not reflect the quantities of chemicals used or dumped at a facility. An understanding of the principles that control the movement and interconversion of organic compounds in the ground and of how sampling and analysis may change a water sample usually help to explain many unexpected results.

The following discussion is divided into three parts. The first part deals with the qualitative aspects of organics in groundwater--primarily with explanations for the occurrence of compounds that are not expected. The second part includes a quantitative discussion: reasons why concentrations of various contaminants may be different from what is expected. The final section includes recommendations that may make monitoring more reliable and facilitate the interpretation of groundwater-quality data.

QUALITATIVE ASPECTS

Many factors can make unexpected organic components appear in a groundwater sample: improper well location; organic compounds introduced during drilling, sampling, and analysis; and organics arising out of biological and chemical transformations.

Management of Toxic and Hazardous Wastes, H. Bhatt, R. Sykes, T. Sweeney (Editors)

Well Location and Depth

It is important to understand the hydrogeology of an area so that the wells monitor the right depth and are located in the right direction (usually downgradient) from the facility being monitored. Incorrect well locations can result in the detection of contamination from a different facility than the one(s) of interest.

Organics Introduced During Well Drilling

Contamination can enter a well during drilling. If a shallow aquifer is contaminated and a deeper one is to be monitored, the shallow, contaminated water can move down the borehole or well casing if the well is not sealed properly. In addition, many drilling techniques require water as part of the drilling fluids, and contaminated drilling water may introduce contaminants to the aquifer. Potable water should always be used where possible for drilling.

Even if there is no zone of contamination above the one to be monitored, and the drilling water is of good quality, contamination can be brought in on drilling equipment if it has not been properly cleaned after a previous job. A variety of oils, greases, and fuels are used in conjunction with drilling; components of these petroleum products may appear in samples from a recently drilled well.

Materials used intentionally during drilling may contaminate water samples. Organic thickeners for drilling fluids can be degraded by bacteria, and may increase the total organic carbon content in water samples. If polyvinyl chloride (PVC) casing is cemented together, residues from the adhesive, including tetrahydrofuran and various ketones, may enter the water. Flush-joint PVC casing and casing secured by screws avoid this potential source of contamination. Another contaminant source may be steel pipe that is coated with organic lubricants.

Organics Introduced During Sampling and Analysis

Winter sampling may require deicers, such as alcohols and glycols, to keep equipment from freezing. Sample bottles for extractable compounds are frequently cleaned with solvents like methylene chloride. The contents of such a sample bottle are not analyzed for methylene chloride. However, the contents of other bottles or vials in the same shipping container may be analyzed for this solvent and become contaminated by vapors from the extractables bottle. If plasticizer compounds (phthalates) are of interest in an investigation, soft plastics containing phthalates should not come in contact with the water samples.

In the laboratory, low levels of methylene chloride, trichlorofluoromethane, acetone, and phthalates may be present.[1] The use of trip and field blanks can frequently help distinguish between actual contamination of the aquifer and contamination intriduced during sampling and analysis. Trip blanks are usually vials of organic-free water that travel to and from the laboratory unopened and are analyzed with the well samples. Field blanks consist of organic-free water that has been exposed to the environment near a well and often run through pumps, bailers, or tubing that will come into contact with actual samples. The water is placed into a clean sample bottle and shipped for analysis. Trip blanks are most often used in conjunction with analyses for volatile organic compounds and field blanks commonly accompany samples to be analyzed for extractable organic compounds.

Trace contamination aside, it is possible for a laboratory to misidentify compounds occurring in high concentration. When gas chromatography is used in the analysis, the retention time of the compound on the chromatographic column is the key to the identification. A complicated sample matrix--a sample containing many chemicals--may alter retention times and lead to compound misidentification. A recent example of this problem involved a facility where high concentrations of benzyl alcohol and benzaldehyde in groundwater were expected near a waste disposal area.[2] The first round of analysis by gas chromatography showed high concentrations of hexachlorobutadiene and/or nitrobenzene and no benzyl alcohol or benzaldehyde. Confirming samples taken 21 months later were analyzed by gas chromatography and mass spectroscopy and showed high levels of only two compounds: benzyl alcohol and benzaldehyde; the "presence" of hexachlorobutadiene or nitrobenzene in the first round of analysis evidently resulted from laboratory misidentification.

Organics Arising from Biological and Chemical Transformations

While there are types of materials like polyethylene that are virtually indestructible in the environment, most consumer, industrial, and agricultural chemicals can be transformed or degraded.[3] Metabolic pathways or chemical environments may not be optimal for rapid degradation, but groundwater moves slowly, typicaly from inches to feet per day. Therfore, even slow reactions can yield detectable concentrations of degradation products near a facility.

Table I gives examples of the kinds of reactions that can occur. Of interest in many investigations is the fate of chlorinated organic compounds. Wood and coworkers have examined the degradation of chlorinated alkanes and alkenes [4] and have found that anaerobic bacteria in groundwater can degrade carbon tetrachloride, tetrachloroethylene, and 1,1,1-trichloroethane to

Table I

Degradation of Organic Coumpounds:
Some Biological and Chemical Transformations

Reaction Type	Examples	
	Starting Compound	Degradation Product(s)
Reductive dehalogenation	DDT[3]	DDD[4]
	carbon tetrachloride[4]	chloroform
	trichloroethylene[4]	dichloroethylene (all forms)
Hydrolysis	phthalate esters[3]	phthalic acid alcohols
	benzylchloride[3]	benzyl alcohol
Oxidation	saturated hydrocarbon (alkane)[5]	fatty acid(s)
	aldicarb[6]	aldicarb sulfoxide aldicarb sulfone

homologous compounds with one or more chlorines removed.

Most of the 113 organic compounds that appear on the U.S. Environmental Protection Agency's list of Priority Pollutants can be transformed to other compounds by microorganisms found in domestic wastewater.[7] Among the few compounds that resist degradation are selected pesticides, ethers, and heavy polychlorinated biphenyls (PCBs).

Chemical reactions not involving organisms can also change the structure of compounds that are dissolved in groundwater. Wastes in lagoons with standing fluid can undergo oxidation and photochemical decomposition before entering the ground. Compounds that are sensitive to hydrolysis like methyl bromide[8] and benzyl chloride can be changed to alcohols without the help of organisms. Organic esters may hydrolyze into their component carboxylic acid and alcohol.

QUANTITATIVE ASPECTS

Even when compounds are identified correctly and the appearance of unexpected compounds is explained, the concentrations that are found may not make sense. Replicate samples may have very different concentrations of contaminants, the concentrations in a particular well may change dramatically from one sampling period to the next, and wells that are close together may yield water of very different qualities. Many of these apparent anomalies can be explained, and changes in sampling or analysis protocols may help to reduce certain types of variability.

Variability Caused by Type of Monitoring

Contaminants in groundwater do not disperse as much as they do in surface water, so that proper placement and depth of wells is very important. Some contaminants like chlorinated solvents are denser than water, and if enough of the material is present to form a separate phase, it may sink through the aquifer, especilly under the influence of a downward hydraulic gradient or along a bore hole. Deep wells would be needed to detect or recover the bulk of this contamination. Other contaminants like some petroleum products are less dense than water, and shallow monitoring may intercept a floating layer of oil. Small amounts of either type of material that are dissolved will not substantially change the density of groundwater. In this case, the depth of monitoring should be chosen for other hydrologic reasons and without special regard to solute density.

The length of screen can dramatically affect the concentration of contaminants in well samples. A well with a short screen that coincides with a thin, heavily contaminated zone will yield heavily contaminated water. A well at the same location with a long screen will usually produce water with lower contaminant concentrations because a substantial part of the screen is open to relatively clean parts of the aquifer.

The concentration of organics can change dramatically as a well is pumped. Production wells that are turned on and off have cones of depression that vary in size. A cone that increases in size during pumping may intercept a zone of contamination after a while, resulting in increasingly contaminated well water. Alternatively, the expanding cone of influence may reach a clean river; the infiltrating river water will dilute the contamination in the aquifer. Finally, even if pumping is held constant, the concentration of contaminants coming out of a well may vary because the contamination is not uniformly distributed in the aquifer.

Variations in Contaminant Concentrations Caused by Attenuation

A waste disposal facility may contribute equal amounts of two organic contaminants, but a monitoring well located some distance away may yield water with very different concentrations of the two substances. Aside from the attenuation resulting from chemical and biological degradation, organic compounds have a wide range of affinities for organic and inorganic solids in the subsurface. Generally, the greater the affinity for solids the lower the affinity for the moving water phase; hence a strongly hydrophobic substance will migrate more slowly in an aquifer than a compound with a relatively high water solubility. PCBs and petroleum hydrocarbons are examples of strongly attenuated substances. In contrast, light alcohols are very soluble, and even industrial solvents like benzene and trichloroethylene are somewhat soluble in water and, therfore, move fairly well.

Probably a better gauge of the movement of organic compounds is a standard measure called the octanol/water partition coefficient. The coefficient is a ratio of the amount of a compound that dissolves in water. A high partition coefficient indicates that the substance preferentially dissolves in octanol, which roughly mimics the adsorptive properties of soil minerals and organic matter. In the environment, a contaminant with a high coefficient would be strongly adsorbed onto soil solids and would not be very mobile. values for the octanol/water partition coefficient are tabulated for many compounds[8] and can be estimated for others.[9] Table II contains values for the aqueous solubility and the partition coefficient for some common chemicals.

Table II.
Aqueous Solubilities and Octanol/Water Partition Coefficients for Selected Compounds[5]

Compound	Water Solubility* (mg/L)	Octanol/Water Partition Coefficient* (unitless ratio)
Benzene	1,800.	130.
Toluene	540.	490.
Aroclor 1221	15.	630.(est)
Aroclor 1260	0.0027	14,000,000.
Methylene chloride	17,000.	18.
1,1,1 Trichloroethane	950.	150.
Trichloroethylene	1,100.	190.
4,4' DDT	0.006	9,500.

* Values are given to two significant figures. The octanol/water partition coefficients are frequently given as logarithmic values.

Although the logarithms of partition coefficients are normally listed, non-logarighmic values ar included in Table II to emphasize that the values range over many orders of magnitude.

Quantitative Problems that Arise During Sampling

Volatile organic compounds, a group frequently monitored for, present special problems during sampling. Volatiles are easily dissipated from a sample if it is agitated, left open, or subjected to a prolonged drop in pressure. No sampling method is perfect, because all methods result in the escape of some contamination. It appears that bailers and peristaltic pumps change sample chemistry the least[10] and should be used for sampling where possible.

Another sampling problem arises when two or more phases are present. In addition to the predominant water phase, there can be solids or other liquid phases, most commonly oils or solvents. An extraction done on such mixtures can result in very high concentrations of contaminants, frequently far in excess of what could have been in the aqueous phase alone. An example is a concentration of 100 mg/L of bis(2-ethylhexyl) phthalate found in a sample collected recently.[2] The aqueous solubility of this compound in 25°C water is 0.4 mg/L. Presumable, one or more nonaqueous phases were present.

Quantitative Problems that Arise in the Laboratory

The laboratory performance problems outlined in this section are not included to cast doubt on the capabilities of industrial and commercial laboratories, but rather to show the formidable problems that arise during the analysis of environmental samples.

The assignment in the field of random or disguised numbers to well samples and replicates is an important element in quality control. Experience on a number of projects has shown that the results for two samples tend to be closer when a laboratory knows that the samples are replicates than when the identity of replicates is disguised. Such results can distort the meaning of the whole sampling program. The cause of this problem probably varies, but may be related to several factors including:

1. The increased likelihood that known replicates are run consecutively through an instrument. The change in calibration or instrument response is probably small during the time that two samples are run, but may be larger if several other samples intervene.
2. Knowing that samples are replicates may bias a technician's procedures and interpretation of the data. Bias can also be introduced when selected replicates are re re-run in

order to better "match" the results. Coding the replicates will eliminate these sources of bias.

Frequently, there is one predominant chemical, and, in the case of chromatography, it may be difficult to detect compounds in lower concentrations that are masked by the predominant one. Special sample preparation or changes in instrument conditions may reveal hidden chromatography peaks.

Samples that are heavily contaminated may leave residues in the instrument and alter its response. Even the use of standard laboratory reference samples (not matrix spikes) may not measure the effect of the matrix of field samples on the instruments. The analyst must be alert to sets of samples that have widely varying concentrations of contaminants and complex matrices. In such cases, instruments may have to be cleaned up or recalibrated frequently. Another analytical problem, especially for soluble organics like alcohols and ketones, is that of poor recovery. Special extraction techniques may be required to analyze for relabively hydrophilic organic compounds such as these.

RECOMMENDATIONS

The following practices should help to minimize the difficulties in monitoring for organic compounds.

1. Contaminants generally flush through an aquifer very slowly. Collect whatever information is available concerning use of chemicals or deposition of waste for as far back in time as possible.
2. The drilling should be watched carefully. The drilling rig and materials used to construct the wells must be clean before use. Glues or cements should be avoided where possible. A source of clean water is needed during drilling. If the quality of the drilling water is not known, a sample of it should be analyzed.
3. The wells should be developed properly: any water added to a bore hole should be removed before sampling. In some cases, considerably more water may have to be pumped out.
4. If chemicals are to be used to clean bailers, tubing and other sampling equipment, cleaning agents should not include compounds that might also be found in the samples.
5. The same equipment, techniques, and personnel (if possible) should be employed each time a well is sampled. Uniform procedures are also important when results from a group of wells are to be compared.
6. The groundwater sample should be examined and described carefully for the presence of sediment, emulsions, and more than one liquid phase.
7. Trip and field blanks are frequently desirable when knowledge about low levels of contamination is important.

8. Samples should not be stored or transported in areas or vehicles with containers of solvents or fuels.
9. Redundant or overlapping analyses can help determine whether sampling and laboratory protocols are adequate. Types of redundancy include:

    a. field replicates (more than one sample taken from one well at one time) coded so the laboratory does not know they are replicates;
    b. total organic carbon (TOC) and chemical oxygen demand (COD) analyses; the relationship between the results for these analyses may vary widely from site to site, but within an area with a certain type of contamination, the results of one analysis may help to confirm the results of the other;
    c. total organic halogen (TOX) analysis (as done by a TOX analyzer) as compared to the results for individual halogenated organic compounds as done by gas chromatography.

10. Results should be verified through a followup round of sampling and analysis, especially if unusual compounds are detected.

REFERENCES

1. Schroeder, R.A., and D.S. Snavely. "Survey of Selected Organic Compounds in Aquifers of New York State Excluding Long Island," U.S. Geological Survey Water Resources Investigations 81-47 (1981).
2. Unpublished project data, Geraghty & Miller, Inc., 1982.
3. Alexander, M. "Biodegradation of Chemicals of Environmental Concern," Science, 211:132-138 (1981).
4. Wood, P.R., F.Z. Parsons, R.F. Lang, and I.L. Payan. "Introductory Study of the Biodegradation of the Chlorinated Methane, Ethane and Ethene compounds," Presented at the American Water Works Association Annual Conference and Exposition, ST. Louis, Missouri, June 7-11, 1981.
5. Morrill, L.G., B.C. Mahilum, and S.H. Mohiuddin. Organic Compounds in Soils: Sorption, Degradation and Persistence. Ann Arbor Science Publishers, Inc., Ann Arbor, Michigan, 1982.
6. Bromilow, R.H. "Breakdown and Fate of Oximecarbamate Nematicides in Crops and Soils," Ann.Appl. Biol., 75:473 (1973).
7. Tabak, H.H., S.A. Quave, C.I. Mashni, and E.F. Barth. "Biodegradability Studies with Organic Priority Pollutant Compounds," Journal WPCF, 53:1503-1518 (1981).

8. U.S. Environmental Protection Agency, Treatability Manual, Office of Research and Development, Washington, D.C., 1981.
9. Lyman, W.J., W.F. Reehl, and D.H. Rosenblatt, eds., Handbook of Chemical Property Estimation Methods: Environmental Behavior of Organic Compounds, McGraw-Hill, New York, 1982.
10. Gibb, J.P., R.M. Schuller, and R.A. Griffen. "Procedures for the Collection of REpresentative Water Quality Data from Monitoring Wells," Illinois State Water Survey and Geological Survey Cooperative Groundwater Report 7, Champaign, 1981.

CHAPTER 12

PRIVATE WELL SAMPLING IN VICINITY OF RE-SOLVE, INC., HAZARDOUS WASTE SITE

Thomas E. Tetreault and
Paul M. Williams
Camp, Dresser & McKee, Inc.

INTRODUCTION AND OBJECTIVE

The Re-Solve, Inc. hazardous waste site is an abandoned solvent reclamation facility located in a rural area of Bristol County, Massachusetts. The site is listed on USEPA's National Priority Site list and is eligible for funding under the Comprehensive Environmental Response, Compensation and Liability Act of 1980 - "Superfund". A site location map and site plan are shown on Figures 1 and 2, respectively.

The site was utilized actively from 1955 to 1980 for reprocessing various waste oils, and several sources of contaminants still remain. Those which represent potential sources of groundwater contamination include four unlined sludge lagoons and a cooling pond which has been filled with sand and other soils which have been contaminated as a result of spills or land disposal.

Concerns were raised that off-site migration of contaminants via groundwater flow could adversely impact potable water quality in nearby private wells, with resulting potential public health impacts. The objective of this paper is to describe the sampling design, protocol and analysis for private drinking water well samples taken from the vicinity of the Re-Solve, Inc. hazardous site, and to evaluate sampling results with respect to the potential for significant public health impacts in the vicinity of the site. A secondary objective of this paper is to describe the private well sampling design in the context of its use in supplementing the groundwater monitoring network designed for the site.

PRIVATE WELL SAMPLING

Existing Data

In December 1981, the Massachusetts Department of Environmental Quality Engineering (DEQE) sampled six private wells in the

Management of Toxic and Hazardous Wastes, H. Bhatt, R. Sykes, T. Sweeney (Editors)

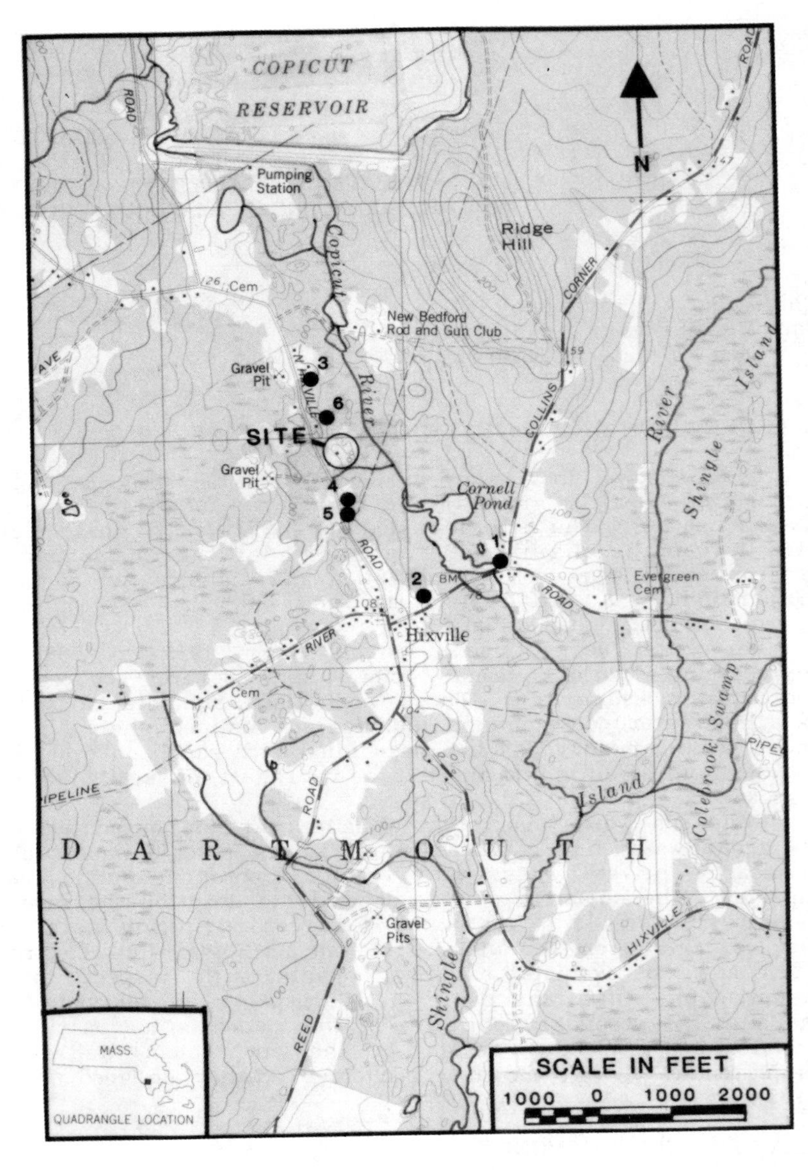

Figure 1. ReSolve, Inc. Site locus map and location of residential well sampling points.

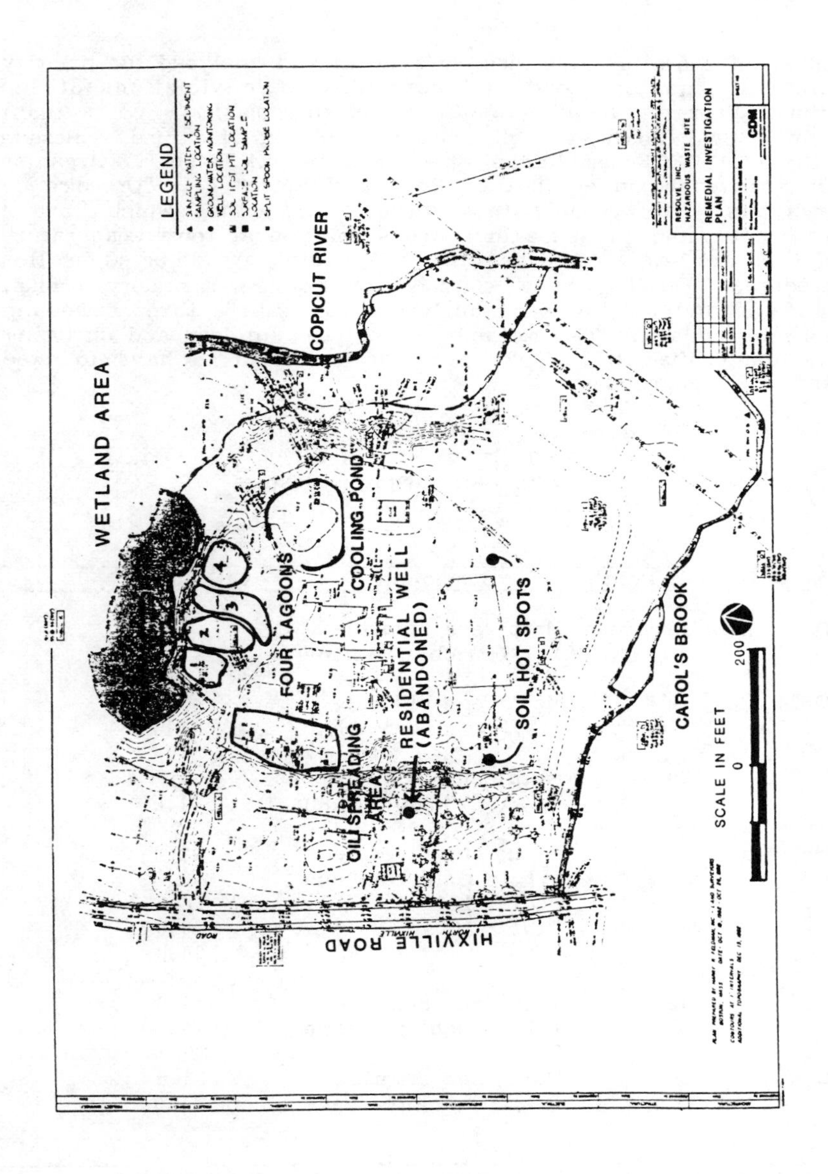

Figure 2. Existing Site Conditions, Re-Sove Inc., December, 1982

vicinity of the site. Samples were taken and analyzed for priority pollutant organics, lead and chromium. No violations of the National Primary Drinking Water Standard's maximum contaminant levels (MCLs) which are enforced by DEQE were indicated. Results of the 1981 DEQE sampling efforts are shown in Table I. Organics were analyzed according to U.S. Procedural Method 624- "Organics by Purge and Trap". Only those organic compounds which have a significant vapor pressure in aqueous solution at room temperature and thus are amenable to partition by purging are detected by this procedure. Quality control consisted of running laboratory blanks, duplicates and spiking each analytical run with a three compound internal standard. The methods of analyses for lead and chromium were unspecified by DEQE, but were assumed to have followed standard methods [1].

Table I
1981 Massachusetts DEQE
Private Well Sampling Results

Sample	Well Depth	Compound	Level (ug/l)
D-1	18'	Methylene chloride	1.1
		1,1,1-trichloroethane	<1.0
D-2	20'	Methylene chloride	<1.0
		1,1,1-trichloroethane	<1.0
D-3	120'	Methylene Chloride	3.7
		1,1,1,-trichloroethane	<1.0
D-4	8'	Methylene chloride	<1.0
		1,1,1-trichloroethane	<1.0
		PCB - 1242	0.03
		PCB - 1254	0.05
		Lead	50.0
D-5	45'	Methylene chloride	1.0
		1,1,1-trichloroethane	<1.0
		Lead	120.0
D-6	24'	Methylene chloride	1.0
		1,1,1-trichloroethane	<1.0

Source: Massachusetts Department of Environmental Quality Engineering, Lawrence Experiment Station (December, 1981)

Sampling Design

After a review of the existing data discussed above, it was determined that re-sampling of all of the private wells which were sampled by DEQE in 1981 would be justified. Home owners at each of the previously sampled residences were contacted for permission to re-sample their well water. All consented, with the exception of the owners at point D-2, who repeatedly could not be contacted. This residence was subsequently dropped from the re-sample design and another, identified as sample point D-7, was added. The "D-7" sample point was selected because it was within 200 feet of the site, and the residents there had registered previously complaints of malodorous water with poor color. These residences were selected for the planned re-sampling program in order to compare the results with previous sampling efforts and thus aid in initially establishing historical trends in potable water quality in the vicinity of the site. Such trends may aid in the determination of potential threats associated with groundwater contamination.

Private Well Sampling Design and Groundwater Monitoring

One aspect of the private well sampling design was the additional information which was provided to supplement the overall hydrogeologic investigations and specifically the groundwater monitoring plan.

The overall objective of the hydrogeologic investigation was evaluation of baseline conditions affected by contamination resulting from activities in and around the site. Data collected from the initial phase of hydrogeologic investigations formed the basis for estimating the extent and rate of contaminated groundwater movement and the type and level of contamination present. Data obtained from private well samples could thus be used to supplement the hydrogeologic investigation, albeit with extreme caution since the investigator has less control over such variables as well as water contamination due to construction, equipment, proximity to septic systems, and the like.

Recognizing these potential limitations, however, an attempt was made to fit the location, depth and other relevant parameters of the private wells into the context of an overall hydrogeologic framework, which is described briefly below.

Hydrogeologic Framework

Local Precipitation in southeastern Massachusetts averages 42 inches per year and is the source of all fresh ground and surface water in the region. Evapo-transpiration is estimated at 22-24 inches per year, thus limiting the amount of water available for runoff or groundwater recharge to 18 to 20 inches per year.

The site lies adjacent to the Copicut River, a small branch of the East Branch, Westport River. The Copicut River arises from the overflow of the Copicut Reservoir approximately 0.75 miles north and upgradient of the site. From the site, the river flows approximately 1/4 mile to Cornell Pond. Overflow from Cornell Pond discharges to a surface drainage network which eventually discharges to Rhode Island Sound.

The geology and groundwater conditions in the vicinity of the Re-Solve, Inc. site have been previously mapped by Willey et al, 1978 [2]. In summary, there are five separate units of interest in the site area; bedrock, glacial till, glacial outwash, fill/recent deposits and swamp deposits. Each of these units and spatial relationships are discussed below.

- Bedrock underlies the entire site area at varying depths ranging from 10' to 40' below land surface. The bedrock consists of a granite or grandiorite, which is a pink or pinkish grey color. The bedrock is fractured to varying degrees at different locations around the site.
- Glacial till underlies the site area, and is variable in thickness and composition. The till is, for the most part, a sand with gravel, cobbles, boulders and silt. The till ranges in thickness from several feet to 20 feet thick.
- Glacial outwash around this site occurs at many locations and ranges in thickness from several feet to several tens of feet. Two separate phases of outwash are generally present. The most predominant outwash type consists of fine to coarse sand and gravel with varying amounts of cobbles and a minor amount of silt. A secondary outwash type consists of a fine to medium sand with minor amounts of silt. Generally, this secondary outwash type (fine to medium sand) overlies the coarser variety and may have been much more predominant prior to the excavation of the site for sand and gravel. All of the outwash encountered at this site is highly stratified into fine and coarse layering of several inches to several feet in thickness. In general, the thicker more productive areas of glacial outwash material occur south of the project area.
- Fill occurs primarily on and immediately adjacent to the Re-Solve site. Some of the fill appears to be of recent vintage while a significant portion has been on site for some time. The fill consists of a mixture of sands, gravels and silts. Recent sedimentary deposits occur primarily along the Copicut River and consist of sands, gravels and silt eroded and reworked from higher basin elevations.
- Swamp deposits occur throughout the study area and consist of black partially decomposed organic matter. In some of the borings, remnant swamp deposits occur beneath fill at depths of up to 11' and mark the original ground surface.

A total of 35 individual monitoring wells were constructed in and around the site at 19 different locations to provide subsurface geologic data, soil samples, groundwater samples and piezometric head. A detailed discussion of this program will be included in a separate paper, but it is significant to note that as many as three types of wells were constructed in "nested" configurations in order to obtain groundwater information from the three major geologic units described above. Known depths of private wells nearby could then be correlated with the more precisely logged groundwater monitoring to provide a cross reference to any quality variations encountered in the private well sampling program. A brief description of the types of monitoring wells and their functions are given below.

- Shallow Wells - This monitoring well type consists of a 5' screen and an appropriate length of riser placed such that the top of the water table is in the screened portion of the well. This type of well was constructed in areas where oils were thought to be floating on the water table.
- Intermediate Wells - This monitoring well type consists of a 5' long screen and appropriate length riser placed such that the well screen is near the bottom of the most permeable section of subsurface materials and/or such that the screen is opposite a soil section which indicate a high level of contaminants. The well screens in these intermediate wells are entirely below the top of the water table and are expected to capture the volatile organic phase of contaminants.
- Deep Wells - This monitoring well type consists of a 2' long screen and appropriate length riser placed such that the screen is installed into the bedrock to capture any possible contaminants in this zone.

Locations of these wells are shown on Figure 2.

Installation of the groundwater monitoring network provided a means of establishing local piezometric surface and gradient with respect to the location of the private wells. Once the monitoring wells had been established using careful survey/topographic control, the value of the private wels becomes evident in that these essentially pre-established non-controlled sample points then become a relatively inexpensive means of contributing to the groundwater monitoring data base while simultaneously serving as a means of establishing any potential direct health impact to residents. As such, these sample points then become a means of supplementing the data base and a means of checking hypotheses developed from the groundwater monitoring network with the actual quality of consumption from pre-established drinking wells. It must be emphasized, however, that the unknown variables associated with each of the private wells sampled places limits upon the conclusions which may be drawn from these data.

The preliminary results of the groundwater monitoring survey established that all of the private wells were located upgradient or laterally gradient of the Re-Solve site, and that none of the residences sampled were directly down gradient of the site. There are no local residences located directly down gradient from the site.

Sampling Protocol

Having briefly established the hydrogeologic framework and prior to actual field sampling, a sampling protocol was defined with the objective of minimizing risk and uncertainty in terms of sampling procedures and analytical results. The remainder of this section describes the protocol utilized for sampling from private residential wells. This protocol includes office, field and laboratory handling procedures for samples from private wells near the site. It does not include laboratory analytical procedures.

The parameters (sample types) analyzed for each of the private wells are shown in Table II along with the appropriate container, volume and preservative.

Table II
Sample Containers and Preservatives

Sample Type	Container	Volume	Preservative
(1) Priority Pollutant Organic	Glass Amber	4000 ml	Cool $4^{o}C$
(2) Volatile Organic	VOA Vial	3x45 ml	Cool $4^{o}C$
(3) Trace Metal Scan	PE Round	1000 ml	HNO_3 pH <2
(4) PCBs	Glass	4000ml	Cool $4^{o}C$

Source: CDM Inc.

Because of the nature of the contaminants present on site, microbiological analyses were determined not to be necessary to assess the potential public health impacts in this case.

In addition to preparation of the necessary sampling containers and preservatives, a complete set of field blanks was prepared. Also prepared prior to actual field sampling were pre-coded sample identification tags and chain-or-custody record sheets.

Sample Collection Procedures

The following provides a generic description of the sample collection procedures utilized for well sampling at the six residences described previously. Minor variations occurred from one sample point to another, primarily due to the individual water system configuration and access considerations. The procedure described was used to collect samples from the residential water supplies for the nonmicrobiological analyses outlined in Table II. The primary objective of these procedures is to collect a sample representative of the groundwater supply and not water standing in the delivery system or well casing.

In a nonpumped well, there will be little or no vertical mixing of the water, and stratification may occur [3]. Water in the screened section will mix with the groundwater due to normal flow patterns, but the well water above the screened section will remain isolated and become stagnant. Stagnant water may contain foreign material inadvertently or deliberately introduced from the surface, resulting in nonrepresentative data and misleading interpretations [4].

In most cases, groundwater samples from existing residential water supplies were obtained from taps or spigots on the existing delivery system. Samples were collected from the tap closest to the well as practical and upstream of any filtration of water treatment device.

Two separate operation steps are required to obtain a representative sample.

- presampling system purging, followed by
- sample collection

Presample Purging

Before any samples are collected, all standing (stagnant) water was purged or removed from the delivery system. The volume of water contained in the well casing, pressure or holding tanks, and other plumbing and appurtenances (pipes, hoses, etc.) was estimated.

Each system was then purged with a minimum of three (3) times the calculated casing volume before sampling commenced. Care was exercised before pumping a well to preclude the possibility of overpumping. Excessive pumping can result in flow entering a well from outside the zone of interest. The purging necessary to obtain a sample representative of the groundwater supply depends on a number of factors:

- pump intake level
- specific capacity of the aquifer
- well efficiency

Information obtained during pumping is required to determine the specific capacity of the aquifer and well efficiency, therefore, the purging volume can only be estimated for a specific well for the initial sampling.

Sampling

After the required volume of water was purged from the delivery system, the sampling tap was shut off. Sample bottles with required preservatives were then brought to the sampling point. The tap was then turned on, and the flow adjusted to about 100 ml/in. Sample bottles were filled as required for specific analyses to be completed (Table II).

Once the tap was shut off, all filters, aerators and treatment systems were reconnected as appropriate. In addition to information normally recorded in the field notebook, the following information was included where available:

- resident's name
- address
- number of household members
- number of children
- any health complaints
- sampling location (specific tap or spigot)
- filtering or treatment systems on delivery system
- Aerator or filter on sampling tap
- well casing diameter (ID)
- water level
- well volume
- pressure on holding tank volume
- appurtenances and other plumbing volume
- total delivery system volume
- purge flow rate
- purge time
- total purge volume

Sampling Results

In January 1983, the five private wells previuosly tested by the DEQE were sampled by CDM along with the "D-7" well located just north of the site on North Hisville Road. In addition, a hand dug well (8.5' - not used as a water supply) at point D-7 and a hand dug well (9.0') at the abandoned house directly west of the site were sampled. The locations of the six residential homes are shown on Figure 2. There were no contaminants detected in the private well water that pose a threat to public health [5]. The findings confirmed the earlier testing completed by the DEQE. Analytical results are presented in Table III.

Table III. Re-Solve Site Findings: Private Wells Organic Compounds

Corresponding DEQE No.	Area	Depth (feet)	Sample		Level	Notes
D-3	PW010	120	001	--	--	--
D-7	PW020	35	002	trichlorothylene 2-hexanone	6.1 LT	H_2S odor
D-4	PW030	8.0	003	--	--	--
D-6	PW040	24.0	004	--	--	--
D-5	PW050	45.0	005	--	--	--
D-1	PW070	--	007	methylene chloride tetrachloroethane	9.7	Field blank
(D-7)	PW100	8.0	001	bis (2-ethylhexyl) phthalate	50	Dug well not used as water supply
-	PW200	9.0	002	--	--	Dug well at abandoned house

Source: U.S. EPA and CDM.

CONCLUSIONS

The sampling methods and protocol described were successful in that they provided verifiable and reproducible results when carefully followed.

After a thorough review of existing analytical data, USEPA and Massachusetts DEQE have determined that the site does not present an immediate public health threatening the potable water supply.

Preliminary results of the private well sampling indicate that conaminated groundwater from the Re-Solve Inc. site has not polluted any of the actively used private wells in excess of established EPA Maximum Contaminant Levels at this time. Streams in the area of the Re-Solve site, i.e. Copicut River and Carol's Brook, have apparantly acted as a hydraulic barrier to contaminant migration, and further field study is attempting to confirm this hypothesis.

There were no contaminants detected in the private well water that pose a threat to public health. These findings confirmed the earlier testing completed by the DEQE in 1981. Since methylene chloride is frequently used in the extraction stage of analysis for organics, the pressence of this compound may be due to laboratory contamination. The presence of bis (2-ethylhexyl) phthalate in sample PW100 (dug well near point D-7) (not used as a water supply) indicates possible contaminant migration from the Re-Solve Site, however, this well is not presently used as a water supply by the residents of this household. Finally, the trace levels of 2-hexanone and 6.1 ug/l of trichloroethene in the sample from the D-7 water supply well do not exceed the established EPA Maximum Contaminant Levels (MCLs) for the given concentrations. Re-sampling of the six private wells will be performed again during the summer of 1983 by a WEPA contractor.

REFERENCES

1. American Public Health Association, American Water Works Association, and Water Pollution Control Federation, "Standard Methods for the Examination of Water and Waste Water," 14th ed., Amer. Pub. Hlth. Assn., Washington, DC (1975).
2. Willey, R.E., J.R. Williams and G.D. Tasker. "Water Resources of the Coastal Drainage Basins of Southeastern Massachusetts, Westport River, Westport to Seekonk." U.S. Geological Survey Hydrogologic Atlas HA-275 (1978).
3. U.S. Environmental Protection Agency, Office of Drinking Water. "Manual of Individual Water Supply System" EPA-570/9-82-004 (1982).
4. Gibson, V.P. and R.D. Singer. "Water Well Manual" Preimer Press, Berkeley, California (1971).
5. Camp Dresser & McKee Inc. "Preliminary Remedial Investigation and Feasibility Study for Re-Solve, Inc. Hazardous Waste Site, Dartmouth, Massachusetts." Report to U.S. EPA dated May 25, 1983 (1983).

SECTION III

TREATMENT

CHAPTER 13

LIQUID HAZARDOUS WASTE TREATMENT DESIGN

T.H. Coughlin, O.A. Clemens, and J. Johnson
Dravo Recovery Systems

INTRODUCTION

A liquid hazardous waste treatment facility capable of treating 360,000 tons per year has been designed for the BKK Corporation by Dravo Corporation. The treatment facility will have the capability of treating at least 80% of the liquid hazardous waste that now enters a California Class I landfill. The facility will be capable of separating the hazardous material from the water phase of the hazardous liquid and produce a high solids material capable of being landfilled in an environmentally safer form. The water phase will be discharged to a publicly owned treatment plant and meet stringent pretreatment standards.

The treatment design is based on existing process technology that has been proven by Dravo in other industrial applications. Experience with effective removal of oils and greases from industrial discharges, along with heavy metal removal from mine discharges and the ability to concentrate the solids into a cake like material was the primary basis for the initial process choice. The integration of known process designs based on proprietary equipment and laboratory testing of composited liquid waste samples resulted in a preliminary design for the treatment facility. The process design was further refined by conducting pilot plant tests on various waste streams. The pilot plant data was representative in that the process equipment used was the same size and capacity that is in the final plant design.

The treatment technology utilizes physiochemical processes that effectively change the hazardous soluble elements into recoverable nonsoluble solids. The solids are further concentrated by a low pressure belt filter press that recovers 95% of the solids. The remaining solids are concentrated in an electroflotation system that results in a final treated effluent of 20 mg/l total suspended solids.

The facility will be capable of decreasing the amount of treatable liquid hazardous waste to be landfilled. The greatly reduced volume of treated liquid hazardous waste can be transported

Management of Toxic and Hazardous Wastes, H. Bhatt, R. Sykes, T. Sweeney (Editors)

to the landfill in a form much less likely to cause major environmental damage due to accidental spills. Furthermore, the material will be a cake-like material that will not pose as serious a hazard in a landfill as would the liquid non-treated material. The prospect exists that the material may be classified non-hazardous due to the higher pH of the solids and the manner in which the heavy metals are encapsulated in the solid material.

DESCRIPTION

The proposed facility would receive bulk loads of liquid hazardous industrial wastes which will be treated using previously tried and proven technologies. The industrial wastes would come (but not limited to) from the following industries:

- Aerospace
- Airlines
- Automotive
- Chemical Companies
- Electric Companies
- Metal Products
- Plating
- Refineries

All the liquid wastes would be held in storage tanks before being processed through the treatment system. The wastes would be checked prior to being stored in order to exclude those liquid wastes which are not treatable, such as radioactive wastes, spent solvents, etc. Transporter manifest records kept on individual liquid waste loads indicate that up to 450,000 tons of liquid hazardous waste are landfilled at one specific location. The portion that can be treated with the Dravo design will represent approximately 80% of the total.

Table I shows the typical categories of waste and their approximate percentage of the treatable total.

Table I. Treatable Waste Categories

Waste Category	%
Acid	24.3
Alkaline Solutions	15.7
Chemical Treatment Solutions	2.0
Paint Sludge	4.6
Drilling Mud	19.9
Oily Wastes	20.8
Tank Bottoms	10.2
Mud and Water	2.5

The environmental acceptability of landfilling liquid hazardous waste is being met with stringent requirements for the owner-operators of landfills. In some states, California for example, the regulatory aim for the near future is to prohibit the landfilling of any hazardous liquids. The need is very apparent for the owner-operator to seek regulatory relief or treatment technologies to continue to be able to receive and landfill liquid hazardous wastes. The treatment option allows the operator to comply with expected future regulations and yields a longer site life by reducing tonnage to be landfilled. Other treatment technologies such as incineration of waste solvents would effectively treat the remaining liquid waste now being landfilled.

WASTE CHARACTERIZATION

Individual liquid wastes needed a very complex process treatment system. Therefore, composite samples of the individual waste streams were collected over a period of time and mixed together in roughly the proportions shown in Table I. Two primary waste streams were then made up for laboratory bench scale test work. These two waste streams were oily waste and neutralized acid plus alkaline waste. Typical ranges of concentrations for the two waste streams and the subnate of the oily waste are shown in Table II.

Table II. Waste Stream Analysis

Parameter	Oily Waste	Oily Waste Subnate	Acid/Alkaline Waste
pH	7.0-10.3	7.0-10.3	8.0-10.5
Settleable Solids	3.0-11.0%	2.0-3.0%	5.0-20.0%
Oil/Grease	3.0-50%	0.5-5.0%	0.1-1.0%
Metals mg/L Total			
Lead	0.1-10.0	0.1-10.0	<.1-170
Nickel	1.0-20.0	1.0-1.5	<.1->1000
Cadmium	<.1-1.0	<.1-1.0	<.1-10
Silver	<.1-0.2	<.1-0.1	<.1-0.4
Copper	0.5-13.0	0.5-3.0	1.0->1000
Chrome	0.1-50.0	<.1-3.3	1.0->1000
Zinc	0.2-8.0	<.1-5.0	1.0->500

The early waste characterization studies indicate that occasional loads of high strength (1%) acids were disposed of by chemical companies. In addition, hydrofluoric acid and chromate waste were received for disposal. These special wastes if allowed to mix with the other composited wastes would present treatment problems. Separate pretreatment systems for these wastes were included in the design, with the product then mixed back into the acid/alkaline waste stream.

The mixing of the various individual streams allowed a generalized physical/chemical treatment process to be used. The process would be able to store the individual wastes in large storage tanks and treat special wastes separately. Chemicals to enhance the recovery of solids and soluble heavy metals would be metered into the system. The use of the belt press to increase the solids content of the waste and the use of the electroflotation system to produce a clean effluent would complete the basic flow sheet. A generalized treatment process would complete the basic flow sheet. A generalized treatment process diagram utilizing these items is shown in Figure 1.

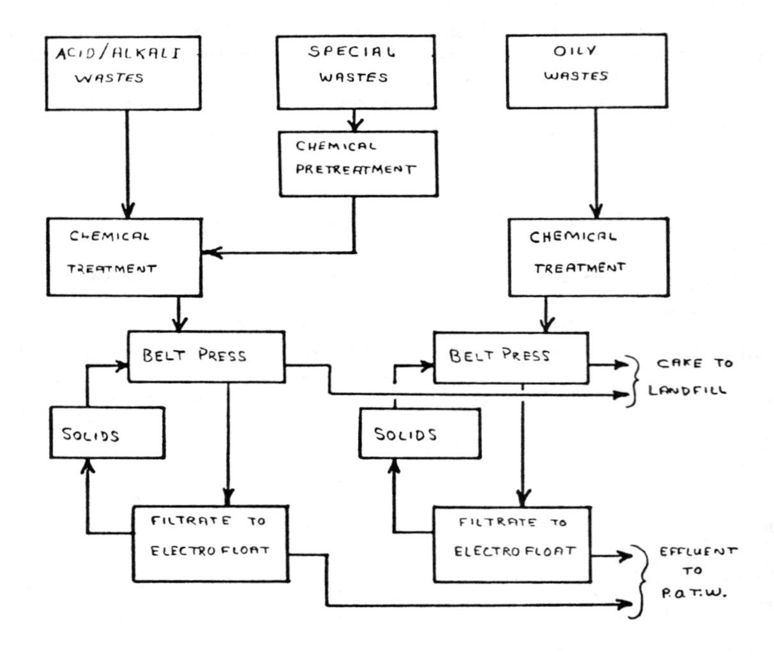

Figure 1. General Process Flow Diagram

LABORATORY AND PILOT PLANT TEST RESULTS

Prior to detailed process design, the composited wastes were treated in the laboratory with a belt press simulator and an electroflotation cylinder. The belt press simulator was capable of yielding preliminary results that indicated whether the waste liquid is pressable. Typical press simulator test results indicate whether the material can produce a friable cake, the percent moisture in the cake, and cake release characteristics from the filter cloth. The electroflotation test results yield information concerning electrical input, chemical dosage, solids production and effluent characteristics. Analytical results for the final treated effluent are summarized in Table III shown below.

Table III. Bench Scale Results - Composited Samples

Parameter mg/L	Acid/ Alkali Feed	Acid/ Alkali Effluent	Oily Waste Feed	Oily Waste Effluent	Combined Effluent
pH	4.7	8.2	6.1	9.2	8.0
COD	5170.	2930.	138000.	2600.	2765.
Oil/Grease	126.	18.	49000.	15.	17.
Sus. Solids	5660.	37.	27330.	137.	87.
Metals Total mg/L					
Cadmium	0.17	N.D.	-	-	0.01
Chromium	115.	N.D.	-	-	0.10
Copper	695.	1.3	-	-	0.70
Iron	555.	0.7	9.4	3.0	1.90
Lead	3.	N.D.	1.2	N.D.	0.10
Nickel	1100.	1.4	-	-	1.20
Zinc	900.	0.3	-	-	0.20

The laboratory test data indicated that the composited waste streams could be treated with results similar to those from other Dravo designed wastewater treatment facilities. To further complement the ongoing design work, a pilot plant test program was initiated to field test larger volumes of composited wastes. The pilot plant was unique in that full scale equipment was utilized to treat the liquid wastewater.

The pilot plant layout, (Figure 2) was comprised of separate storage tanks and trailer mounted mobile treatment units. The mobile belt press unit was comprised of a 2 meter press, chemical mixing tanks, air compressor, chemical feed pumps and conditioning

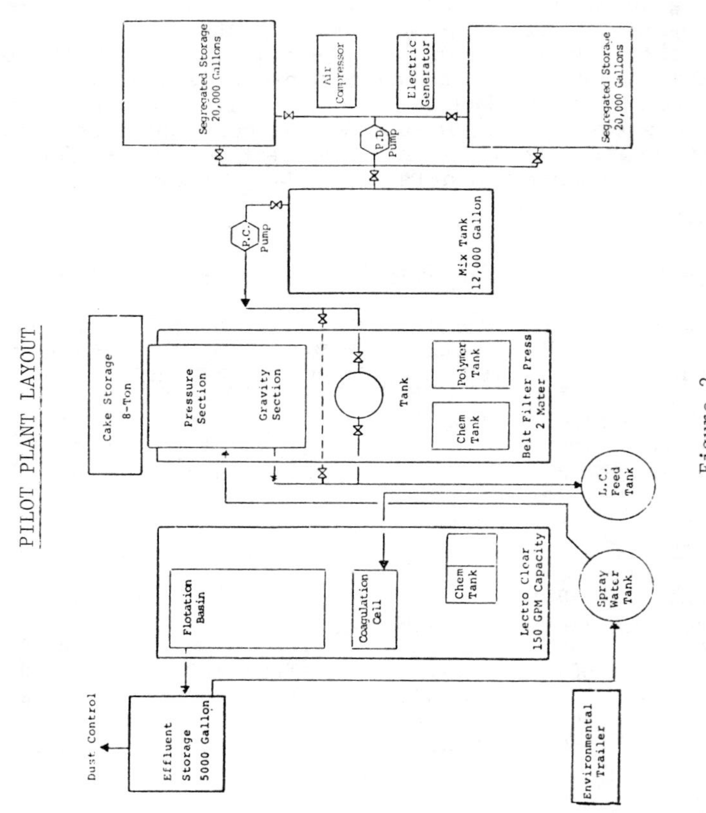

PILOT PLANT LAYOUT
Segregated Storage 20,000 Gallons
Air Compressor
Electric Generator
Segregated Storage 20,000 Gallons
P.D. Pump
P.C. Pump
Mix Tank 12,000 Gallon
Cake Storage 8-Ton
Pressure Section
Gravity Section
Tank
Polymer Tank
Chem Tank
Belt Filter Press 2 Mater
L.C. Feed Tank
Flotation Basin
Coagulation Cell
Chem Tank
Lectro Clear 150 GPM Capacity
Spray Water Tank
Dust Control
Effluent Storage 5000 Gallon
Environmental Trailer

Figure 2

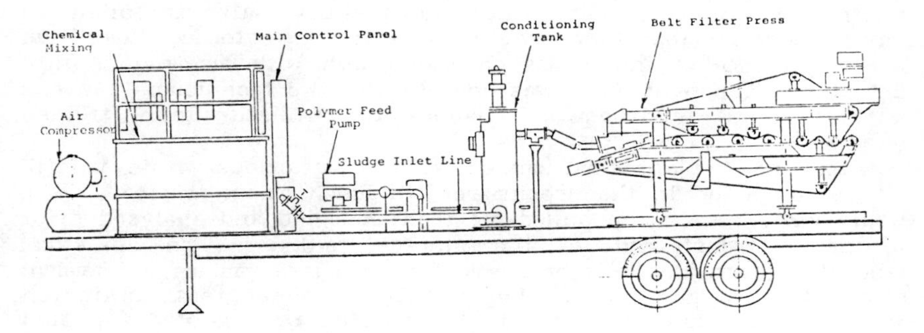

Figure 3. Mobile Belt Press Unit

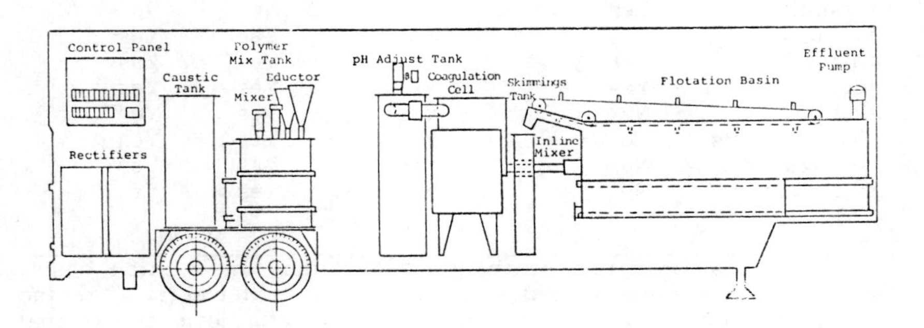

Figure 4. Mobile Electroflotation Unit

tank (Figure 3). The mobile electroflotation unit was comprised of chemical mixing tanks, rectifiers, pH adjustment tank and flotation basin. (Figure 4) The pilot plant set up also included an environmental laboratory equipped with an atomic absorption spectrophotometer and a gas chromotograph with electron capture detector. The pilot plant was operated by two technicians and one advisor, the environmental laboratory utilized four trained chemists.

The pilot plant was located at the hazardous waste landfill site which aided in the procurement of fresh waste to treat. The waste loads were to be pulled off road, sampled and analyzed prior to being transferred to the storage tanks. The analytical laboratory performed a spot check for phenols, cyanides, hexavalent chromate, pH and odor. The results of these tests indicated whether the waste load was to be rejected or accepted for pilot plant treatment. If the waste was accepted, then further chemical analysis of the waste was begun.

Table IV outlined the schedule of analytical testing for the waste loads and for key steps within the pilot plant system.

Table IV. Pilot Plant Laboratory Analysis

Parameter	Waste Loads	Waste Feed	Cake	Effluent
Odor	Yes	No	No	No
Cyanides	Yes	No	No	No
Phenols	Yes	Yes	Yes	Yes
Chromate	Yes	Yes	Yes	Yes
pH	Yes	Yes	Yes	Yes
Solids	No	Yes	Yes	Yes
Heavy Metals	No	Yes	Yes	Yes
Oil/Grease	No	Yes	Yes	Yes
Chlorinated Solvents	No	Yes	Yes	Yes

The actual operation of the pilot was done on a batch basis, that is, once a storage tank was filled, the material was treated through the process. The correct dosage of chemicals to pretreat the liquid waste was determined in the laboratory. This resulted in the soluble elements in the liquid forming insoluble precipitates. The waste was further treated in the laboratory using the belt press simulator and electroflotation cylinder. Approximate product characteristics such as metal content of the effluent and cake percent solids could be determined at this point.

The pilot plant treated three major types of composited liquid wastes. The analytical information shown in Table V summarizes the average metal values for feed into the treatment system and effluent out of the system.

Table V. Analytical Data for Different Composites

Metal-Total mg/L	Metal Plating In	Metal Plating Out	Oily Water In	Oily Water Out	Acid/Alkali In	Acid/Alkali Out
Cadmium	96.	<0.1	0.3	0.09	53.	0.1
Chrome	1248.	2.4	4.6	0.01	804.	2.2
Copper	1002.	1.4	2.4	0.13	640.	1.6
Lead	21.	0.5	3.5	1.02	17.5	0.6
Nickel	1492.	0.3	7.5	2.55	354.	1.1
Silver	0.5	<0.1	<0.1	0.06	0.2	0.1
Zinc	54.	0.1	10.1	0.66	127.	0.7
Sus. Solids	13.5%	28.0	900.	N.D.	3.3%	20.
Oil/Grease	N.D.	N.D.	0.3%	N.D.	N.D.	N.D.

Table V data presents an average of all data for all tests that were run. This included runs that were for marketing and promotional demonstrations. Table VI shows test results from 10 different runs where effluent quality was a primary concern. In addition, test run #6 also incorporated hexavalent chrome pretreatment.

Table VI. Analytical Results of Metal Plating Tests

Metals Total Mg/L	1	2	3	4	5	6	7	8	9	10
Cadmium	<0.05	<0.05	<.1	<.1	<.1	<.1	<.1	<.1	<.1	<.1
Chrome	-	-	-	-	-	.3	-	-	-	-
Copper	-	-	-	-	-	.1	.1.	.1	.2	.2
Lead	.3	.2	.3	.1	.3	<.1	<.1	.4	.2	.2
Nickel	.3	.5	.2	.1	.3	<.2	<.2	.3	.2	.1
Silver	<.1	<.1	<.1	<.1	<.1	<.1	<.1	<.1	<.1	<.1
Zinc	.1	.1	<.1	<.1	.1	.1	<.1	.1	.1	.1

Table VII presents data for metal recovery in the belt press cake and the total metal reduction from feed to final effluent.

The primary result of the pilot plant test was to show that the equipment was capable of producing good quality effluent and reducing the total tonnage reporting to the landfill. The removal of heavy metals produced the desired results and indicated that the equipment was capable of treating the volume needed to attain engineering design throughput. The products from the pilot plant were returned to the landfill operation. The treated effluent was used for dust control throughout the landfill while the filter cake was transported to the disposal area. The pilot plant operated approximately 3 months before being dismantled and set up at

Table VII. Typical Metal Summary for Acid/Alkali Composite

Total mg/L	Composite Feed	Press Cake	Press Filtrate	Skim Float	Float Effluent	Metal Reduc
Silver	0.17	0.43	0.03	0.58	0.01	41%
Cadmium	2.59	16.8	0.11	100.	0.08	96%
Chrome	247.	1878.	0.20	1351.	0.2	99%
Copper	318.	1684.	0.24	982.	0.2	99%
Nickel	13.	141.6	1.17	405.	0.5	96%
Lead	36.7	195.5	0.62	30.	0.6	98%
Zinc	132.	628.2	0.28	166.9	0.1	99%
Sus. Solids	8.64%	40.%	1%	6%	0.01%	99%

another installation for production treatment of metal plating wastes.

PLANT DESIGN

The design of the 360,000 ton liquid hazardous waste treatment facility is based primarily on the concept that the liquid waste will yield a good quality effluent and a high solids metal bearing filter cake. The engineering problem is to integrate the information that the process engineer has with that of the actual plant designer. The concept of compositing the wastes in two major streams, oily waste and acid/alkali, was proven workable by laboratory test work. The oily waste stream process design is comprised of treatment processes that have been previously shown to be effective in other Dravo designed plants. THe oily waste characteristics and treatment products from laboratory work lead support to the use of previous process design.

The major design criteria is to assure that the proportion of each waste stream to the whole and the characteristics of each waste stream are correctly defined. Waste categories listed in Table I show that an approximately 60% oily waste material and 40% acid/alkali material will comprise the total. The chemistry of the wastes shown in Table II helped to set the criteria for materials of construction, effluent criteria, press cake production, and equipment selection.

The treatment process of incoming wastes to the treatment facility begins with the waste being pretested to note whether the waste is treatable. If the waste is treatable, i.e., found not to contain solvents, phenols, cyanides, hexavalent chrome or highly odiferous materials, the waste load is directed to a 10,000 gallon truck dump station.

The waste liquid is transferred from the truck dump station to one of three 125,000 gallon capacity oily waste receiving tanks. The tanks are equipped with mixers to ensure a homogenous solution. The contents of a full tank are tested in the laboratory to determine chemical feed rates to properly treat the waste. Once the chemical pre-treatment controls are set, the liquid waste is pumped from the storage tanks to a cyclone separator. The suspended solid materials will be centrifugally forced into the center of the cyclone and discharged from the bottom. The remaining liquid will be discharged from the top of the cyclone and be sent to an A.P.I. separator.

The A.P.I. separator's primary function is to remove any oily material that can float to the surface. Settable solids will be gravity discharged and blended with the cyclone underflow. The discharged oil is resold or used as a fuel for flume air incineration. the treated A.P.I. effluent will be gravity discharged to two pH adjustment tanks. The tanks are equipped with mixers and pH monitor/controller units. The waste pH leaving these tanks will be between 6 and 10 depending on the heavy metal waste load.

The wastewater then enters an electrocoagulation unit where low voltage DC power is applied. The voltage comes from a series of electrodes installed in the coagulation basin. The wastewater flows by gravity to the electroflotation basin where polyelectrolytes are added to the wastewater to aid in solid agglomeration. The electroflotation basin produces millions of micron sized bubbles which attach to the agglomerated solids lifting them to the surface where they are removed by surface skimmers. The skim product is stored in a tank until a level activated pump transfers the skim to a solids holding tank. The effluent is discharged to a clearwell from where it can either be recycled as service water, impounded in a 125,000 gallon holding tank (if out of specifications,) or discharged to a flow recording flume prior to final discharge to the sanitary sewer.

The settled solids from the cyclone and A.P.I. separators are pumped to one of two 35,000 gallon sludge holding tanks where the flotation solids are also discharged. These two tanks will be equipped with mixers, and have provisions to add bulk binders (such as saw dust) in case of low solids containing wastes. The sludge is pumped from the holding tanks to a conditioning tank where an operator can visually note the effectiveness of pretreatment polymers and chemicals prior to discharge onto the belt press. The conditioning tank also contains a variable speed mixer to enhance polymer blending and floc growth. The conditioned material flows by gravity to a 2 meter belt press for dewatering. The material first flows onto a gravity drainage section, where most of free water is drained and collected, and then continues to a medium pressure zone and finally to a higher pressure zone for additional dewatering. The final cake that is discharged will contain 40 to 45% solids. The dried cakle is collected and hauled to a landfill.

Highly acidic materials which need neutralization prior to

entering the main treatment system are to be stored in one of three 12,500 gallon tanks. Chromate material is received and stored in a 15,000 galon tank prior to being reduced to the trivalent chrome form. The products from both of these separate treatments are discharged to the 125,000 galon storage tanks or, if the solids content is high enough, to the belt press sludge holding tanks.

The major tanks are enclosed and vented a hydrogen peroxide solution to destroy any odors that are present. The scrubber exhaust is fed to an incinerator which has as main fuel supply the collected waste oil from the A.P.I. separator. The tanks are provided with diking to contain any fluids from a tank rupture or line failure. The diking is sized to contain 110% of the total volume of the set of tanks within each area.

ECONOMICS

An economic analysis was performed on the proposed plant using worst case situations for specific parameters. The evaluation indicated that five independent variables could be adjusted to affect the final discounted cash flow - return on investment 9DCF-ROI). The preliminary financial analysis indicated an initial DCF-ROI of over 30% and, by adjusting the independent variables, the finalized expected DCF-ROI was found to approach 40%.

The five sensitivities reviewed were the following:

-	Volume	For every 1% change in the total volume the project DCF-ROI changed by 0.15%.
-	Price	For every 1% change in the price charged for treating waste the DCF-ROI changed by 0.3%.
-	Solids Content	A 50% reduction in solids influent loading increased the DCF-ROI by 7.0%.
-	Cake Disposal	Alternate site disposal to other Class 1 landfills reduced the DCF-ROI by 5.2%.
-	Debt Interest Ratio	DCF-ROI was not materially affected by small 2% changes in interest charges.

Income for the operation of the plant is based on 6 separate areas that would generate revenues. THese areas are their projected percentage of the total revenue are shown below:

-	Acid/Alkali	37.0%
-	Oily Waste	39.0%
-	High Acidic	9.6%
-	Truck Washes	4.2%
-	Oil Recovery	5.3%
-	Lab Analysis	4.5%

Operating expenses for the proposed facility are presented

below with their projected percentage of the total.

- Labor Costs 19 employees	7.3%
- Management	16.5%
- Chemicals	7.2%
- Electrical	1.5%
- Cake Disposal	26.7%
- Taxes and Insurance	14.3%
- Sewer Fees	0.5%
- Maintenance	3.9%
- Depreciation	15.1%
- Interest	6.9%

Finally, contribution of each major component to the overall facility capital cost is shown below.

- Land	15.3%
- Offsite Development	12.2%
- Building	5.1%
- Proprietary Equipment	6.1%
- Equipment	44.0%

CONCLUSIONS

The onset of stringent environmental regulations dealing with the disposal of liquid hazardous waste has prompted B.K.K. Corporation to begin the design and development of a liquid hazardous waste treatment facility. A waste characterization and laboratory treatment program determined that the majority of liquid hazardous waste could be composited into two treatable streams. The design of a treatment facility was then based on previous experience gained from other operating facilities, and further verified by the installation of a pilot plant utilizing actual plant size equipment.

The net results of the treatment process is that 75% of the total treatable waste will not be landfilled, but discharged into a sanitary sewer system after meeting stringent pretreatment standards. The cake product will be landfilled in an environmentaly much safer form than the original hazardous liquid. The landfills active life will be extended and the potential problem of liquid leachate caused by hazardous liquid implacement restricted.

CHAPTER 14

IN SITU STABILIZATION AND CLOSURE OF AN OILY SLUDGE LAGOON

J.W. Thorsen, M.F. Coia, and A.A. Metry
Roy F. Weston, Inc.

GENERAL BACKGROUND

The inactive lagoon site occupies over four acres in western Pennsylvania as seen in Figures 1 and 2. The focal point of the site is the open lagoon, an earthen diked lagoon of about one acre, containing approximately 30,000 cubic yards of asphaltic sludge and 200,000 gallons of acidic liquid supernatant. Operation of the site began in the 1930's when the oil company used the lagoon for the disposal of white oil production wastes. For a period of over 40 years, the lagoon was used for the disposal of sludge residues.

The waste material consists of white oil production wastes, residue from waste motor oil re-refining, coal fines, and fly ash. In 1968, when a spill of an estimated 3,000 gallons occurred, the Allegheny River was drastically impacted, killing an estimated 4,000,000 fish and resulting in the shutdown of water supplies.

Acid sludges, similar to the one at this lagoon, are typically 30 to 35 percent hydrocarbons, and 65 to 70 percent sulfur, green sulfonic acid, sulfuric acid and acid esters, and sodium sulfate. The supernatant has a low pH, because of the highly soluble nature of sulfuric acid, sulfonic acid, and sodium sulfate in water. This is consistent with the materials found in this lagoon. Alkyl benzene sulfonate (ABS), a well known surfactant or detergent is also present in the lagoon sludge and is an end product of the sulfonic acid reaction.

FIELD INVESTIGATIONS

The field investigations at the lagoon area incorporate the following tasks:

- topographic site survey;
- site geophysical survey;
- air sampling;
- lagoon sludge and supernatant sampling;
- monitoring well and drinking water well sampling;

Management of Toxic and Hazardous Wastes, H. Bhatt, R. Sykes, T. Sweeney (Editors)

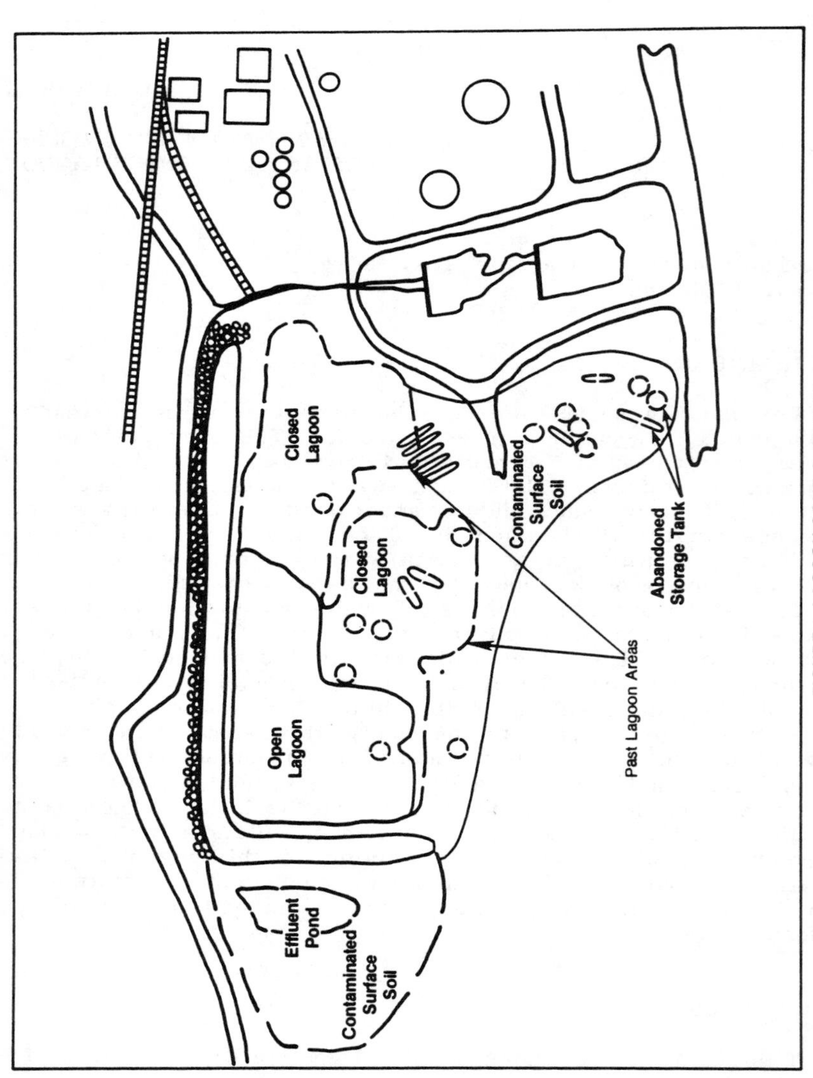

FIGURE 1 LAGOON CONFIGURATION

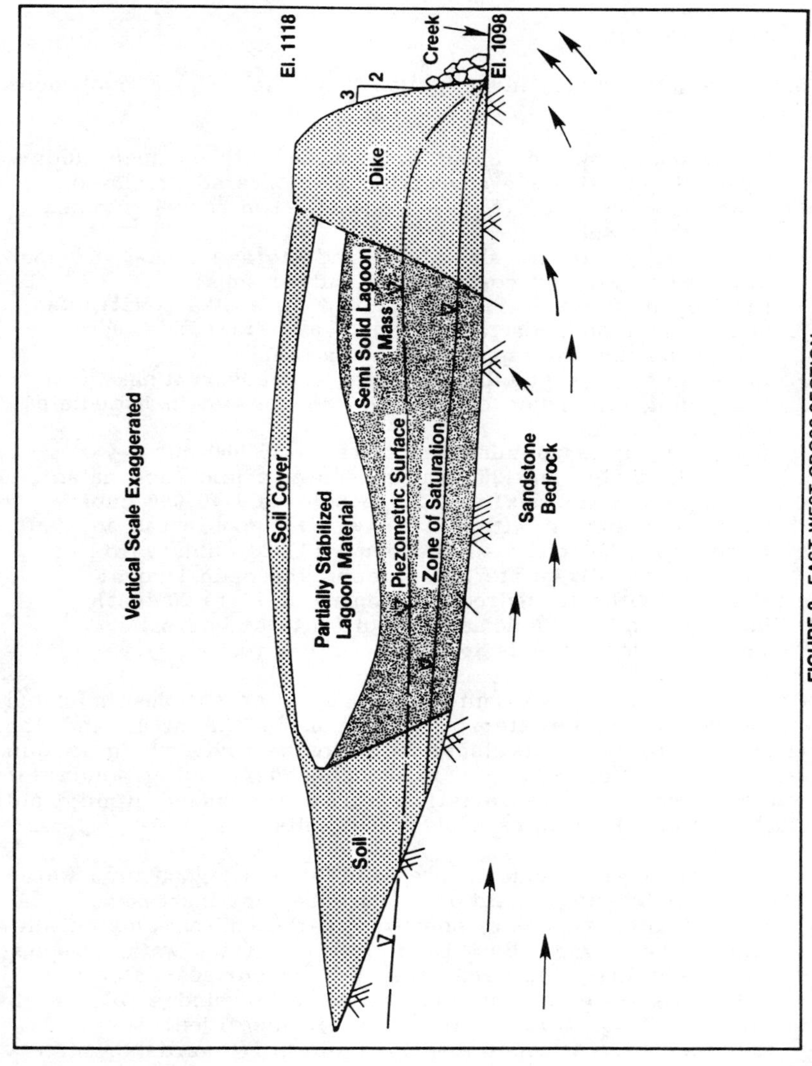

FIGURE 2 EAST-WEST CROSS SECTION THROUGH CLOSED LAGOON

- surface water sampling;
- soil sampling;
- tank sampling.

Based on the field investigation, the following conclusions were made:

- the lagoon comprises about three acres with an open sludge lagoon (approximately one acre), and covered or closed sludge lagoon areas (between one and two acres) perched above the creek;
- at the site there are also abandoned storage tanks, a lagoon overflow pond, and contaminated suface soils;
- the lagoon is contained by a 22-foot high dike constructed of silty clay and shale. This dike has 1.5:1 side slopes and is heavily rip-rapped along the creek;
- Based upon dike stability analyses and observations of no soil piping or tension cracks, the dike is not in immediate danger of failure;
- the open lagoon contains an estimated 35,000 cubic yards of asphaltic sludge and 200,000 gallons of acidic supernatant;
- a sludge and soil mixture of approximately 40,000 cubic yards is contained within the covered lagoon areas adjacent to the open lagoon. An additional 13,000 cubic yards of contaminated dike material surround the open lagoon;
- shallow depths to bedrock (maximum of 15 to 20 feet) indicated the probable lagoon bottom is at bedrock. Contamination extends to the top of bedrock.

Chemical tests of the monitoring well water samples indicated the presence of groundwater contamination in the area, and the presence of increased contamination due to the acidic sludge within the lagoon. The following general conclusions about groundwater and surface water characteristics were made based upon field investigations and laboratory analytical results:

- the groundwater and surface water in the lagoon area were contaminated upgradient from the site, but increased contaminant levels were observed in the soil samples on-site and in the downgradient ground- and surface-water samples;
- the predominant inorganic ground- and surface-water pollutants resulted from the extremely low sludge pH. High levels of TOC, COD, and MBAS were identified;
- bedrock surface slopes steeply downward towards the creek. Groundwater flows in the same direction;
- two different groundwater flow regimes were identified beneath the lagoon site: a perched water table in the soil overburden and a semiconfined aquifer in the fractured sandstone bedrock;
- infiltration recharges the shallow water table creating an

increased potential for groundwater contamination beneath the open lagoon;
- surface runoff, subsurface seepage, and groundwater inflow are identified as potential mechanisms for contaminant migration to the creek.

DEVELOPMENT AND SCREENING OF REMEDIAL ACTION STRATEGIES

Remedial actions at industrial waste disposal sites include a wide spectrum of options to manage the wastes and the potential or actual contamination of groundwater, surface water, and soils. Previous remedial action experiences have demonstrated the site-specific nature of the various options. No two sites are alike in their waste types and quantities, or in their hydrogeologic environments. The selected remedial action strategies must reflect the existing site-specific constraints.

Basic information must be collected to identify and evaluate potential remedial action strategies. This information includes:

- a characterization of the site's hydrogeologic conditions, including soil types, groundwater flow patterns and quality, surface water quality, and climatic conditions;
- knowledge of the waste characteristics, including waste types, compositions, quantities, and past disposal practices;
- understanding of the nature of the potential and actual environmental impacts associated with the waste site, and evaluation of the potential impacts of the remedial actions;
- evaluation of the various remedial action technologies to assess their technical feasibility and cost-effectiveness at the abandoned site.

The purpose of this section is to outline the types of remedial action technologies that are available and to identify potential strategies for implementing remedial action at the lagoon. A list of technology options are outlined in Table I. After preliminary screening, five major strategies were evaluated in detail; these include:

- on-site containment;
- on-site fixation;
- on-site encapsulation;
- off-site processing;
- off-site secure landfilling.

Elements of these concepts are listed in Table II.

WASTE CONTAINMENT

After detailed evaluation of various remedial action strategies, waste containment was determined to be the most cost-effective, environmentally-acceptable alternative. The waste containment strategy incorporates surface management and lagoon stabilization techniques to control the wastes at the lagoon site. The primary goal of waste containment is the control of infiltration into contaminated areas. This will both minimize leachate generation and dispersion and will control groundwater flow. Table III summarizes the remedial action components which may be implemented as part of this strategy. The various components of the waste containment strategy are shown schematically in Figure 3.

Technical Feasibility

The waste containment strategy is designed to accomplish in situ isolation of the contaminated lagoon areas. This subsection discusses the technical feasibility of the modules which comprise the waste containment concept.

General Site Management

Site management includes:

- security;
- monitoring;
- site cleanup;
- surface runoff controls.

Security would be required both during and after remedial action activities. This can be accomplished with a fence surrounding the site and possibly the use of a guard during the construction period.

Site cleanup includes the removal and off-site disposal of abandoned storage tanks which are present, in varying degrees of decay, on the lagoon site. Many of the tanks are completely dismantled and have only scrap steel value, whereas some tanks remain intact and could be reused. Handling and transportaion problems will dictate applicable tank disposal methodolgies. Transportation size constraints may require complete dismantling prior to removal. The scrap tankage will be taken off-site for disposal as a nonhazardous waste. Wastes from the tanks will be placed in the open lagoon and will be managed with the lagoon sludges.

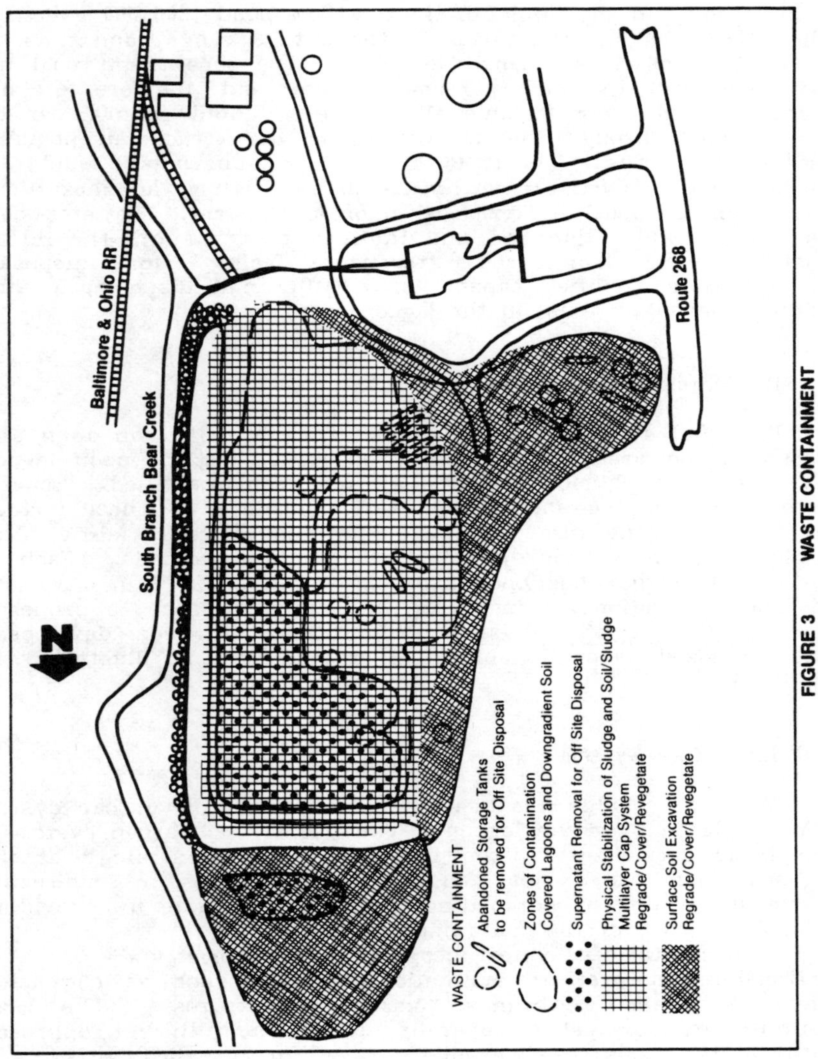

FIGURE 3 WASTE CONTAINMENT

Supernatant Removal

Supernatant refers to the standing water on the surface of the open lagoon and the liquid in the overflow pond (200,000 gallons). The pH of the supernatant is in the 2 to 3 range, and thus, it should be handled as a corrosive liquid. The supernatant could not be properly disposed of or reused on-site; and, therfore, off-site disposal is the only feasible alternative. Liquid removal can be accomplished through the use of vacuum tank trucks or portable centrifugal pumps. As it is loaded, the supernatant would be sampled and analyzed to verify its characteristics; and hazardous waste manifest would be completed prior to shipment. Haulers would be appropriately licensed and insured to transport the liquid wastes to an approved treatment facility for disposal. Neutralization facilities capable of treating and disposing of the acidic supernatant exist in the lagoon area.

In-Situ Physical Stabilization

Physical stabilization of the sludges in both the open and closed lagoon areas is necessary, especially for the open lagoon where existing physical stability has not been quantified. Several methods for sludge stabilization were considered. These include stabilization using lime, fly ash, kiln dust or soil, fabric filter application, and various combinations of these methods. Each of these technologies has been implimented successfully in a variety of field applications. Based on previous experience, a proposed strategy for in-situ physical stabilization has been developed. The proposed lagoon stabilization methodology is illustrated in Figure 4.

Multilayer Cap System

When the sludge has been stabilized, a multilayer cap system will be placed over the lagoon. The use of such a cap system is the basic component in the infiltration control strategy at the lagoon. The cap system can accomplish the task of minimizing direct infiltration of percolating water into the site by providing a surface seal over the contaminated area.

A successful cap system incorporates the use of low-permeability natural or synthetic materials, such as compacted clays, synthetic membranes, or soil admixtures. The most appropriate cap system generally utilizes a multilayer approach with native soils, well-graded gravel or crushed rock, and compacted clays. A multilayer cap system such as the one shown in Figure 5 has been considered for the containment strategy at the lagoon site, and it consits of the following three basic layers:

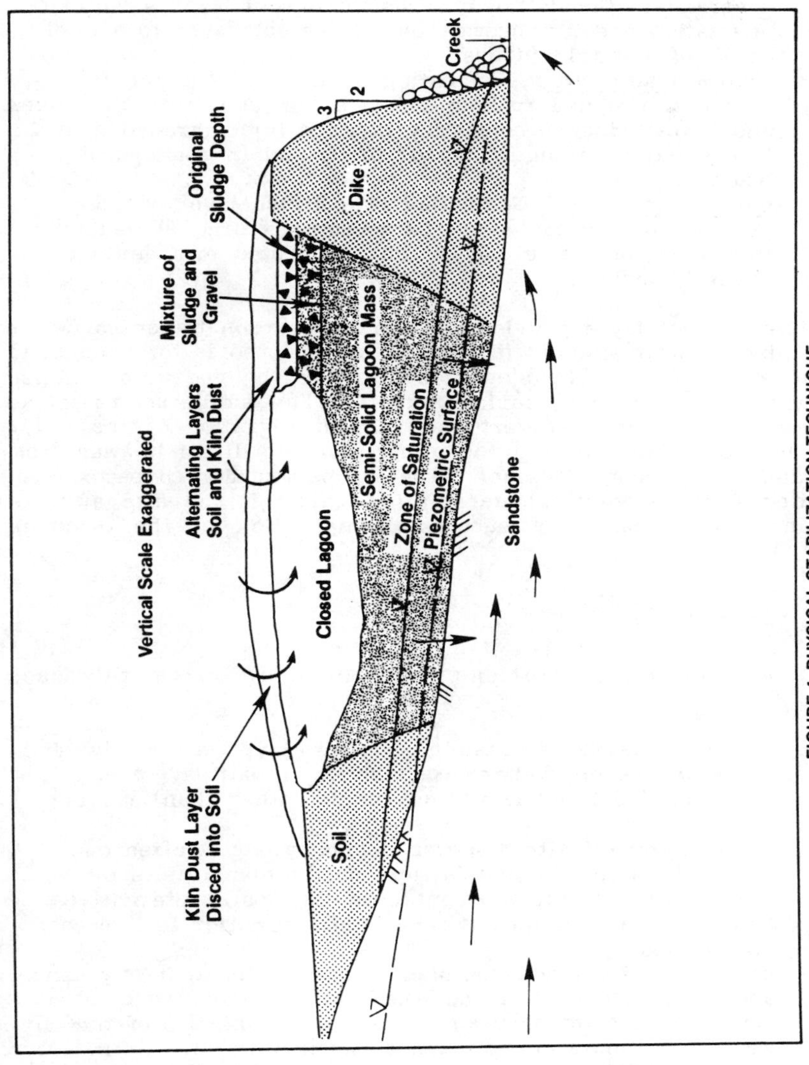

FIGURE 4 PHYSICAL STABILIZATION TECHNIQUE

- coversoils: Consisting of a 6-inch topsoil layer to support vegetation and a noncompacted native soil layer to a total depth of about 18 inches;
- middle drain layer: Consisting of an 18-inch layer of gravel or crushed rock to act as a flow zone with the upper and lower 3 inches consisting of more finely graded gravel to prevent plugging of the drain by soil from above or below;
- bottom cap layer: Consisting of a 6-inch layer of a low-permeability compacted bentonite-soil mixture followed by a final layer of native clayey soils compacted to a depth of about 12 inches;

The surface layer provides support vegetation to control water and wind erosion, and it serves as a media for enhanced evapotranspiration. The sloped terrain and the vegetation reduce infiltration by enhancing surface runoff. The middle layers act as a porous flow zone to divert the vertical migration of the water that does percolate through the upper layer to direct it away from the underlying sludge lagoons. The low-permeability compacted clay impedes further vertical percolation through the site and its presence is crucial to a successful application of the drainage layer.

CONCLUSION

The following conclusions could be drawn from this case history:

- oily waste lagoons containing asphaltic-type waste may tend to be self-sealing; therefore, the major pathway of contamination is overflow and runoff rather than leaching and seepage;
- removal and off-site disposal of such waste is often one order of magnitude that is more costly than in situ or on-site management and containment. Costly alternatives, however, may not be environmentally superior to less costly approaches;
- utilization of common pozzolana (cement kiln dust, fly ash and lime, cement, soil and lime, etc.) is effective in improving the physical and structural properties of the oily waste in preparation for lagoon closure;
- in-situ solidification and closure of inactive oily waste lagoons is technically feasible and cost-effective;

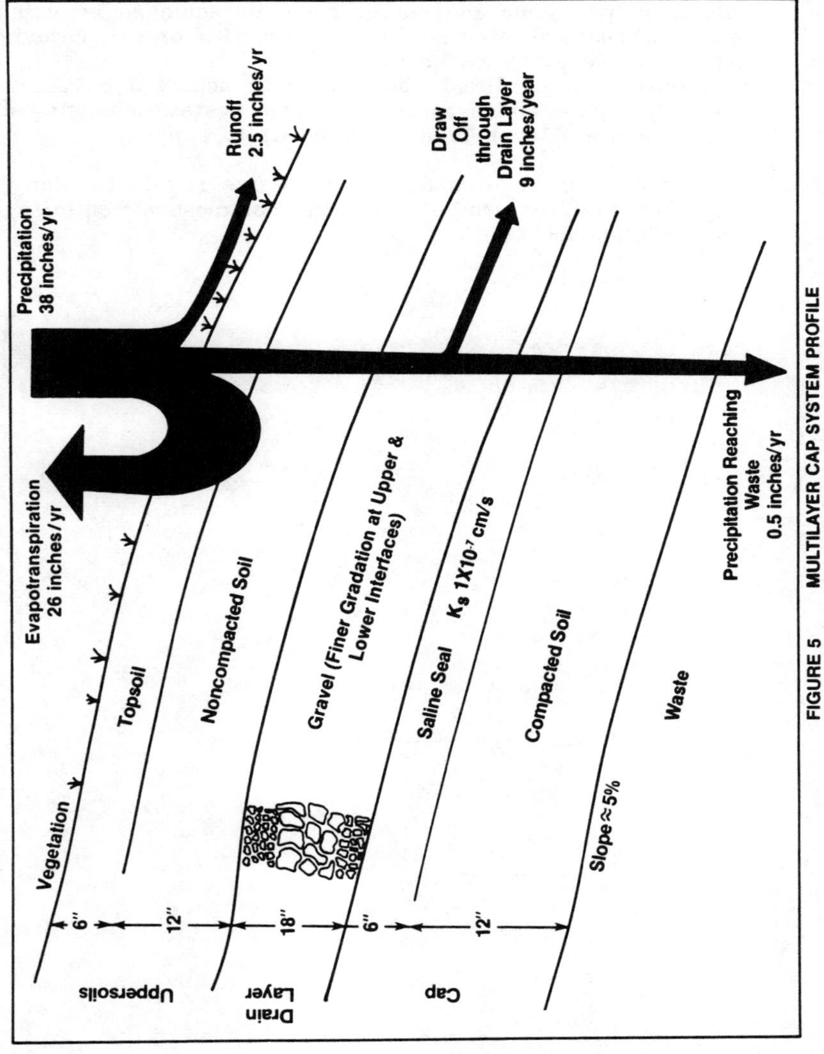

FIGURE 5 MULTILAYER CAP SYSTEM PROFILE

- mixing of pozzolana and sludge could be achieved by using either mechanical mixing (e.g., a pub mill) or earth-moving equipment (e.g., a backhoe);
- containment of solidified waste could be achieved by using perimeter dikes and a multilayer cover system consisting of a cap, a drain layer, and a soil cover to support vegetation;
- this in-place closure concept is a passive remedial action approach that requires only minimal postclosure monitoring and maintenance.

Table I. Potential Remedial Action Strategies at the Lagoon

On-Site

1. Site Management
   - o Waste Stabilization
   - o Grading
   - o Surface Water Diversion
   - o Revegetation
   - o Monitoring and Maintenance
2. Infiltration Controls (Cap Systems)
   - o Synthetic Membrane
   - o Clay
   - o Soil Admixtures
   - o Site Management (see 1 above)
3. Waste Fixation
   - o In Situ
   - o Removal/Fix/Replace
   - o Site Management (see 1 above)
4. Waste Encapsulation
   - o Cap Systems (see 2 above)
   - o Liner Systems
     - Compacted Clays
     - Synthetic Membranes
     - Asphaltic Liners
     - Concrete Mixtures
     - Soil Admixtures
   - o Site Management (see 1 above)
5. Incineration
6. Passive Groundwater Controls
   - o Bottom Sealing (liner)
   - o Interception Trenching
   - o Cutoff Walls
     - Slurry Trench
     - Grout Curtain
     - Synthetic Membranes
     - Composted Clay
7. Active Groundwater Controls
   - o Extraction
   - o Injection
8. Groundwater Treatment (physical/chemical/ biological) Matrix Available
   - o Destruction
   - o Volume Reduction
   - o Stabilization
9. Land Treatment of Groundwater and Waste Off-Site
   - o Excavation with Off-Site Disposal
     - Disposal as Hazardous Waste
     - Fix and Dispose as Solid Waste
   - o Incineration
   - o Treatment and Utilization for Energy Recovery

Table II. Component Technologies of Remedial Action Strategies

Remedial Action Components	On-site Containment	On-site Fixation	On-site Encapsulation	Off-site Processing	Off-site Secure Landfill
General Site Cleanup	•	•	•	•	•
Supernatant Removal and Disposal	•	•	•	•	•
Excavate Contaminated Soil and Disposal	•	•	•	•	•
Physical Stabilization (*In Situ*)	•				
Multilayer Cap System	•		•		
Cover/Regrade/Revegetate	•	•	•	•	•
Site Security	•	•	•	•	•
Site Monitoring/Maintenance	•	•	•	•	•
Waste Excavation		•	•	•	•
Waste Handling/Staging/Storage		•	•	•	•
Waste or Soil Mixture Reburial		•	•		
Waste Mixing/Fixation		•			
Clay Cap		•			
Off-Site Waste Transport				•	•
Off-Site Waste Disposal				•	•
Off-Site Waste Processing				•	

Table III. Waste Containment Strategy Summary[1]

Open Lagoon	Closed Lagoon	Effluent Pond	Site Management
Supernatant removal and off-site disposal	Physical surface stabilization (if necessary)	Liquid removal and off-site disposal	General site cleanup
Physical stabilization of sludge in place	Application of multilayer cap system	Excavation of contaminated surface soils and disposal in open lagoon	Security
Application of multilayer cap system	Cover/revegetate	Backfill/regrade pond	Monitoring
Cover/revegetate		Cover/revegetate	Surface water diversions

[1]Passive and/or active groundwater controls may be implemented in conjunction with the scenario.

CHAPTER 15

HAZARDOUS WASTE REDUCTION THROUGH IN-PROCESS CONTROLS, PROCESS SUBSTITUTIONS, AND RECOVERY/RECYCLING TECHNIQUES

Dr. John A. Gurklis
Battelle Columbus Laboratores

INTRODUCTION

Compliance with effluent guidelinies (for waters discharged to streams), pretreatment standards [for waters discharged to publicly owned treatment works (POTW)], and with RCRA (Resource Conservation and Recovery Act) regulations has resulted or will result in significant cost increases for firms carrying out electroplating and metal finishing operations [1]. To achieve suitable aqueous effluents, many firms have or will have to install waste treatment facilities for the first time and others have or will have to upgrade existing facilities. Accordingly, this is a good time for the managers of these plants to explore the use of in-process controls to cut down on processing solution dragout and rinsewater volume, and also the use of recovery/recycling techniques for conserving plating and processing chemicals. Implementation of in-process controls, process substitutions, and recovery/recycling techniques may be amongst the more effective means of minimizing the size of the waste treatment facility and/or the amount of hazardous wastes to be treated and disposed under RCRA.

Currently, recovery of metal values from mixed wastewater treatment sludges is generally not practical for technical or economic reasons [1]. The recovery problems relate to the fact that metals in mixed metal sludges are generally difficult to separate and may involve relatively complex and costly chemical and or metallurgical procedures. Accordingly, in-process recovery and/or recycling of plating and other processing bath chemicals contained in rinsewaters appear to be promising approaches to reduce processing chemicals costs as well as to minimize the quantity of metal-bearing sludges generated. Amongst the more promising and more widely used techniques employed in electroplating and metal finishing plants for economical recovery/recycling of valuable plating chemicals now going to waste treatment are evaporation, reverse osmosis, and "save rinses". A discussion of these technologies together with detailed costing of representative recovery/recycling operations by the use of

Management of Toxic and Hazardous Wastes, H. Bhatt, R. Sykes, T. Sweeney (Editors)

evaporation and reverse osmosis are presented below.

REPRESENTATIVE ELECTROPLATING OPERATIONS AND WASTEWATER TREATMENT

Most captive or job shops operate several lines that contain different types of cleaning, activating, and plating solutions for electrodepositing a variety of metallic coatings on many different substrate materials [1,2,3,4]. Chemicals, such as nickel sulfate, nickel chloride, nickel sulfamate, zinc cyanide, copper cyanide, cadmium cyanide, chromic acid, lead fluoborate, and tin fluoborate, are used in the preparation of a wide variety of solutions that are employed in the electroplating and metal finishing industry to produce protective or decorative finishes on metal and plastic products. In addition, anodizing of aluminum, and the application of chromate and phosphate conversion coatings on various metals are frequently carried out in electroplating and metal finishing plants. The interrelationships among dragout, rinsing, wastewater treatment, and sludge disposal are discussed below.

Dragout and Product Flow in Electroplating

Product or workpiece flow, solution dragin and dragout,* and rinsewater flow for an electroplating operation with a two-tank countercurrent rinsing arrangement are shown in Figure 1. Rinsing operations are required in electroplating and metal finishing operations in order to remove the plating or processing solutions (dragout) adhering to the workpieces in order to provide a satisfactorily clean surface for subsequent processing or treatment. The rinsewaters contain the plating or processing chemicals, which, if not recovered, are sent to wastewater treatment. Thus, the principal source of water borne wastes going to the wastewater treatment plant are those resulting from the rinsing of dragout solutions adhering to workpieces.

Typical Wastewater Treatment Operations in Electroplating and Metal Finishing

---

* Dragout refers to the solution that adheres to the parts and associated fixtures removed from a bath. More precisely, it is defined as that solution which is carried past the edge of the tank.

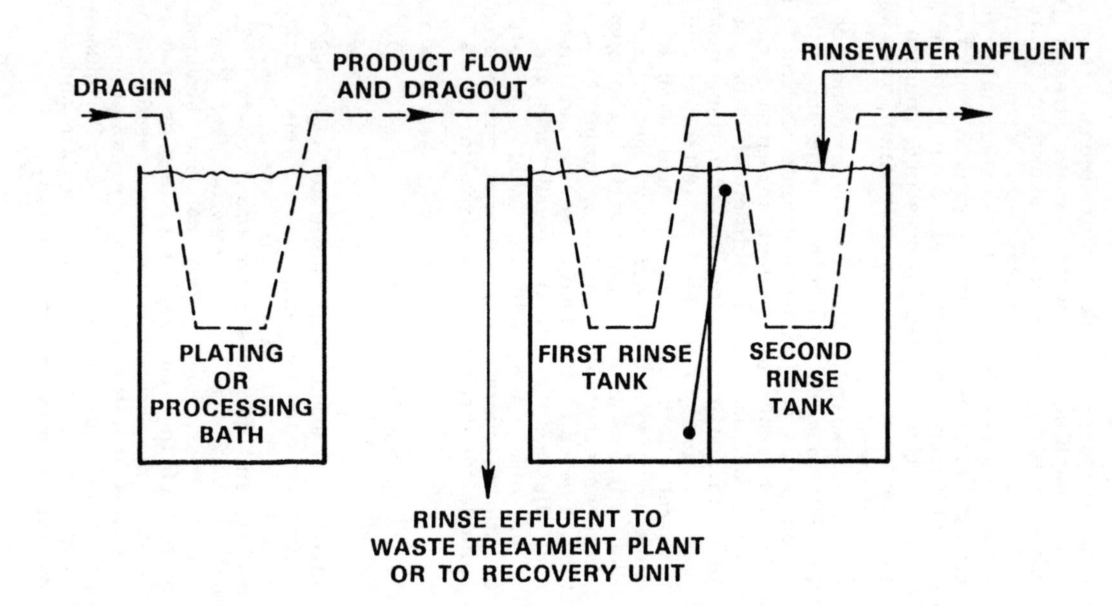

Figure 1. Product and Rinsewater Flows in Electroplating and Metal Finishing Operations

A representative system for the conventional treatment generally employed for handling segregated waste streams from electroplating and metal finishing operations is shown in Figure 2. As indicated in Figure 2, the rinsewaters from various electroplating operations generally are segregated into three principal waste-bearing streams, namely cyanide, hexavalent chromium (chrome), and mixed acid-alkali. In some instances, at periodic intervals spent concentrated processing solution are slowly introduced into the appropriate relatively dilute rinsewater streams for treatment and disposal.

Cyanide destruction is generally achieved by oxidation of the cyanide in an alkaline environment to carbon dioxide and nitrogen by the addition of chlorine gas and sodium hydroxide or lime. THe reduction of hexavalent chromium to trivalent chromium is usually carried out in an acidic medium by the addition of either sulfur dioxide or sodium bisulfite. After cyanide oxidation and chromium reduction, these two streams go to the neutralization and pH adjustment tank or vessel where they join the acid-alkali stream. The trivalent chromium along with the other heavy metals from the cyanide and acid-alkali streams are precipitated with the final pH in the tank being between about 8.5 and 9.0 in order to achieve maximum precipitation of most of the heavy metallic ions present. The precipitation and adjustment of pH are usually achieved using sodium hydroxide or lime and sulfuric acid. Settling, clarification, and dewatering operations are carried out to produce a treated effluent suitable for discharge to a stream or to a POTW.

Dragout and Rinsing Operations in Electroplating and Metals Finishing

As mentioned above, the amounts of rinsewater employed after processing steps in electroplating and other metal finishing operations determine to a great extent the size of the on-site waste treatment facility and the waste treatment costs. Thus, minimizing the amount of dragout and the rinsewater flow after each operating step is highly desirable. Practical and economically attractive implementation of the recovery of plating or processing chemicals from rinsewaters by evaporation, reverse osmosis or other methods generally requires that the amount of rinsewater employed in a particular operation be relatively small. Small volumes of rinsewaters are desirable in order to minimize the size and cost of the units and the expenditures of energy required.

Rinsing Operations and Calculations

The determination of what degree of rinsing is adequate is a very important consideration in electroplating and metal finishing operations. Specific data on optimal concentration levels in

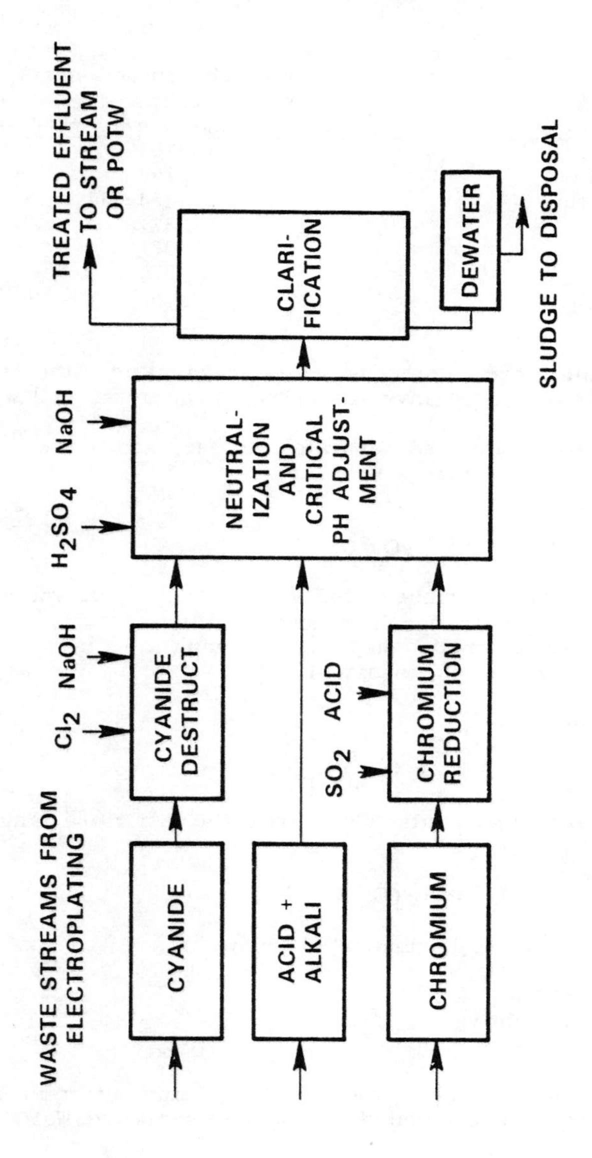

Figure 2. Representative Treatment of Electroplating Wastewaters

rinsewater following various processing steps are relatively scarce. However, the following tabulation gives a set of representative maximum concentration levels suggested for the rinses following typical processing steps [6].

Rinses Following These Processes	Suggested Maximum Concentration of Dissolved Solids in Final Rinse, mg/1
Alkaline cleaning	750
Acid cleaning, dips	750
Chromate passivating	350-750
Cyanide plating	37
Copper plating	37
Nickel plating	37
Chromium plating	15

The following are formulas for calculating theoretical rinsewater amounts for alternative arrangements of rinse tanks [7]:

1. Series rinse (water fed separately to each tank)

$$r = n(C_o/C_f)^{1/n}$$

where r = ratio of rinse water volume to dragout volume
C_o = concentration in process solution
C_f = concentration in last rinse tank
n = number of rinse tanks

2. Single tank rinse

$$r = C_o/C_f$$

3. Counterflow rinse (water flows from the last rinse tank to the first)

$$r = (C_o/C_f)^{1/n}$$

The use of these formulas is illustrated below.

Effectiveness of Alternative Rinse Tank Arrangements

The amounts of rinsewater required with four different rinse tank arrangements were calculated for a representative Watts-type nickel plating bath containing nickel sulfate, nickel chloride, and boric acid (total solids content = 250 g/l) to illustrate their relative effectiveness. The following data were assumed to apply to the nickel-plating step:

C_o = concentration in processing bath = 250 g/l or 250,000 mg/l
C_f = concentration in last rinse tank = 0.037 g/l or 37 mg/l
DO = processing solution dragout = 1.0 gal/hr
r = ratio of rinsewater volume to dragout volume
RE = rinsing efficiency = 70 percent (assumed).

The following equation was used to determine rinsewater flow rate (RF):

$$RF = (r)\ (DO)\ (RE).$$

For the four different rinse flow arrangements, the rinse flow rates were calculated as shown below:

1. Single tank rinse

   $RF = (250{,}000/37)\ (1.0)\ (1/0.70) = 9653$ gal/hr

2. Two-series rinse tanks

   $RF = 2\ (250{,}000/37)^{1/2}(1.0)\ (1/0.70) = 235$ gal/hr

3. Two-counterflow rinse tanks

   $RF = (250{,}000/37)^{1/2}(1.0)\ (1/0.70) = 117$ gal/hr

4. Three-counterflow rinse tanks

   $RF = (250{,}000/37)^{1/3}(1.0)\ (1/0.70) = 27$ gal/hr.

As can be seen from the above calculations, the two-counterflow rinse tanks system uses ony one half of the water required in the two-series rinse tanks system, and one to two orders of magnitude less water than the single rinse tank system. These calculations indicate that two or more counterflow rinse tanks should be employed wherever possible to conserve rinsewater. For recovery of dragout solutions using evaporation or reverse osmosis techniques, the use of three or more counterflow reverse tanks would be preferred to reduce the amount of rinsewaters to be processed and thus cut down on the size of the recovery unit required.

CUTTING WASTE GENERATION BY DRAGOUT REDUCTION AND PROCESS SUBSTITUTIONS

Reduction of dragout and use of process substitutions are effective methods of reducing the amounts of hazardous wastes generated in the electroplating and metal finishing industry. The application of these methods, for reducing hazardous waste generation and the subsequent waste treatment and disposal, is discussed below.

Ways to Reduce Dragout

Reduction of the amount of dragout solution on parts and fixtures will: (1) decrease tha amount of rinsewater required, (2) cut down the amount of waste treatment and sludge disposal required, and (3) reduce the costs associated with the replacement of plating or processing chemicals. The following are suggested techniques for decreasing solution dragout in electroplating and metal finishing operations:

- Reduction of the velocity of rack or barrel withdrawal from the bath and prolongation of the drain time over the bath
- Better design and arrangement of parts on racks to promote solution drainage
- Use of drip boards
- Use of more dilute baths, so that for a given volume of dragout solution the chemicals content would be lower
- Addition of wetting agents to baths to lessen surface tension and thereby decrease the amount of solution adhering to parts and racks
- Use of higher operating temperatures, which generally lower solution dragout on parts and racks occurs
- With baths that have some evaporation because of normal operation at somewhat elevated temperatures (e.g., 100 to 150 F or higher), the use of controlled fog or spray rinses over the plating bath to return part of the dragout solution directly to the bath.

The use of more dilute baths, wetting-agent additions, or higher solution temperatures must be evaluated critically so that their use does not interfere with the normal operation of the baths and the production of quality parts.

Process Substitutions

Process substitutions offer opportunities for cutting down on some of the hazardous waste pollution load generated by the electroplating and metals finishing industry. Several examples of process substitutions, that have been used successfully to cut down on hazardous waste generation and the needs for its treatment and disposal, are cited below:

- Substitution of an acidic copper sulfate bath for an alkaline copper cyanide bath is successful for some electroplating applications. The acidic copper bath chemicals cost substantially less than the cyanide bath chemicals. Also significant is that, by this substitution, one eliminates the need for the expensive wastewater treatment operation required for the destruction of cyanide. For example, it costs about $3.00 for the chlorine and sodium hydroxide required for the destruction of 1.0 pound of cyanide.
- Vacuum deposition of aluminum coatings on some aircraft parts has been successfully employed as a substitute for a cadmium coating electroplated from a cyanide bath. Such a process substitution results in eliminating the need for cyanide destruction and the subsequent disposal of the cadmium-bearing sludge.
- For some applications, mechanical or peen plating of cadmium onto steel parts, using a noncyanide solution containing cadmium powder, can be employed as a substitute process for the electroplating of cadmium from a cyanide bath.
- During the last few years, there has been a significant increase in the use of trivalent baths for electroplating chromium for some applications. The substitution of the trivalent bath for the hexavalent chromium bath eliminates the need for the relatively expensive chromium reduction step in the wastewater treatment plant.

Before one adopts any of the process substitutions mentioned above as a means of cutting down on hazardous waste generation, they should assure themselves that the substitute processes can economically produce quality coated parts that comply with all existing specifications [8].

WASTE REDUCTION BY USE OF RECOVERY/RECYCLING TECHNIQUES

As mentioned earlier, evaporation, reverse osmosis, and "save rinses" are amongst the more promising and more widely used recovery/recycling (R/R) techniques employed in electroplating and metal finishing plants to conserve valuable chemicals and thereby reduce waste generation. an assessment of these R/R technologies and their economic aspects is presented below.

Evaporation

Evaporation is probably the most widely used process for the recovery of plating or processing chemicals from rinsewater streams. As indicated earlier, adequate rinsing of the plated parts must be provided in order to remove excess plating chemicals and contaminants to prevent staining or spotting of the product or to promote satisfactory subsequent processing of the plated product. Evaporation is an established process that has been proven successful on many types of plating baths [1,3,4,5,9,10]. Economical recovery of valuable chemicals from many plating baths, including cadmium cyanide, zinc cyanide, copper cyanide, nickel, chromium, and fluoborate has been achieved. It is estimated that currently several hundred units are in operation for recovery of plating or metal processing solutions from rinse streams [4].

Several types of units are employed: atmospheric evaporators, submerged-tube evaporaor, climbing-film evaporators, and flash evaporators. All types perform essentially the same function, i.e., concentrating the chemicals in the rinsewater solution by evaporation of the water; some also recover water condensate suitable for return to the rinse tanks on the plating line. Evaporation is frequently carried out under vacuum to prevent thermal degradation of plating bath constituents or additives and to reduce the amount of energy required for the process. Multiple-effect evaporators provide greater thermal efficiency but at greater capital cost than single-effect units.

The general operations involved in the application of evaporation in electroplating and metal finishing plants for recovery/recycling of plating or processing chemicals from rinsewaters are illustrated in Figure 3 [9]. As shown in Figure 3, evaporation is used to recover chromium plating chemicals from rinsewater discharged from the first rinse tank. As indicated earlier, th use of a multistage counterflow rinse system is generally required to reduce the volume of rinsewater to a quantity that can be processed economically by recovery/recycling techniques. The closed-loop evaorative recovery syustem, shown in Figure 3, employs a three-tank countercurrent rinse. Essentially all of the chromium plating chemicals dragged out from the bath are recovered, so that fresh chromic acid, corresponding only to the amount of chromium plated on the workpieces plus the small amount of trivalent chromium (formed in the bath during normal plating operations) removed in the cation exchanger must be added to keep the bath operating properly. Water consumption is reduced to the water lost to surface evaporation from the chromium plating bath. In addition to removing trivalent chromium from the rinsewater, the cation exchange column also removes dissolved iron and other metallic ions (arising from the workpieces processed) to prevent the build-up in the bath of metallic impurities usually associated with closed-loop operation.

Generally similar techniques to those described above for chromium chemicals recovery can be used for the recovery/recycling of plating chemicals from a variety of different plating processes.

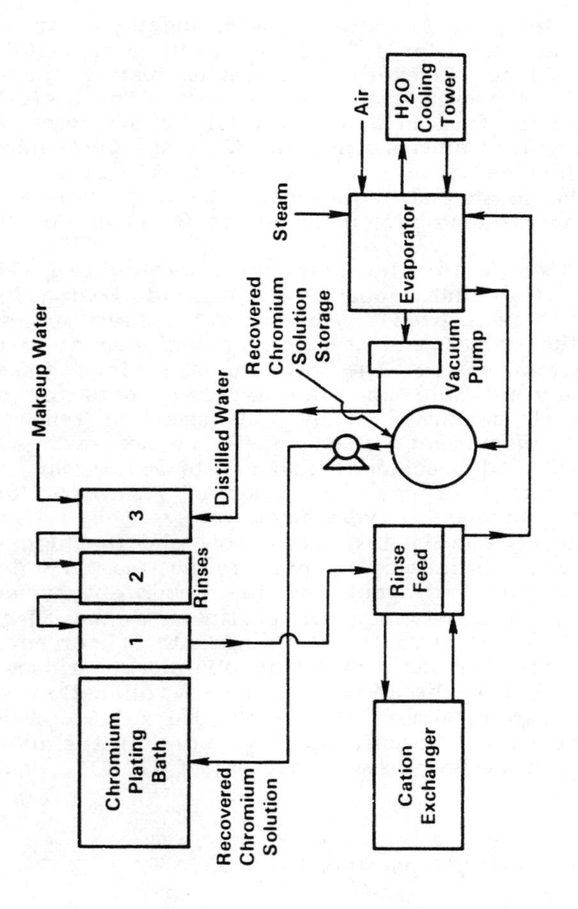

Figure 3. Schematic of the Recovery of Chromium Plating Solution by Evaporation

Economics of Evaporative Recovery

An economic analysis of two different-sized evaporator systems for recovery of chromium plating chemicals from rinsewaters is presented below.

Twenty Gallon/Hour Evaporator System

Detailed estimated total investment costs, together with annual fixed and operating costs for a 20-gal/hr evaporator installation (shown in Figure 3) for recovery of chromium plating chemicals* from rinsewaters, are presented in Table I [9]. The costs listed in Table I were derived from an EPA report [9] by conversion of Mid -1978 dollars to Dec-1982 dollars using the CE Plant Cost Index for the time period involved, i.e., a factor of 1.435 [11]. It was assumed that the costs of equipment, labor, utilities, and materials, etc., increased in proportion to the CE Plant Cost Index for the period.

The installed costs for the evaporator system are $51,250 (Dec-1982 dolars); the total annual operating and fixed costs are $13,130 and $18,770, respectively. The costs for steam and system depreciation are the major costs of operation and amount to about 51 percent of the total costs. The total annual savings, based on the recovered chromic acid and the avoided costs for waste treatment and disposal amount to $21,370. As shown in Table I, the cash flow from the investment (which equals the net savings after taxes of 48 percent + depreciation) amounts to $6540/year. This cash flow corresponds to a payback period of 7.9 years for the evaporator system, which is not very favorable.

Because of the high initial investment cost and the high steam consumption cost, the economics of evaporative recovery depend greatly on the volume of dragout and the amount of rinsewater required for adequate rinsing. For example, if the 20-gal/hr evaporator unit (Table I) were fed a stream with 50 percent more dragout chemicals for the same 20-gal/hr quantity of rinsewater, the annual savings for the overall recovery operation would increase by 50 percent to about $32,100 ($21,370 x 1.5). The net savings would increase to about $13,250, so that the payback period, based on a cash flow from investment of $12,010, would be reduced to 4.3 years.

Seventy-Five Gallon/Hour Evaporation System

---

* The chromium plating chemicals are essentially chromic acid and sulfuric acid present in a ratio of about 100 CrO_3 to 1 H_2SO_4.

Table I.
Economics of a 20-gal/hr Evaporator System for Recovery of Chromium Plating Chemicals, Operating 5,000 hr/yr [a]

Item	Amount[b] (Dec-82 $)
Installed cost, 20-gal/h evaporator ($):	
Equipment:	
Evaporator	27,270
Tanks	2,380
Pumps	1,080
Cation exchanger	8,900
Piping	3,590
Miscellaneous	650
Subtotal	43,870
Installation, labor and materials:	
Site preparation	400
Plumbing	4,100
Electrical	1,770
Equipment erection	390
Miscellaneous	720
Subtotal	7,380
Total installed cost	51,250
Annual operating cost ($/yr):	
Labor, 100 hr/yr at $10/hr	1,000
Supervision	(c)
Maintenance, 6% of investment	3,140
General plant overhead	1,260
Raw materials, cation exchanger:	
H_2SO_4, 3,500 lb/yr at $0.040/lb	140
NaOH, 3,000 lb/yr at $0.115/lb	340
Utilities:	
Electricity, $0.065/kWh	2,010
Cooling water, 1,000 gal/hr/1,000 gal	720
Steam, $4.30/10^6 Btu	4,520
Total operating cost	13,130

Table I. (Continued)

Item	Amount[b] (Dec-82 $)
Annual fixed costs ($/yr):	
Depreciation, 10% of investment	5,120
Taxes and insurances, 1% of investment	520
Total fixed costs	5,640
Total costs of operation	18,770
Annual savings ($/yr):[d]	
Recovered plating chemicals (1.875 lb/hr H_2CrO_4)	10,500
Water treatment chemicals	6,460
Sludge disposal	4,310
Water use, 13 gal/hr at $1.58/1,000 gal	100
Total annual savings	21,370
Net savings = annual savings - (operating cost + fixed cost) ($/yr)	2,570
Net savings after taxes, 48% tax rate ($/yr)	1,330
Average ROI[e] = (net savings after taxes/total investment) x 100 (%)	2.6
Cash flow from investment = net savings after taxes + depreciation ($/yr)	6,450
Payback period = total investment/cash flow (yr)	7.9

(a) Data were taken or derived from Reference 9.

(b) To update the costs in the source to Dec-1982 values, the mid-1978 costs of equipment, labor, utilities, and materials, etc., were assumed to have increased in proportion to the CE Plant Cost Index. The CE index factor used was 1.435 [11].

(c) Minimal.

(d) Based on a 90% operating factor.

(e) ROI = Return on investment.

The economics, based on a case study of a 75-gal/hr rising film evaporator at the Phillips Plating Company (Phillips, Wisconsin), for concentrating the chromium plating dragout in the rinse stream for recycling to the plating bath are summarized in Table II [5]. Installation of this closed-loop recovery system reduced the required addition of anhydrous chromic acid (CrO_3) to the plating bath by approximately 4 lb/hr.

The total cost for the installed evaporator system was about $79,000 in Dec-1982 dollars. The costs listed in Table II were derived by updating costs expressed in 1979 dollars to costs in Dec-1982 dollars using the CE Plant Cost Index for the time period involved; i.e. a factor of 1.315 was used [11]. The total annual capital and operating and maintenance costs for the evaporator system, based on operating 6000 hr/yr, amounted to $40,310. The annual savings on chromic acid plating chemicals amount to $28,400. Thus, based only on the value of the recovered chromium plating chemicals the evaporator operation would appear to lose about $11,900 per year. However, when one includes the avoided costs of waste treatment and sludge disposal* that would have been required for the chromium plating chemicals if they had not been recovered, the annual savings for the Phillips evaporator system amount to $65,750. With this savings, the payback period after taxes (at 48 percent) and depreciation is 3.8 years. This 3.8-year payback for the 75-gal/hr unit was only about 1/2 that of the 20-gal/hr unit (Table I). To a significant extent, the economies of scale were operative, in that the 3.5 times larger capacity unit cost only 1.54 times more than the smaller one, so that the annual capital charges were also only 1.54 times as large.

Reverse Osmosis

In reverse osmosis (RO), pressure is applied to the more concentrated side of a semipermeable membrane, causing the permeate (mainly water) to diffuse through the membrane in the direction opposite to the osmotic pressure, that is, from the concentrated to the dilute side. The concentrated solution of dissolved solids (concentrate) left behind can be further treated to achieve a more concentrated solution or can be returned to the original plating or processing tank. The recovery of plating solution dragout by the application of reverse osmosis is illustrated in Figure 4 [9]. The feed to the RO unit is taken from the first rinse tank of a counterflow rinse system. The feed is separated under pressure (generally, about 400 to 800 psi) into a purified permeate stream and a concentrate stream by the plating bath, replacing evaporated water and dragged out chemicals. The permeate goes to the last

---

* Phillips uses a Sulfex^{TM} insoluble sulfide treatment system.

Table II.
Economic Evaluation of a 75-gal/hr Evaporator System for Recovery of Chromium Plating Chemicals[a]

Item	Amount[b] (Dec-82 $)
Installed cost for 75-gal/h evaporator	$78,900
Annual costs at 6,000 hr/yr:	
Depreciation (10-yr life)	7,890/yr
Taxes and insurance	790/yr
Maintenance	4,730/yr
Labor (1/2 hr/shift at $7.90/hr)	2,960/yr
Utilities:	
Steam (at $4.60/$10^6$ Btu)	19,730/yr
Electricity	790/yr
General plant overhead	3,420/yr
Total annual cost of operation	40,310/yr
Annual savings:	
Replacement CrO_3	28,400/yr
Waste treatment reagents	20,250/yr
Sludge disposal	7,100/yr
Total annual savings	65,750/yr
Net savings after taxes (NSAT) at 48%	13,230/yr
Average ROI[c] = NSAT/total investment	16.8%
Cash flow for investment = NSAT + depreciation	21,120/yr
Payback period = total investment/cash flow	3.8/yr

(a) Data were taken or derived from Reference 5.

(b) The 1979 costs used in the source were updated to Dec-1982 costs using the CE Plant Cost Index; the adjustment factor used was 1.315 [11].

(c) ROI = Return on investment.

rinse tank of a counterflow system, providing water for the rinsing operation.

The performance of RO units is defined by flux--the rate of passage or purified rinsewater through the membrane per unit of surface area--and the percent rejection of a dissolved constituent in the rinsewater, which relates to the membrane's ability to restrict that constituent from entering the permeate stream [9,12]. Percent rejection is defined by:

$$\text{Percent rejection} = \frac{C_f - C_p}{C_f} \times 100$$

where C_f = concentration in feed stream
C_p = concentration in permeate stream.

Typical percent rejection values for various cations and anions are shown in Table III [9].

TABLE III. Reverse Osmosis Operating Parameters[a]

Feed solution	Maximum Concentration of Concentrate, percent	Rejection, percent
Ni^{+2}	10-20	98-99
Cu^{+2}	10-20	98-99
Cd^{+2}	10-20	96-98
CrO_4^{-2} [b]	10-12	90-98
CN^{-1} [b]	4-12	90-95
Zn^{+2}	10-20	98-99

(a) Source: Reference 9.
(b) Performance depends greatly on pH of solution.

Currently, the principal applications of RO systems in electroplating involve mostly recovery of plating chemicals from nickel baths and from a limited number of acidic zinc, copper, and chromium baths.

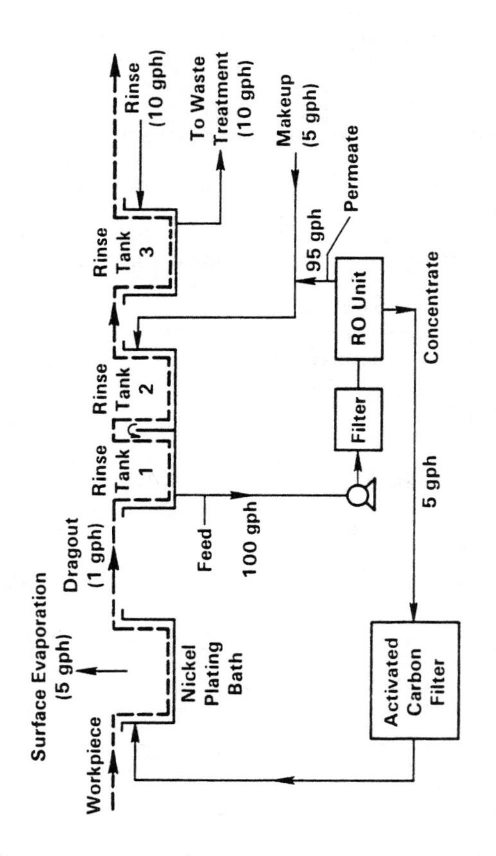

Figure 4. Reverse Osmosis for Recovery of Nickel Plating Dragout

Economics of Recovery of Nickel Plating Chemicals by Reverse Osmosis

The use of reverse osmosis for opent-loop recovery and recycling of nickel plating chemicals from rinsewaters (100 gal/hr flow) is shown in Figure 4 [9]. The rinsing system consists of two counterflow rinses and a third running rinse, whose rinsewaters go to waste treatment. A 50-u filter is used in the system to prevent blinding of the membrane by solid particles. The RO unit, which is preassembled by the manufacturer, consists of a high pressure centrifugal pump and six membrane modules; installation requires mainly piping and electrical connections. An activated carbon filter is used to avoid organic contaminant build-up in the plating bath.

The economics of the RO operation, shown in Figure 4, for recovering nickel plating chemicals are summarized in Table IV [9]. Total installed equipment costs, together with annual operating and fixed costs, are given in the table in Dec-1982 dollars, calculated by updating the published costs by the CE Plant Cost Index factor of 1.435. The total installed costs for th RO system are $28,000 (Dec-1982 dollars); the total annual operating and fixed costs are $6,460 and $3,090 respectively. The annual savings on nickel plating chemicals, together with the savings resulting from the avoided costs of wate treatment and sludge disposal for the nickel plating rinsewaters, amount to a total of $16,590. The average ROI (return on investment) for the system was 13.1 percent. The cash flow from the investment (which equals net savings after taxes of 48 percent + depreciation) amounts to $6,460 per year. This cash flow corresponds to a payback period of 4.3 years for the RO system, which is economically attractive.

Save Rinses for Dragout Recovery

The use of "save" or "still" rinses provides a relatively simple, low-cost, and effective means of direct recovery and recycling of plating dragout [1,3,4,13]. The "save" rinse tank following the plating tank serves to collect a significant portion of the plating solution carried in on the parts, racks, or barrels. Periodically, part of the strong solution in the "save" rinse tank is returned to the plating tank.

The volume that can be returned to the plating tank will be governed by the amount of evaporation that occurs in the process tank. In some instances, part of the flow from subsequent rinse tanks can be counterflowed to the "save" rinse tank and the efficiency of the recovery process improved. As with some other recovery techniques, it is important to obtain maximum rinsing efficiency with a minimum amount of water. Were one to put in about four or five counterflow rinse tanks in conjunction with a "save" rinse tank, one could recover almost the entire dragout

Table IV.
Economics of Reverse Osmosis System for Nickel Plating Chemicals Recovery, Operating 5,000 hr/yr[a]

Item	Amount[b] (Dec-82 $)
Installed cost, 330-ft^2 unit ($):	
Equipment:	
RO module including 50-μm filter, pump, and 6 membrane modules at 55 ft^2 per module	21,530
Activated carbon filter	2,870
Piping	720
Miscellaneous	720
Subtotal	25,840
Installation, labor and material:	
Site preparation	290
Plumbing	430
Electrical	720
Miscellaneous	720
Subtotal	2,160
Total installed cost	28,000
Annual operating cost ($/yr):	
Labor, 100 hr/yr at $10.00/hr	1,000
Supervision[c]	(c)
Maintenance	1,680
General plant overhead	950
Raw materials:	
Module replacement, 2-yr life (6 x $460)/(module x 0.5 yr)	1,380
Resin for carbon filter	720
Utilities, electricity (0.065/kWh)	720
Total operating cost	6,460
Annual fixed costs ($/yr):	
Depreciation, 10% of investment	2,800
Taxes and insurance, 1% of investment	290
Total fixed costs	3,090
Total cost of operation	9,550

Table IV. (Continued)

Item	Amount[b] (Dec-82 $)
Annual savings[d] ($/yr):	
Plating chemicals:	
1.65 lb/hr $NiSO_4$	8,090
0.34 lb/hr $NiCl_2$	2,280
Water treatment chemicals	3,620
Sludge disposal cost	2,600
Total annual savings	16,590
Net savings - annual savings - (operating cost + fixed cost) ($/yr)	7,040
Net savings after taxes (48% tax rate) $7,040 x 0.52 ($/yr)	3,660
Average ROI[e] = (net savings after taxes/total investment) x 100 (%)	13.1
Cash flow from investment = net savings after taxes + depreciation ($/yr)	6,460
Payback period = total investment/cash flow (yr)	4.3

(a) Data were taken or derived from Reference 9.

(b) The mid-1978 costs used in the source were updated to Dec-1982 costs using the CE Plant Cost Index; the adjustment factor used was 1.435.

(c) Minimal.

(d) Based on a 90% operating factor.

(e) ROI = Return on investment.

flow. However, in practice, because of space limitations, costs of additional tanks, and rinsing efficiency considerations, etc., the use of systems with more than 3 counterflow tanks in conjunction with a save rinse tank is rather limited.

Recoveries can range from about 50 to 90 percent or more depending upon the rinse and evaporation rates of the particular plating operation involved. The principal costs involved in the implementation of "save" rinse tanks are the costs of the tanks and their installation in selected lines. The latter can be costly in some installations, as the needed space in a line may not be readily available and expensive modifications to the line might be required to provide the needed space.

Selection for Candidate Plating Operations for Recovery/Recycling Systems

The following is a suggested procedure for identifying candidate waste rinsewater streams from plating operations that offer promise for the recovery/recycling of the dragged out plating chemicals:

1. Select plating operations with high production of plated parts (i.e., especially those with high dragouts of plating chemicals going to the rinse tanks).
2. Further, select those baths whose dragout chemicals result in high wastewater treatment and sludge disposal costs. For example, the treatment and disposal costs associated with cyanide baths and hexavalent chromium baths are much higher than those for most other plating baths.
4. Carry out preliminary calculations using measured or estimated dragout values, together with measured or estimated rinsewater rates (or anticipated rates with the installation of a counterflow rinse tank system, if not slready in place), to estimate recovery operation costs along the lines illustrated earlier in Tables I, II, and IV. For these calculations, it is important to use current actual prices for plating bath and treatment chemicals (i.e., data on prices paid on recent purchases from the plant's accounting office or from prices quoted by the local suppliers of chemicals) as well as utilities and other items.
5. Estimate investment and operating costs for recovery operations, along with the potential savings (avoidance of treatment and disposal costs, and costs for virgin materials) resulting from the implementation of recovery, using data and information available from the literature (pertinent U.S. EPA reports and other sources) and from trade literature from recovery/recycling equipment manufacturers.

6. Once specific plating rinsewater streams have been identified and evaluated as promising candidates for recovery/recycling using specific technologies, obtain bids from several equipment manufacturers or suppliers for use in making final decisions regarding the desirability of setting up one or more R/R operations in one's plant.

REFERENCES

1. Gurklis, J.A., "Handling and Disposal of Special Wastes", Plating and Surface Finishing, 65, 10, 12-18, October 1978.
2. "Assessment of Industrial Hazardous Waste Practices-Electroplating and Metal Finishing Industries--Job Shops", prepared for U.S. EPA Hazardous Waste Management Division by Battelle Columbus Laboratories, PB-264 349, September 1976.
3. U.S. EPA, "Development Document for Interim Final Effluent Limitations and New Source Performance Standards for the Metal Finishing Segment of the Electroplating Manufacturing Point Source Category", U.S. EPA Effluent Guidelines Division, Office of Water and Hazardous Materials, EPa-440/1-75/040-a, April 1975.
4. U.S. EPA, "Development Document for Proposed Existing Source Pretreatment Standards for the Electroplating Point Source Category", Effluent Guidelines Division, Office of Water and Hazardous Materials, EPA 440/1-78/085, Washington, D.C., February 1978.
5. U.S. EPA, Office of Water Planning and Standards, and Center of Environmental Research Management, "Environmental Regulations and Technology--The Electroplating Industry", EPA 625/10-80-001, Washington, D.C. and Cincinnati, Ohio, August 1980.
6. Novotny, C.J., "Water Use and Recovery", Finishers' Management, 18, 2, pp. 43-46, 50, February 1973.
7. Graham, A.K., Electroplating Engineering Handbook, Third Edition, Chapter 34, Van Nostrand Reinhold Co., New York, New York, pp. 754-771, 1971.
8. Vaaler, L.E. and Gurklis, J.A., "Cadmium Plating with Environmental Restrictions", Proceedings of the Workshop on Alternatives for Cadmium Electroplting, EPA-560/2-79-003, March 1979.
9. U.S. EPA, Industrial Environmental Research Laboratory, "Environmental Pollution Control Alternatives: Economics of Wastewater Treatment Alternatives for the Electroplating Industry", EPA 625/5-79-016, Cincinnati, Ohio June 1979.
10. U.S. EPA, Industrial Environmental Research Laboratory, "Summary Report on Control Technology for the Metal Finishing Industry--Evaporators", EPA-625/8-79-002, Cincinnati, Ohio, June 1979.

11. "Economic Indicators", Chemical Engineering, 90,5,7, March 7, 1983.
12. Spatz, D.D., "Reverse Osmosis/Ultrafiltration Application to Water Reuse and Material Reclamation", Paper #29, Osmonics Inc., Hopkins, Minnesota, May 1975.
13. Ry, C., "Methods and Technologies for Reducing the Generation of Electroplating Sludges", Second Conference on Advanced Pollution Control for the Metal Finishing Industry, EPA, 600/8-79-014, Cincinnati, Ohio, June 1979.

SECTION IV

WASTE RECYCLE

CHAPTER 16

NEW YORK STATE INDUSTRIAL MATERIALS RECYCLING PROGRAM

Pickett T. Simpson, P.E.
New York State Environmental
Facilities Corporation

INTRODUCTION

As all are well aware, New York state is the scene of one of the infamous hazardous waste disposal sites that have had a serious economic, political, social and environmental impact on not only the citizens of New York State but on the country as a whole. The site was Love Canel. As a result of all this activity, the New York State Legislature began, in 1978, to enact necessary legislation to ensure the citizenry against such effects and to promote activities in the hazardous waste management field that would protect the environment, the general public health and provide an economic and environmentally sound climate for all interests in New York State. Many efforts were made in the field of regulation of hazardous waste management facilities and, to complete the overall concept of a sound, total program, the Governor signed into law on July 31, 1981 a bill entitled the New York State Industrial Materials Recycling Act.

This is a major program to encourage industries to reduce, recycle and reuse industrial materials including industrial solid waste and industrial hazardous waste. A many faceted program, it not only initiated activities in the waste exchange field but encouraged the development of information exchange, technology transfer and technical assistance.

The law, which is an amendment to the Public Authorities Law of the State of New York, specifically mandates the New York State Environmental Facilities Corporation, a public benefit corporation, to establish a comprehensive program to assist industries that generate industrial materials by:

1. encouraging the reduction, recovery and recycling of these materials;
2. providing industries with technical information and assistance; and,
3. encouraging the exchange of materials.

Management of Toxic and Hazardous Wastes, H. Bhatt, R. Sykes, T. Sweeney (Editors)

The legislation further mandates the kinds of assistance to be provided and this includes:

1. review and compilation of research and development information on methods, technologies, reduction, recycling and disposing of materials;
2. research into available markets for recycled materials and preparation of a list of these materials;
3. maintenance of data on existing and projected production of industrial materials;
4. development of technical reference information on methods and economic means to reduce and recycle materials;
5. preparation of a handbook on recovery and recycling; and,
6. establishment and maintenance of an information clearinghouse for industrial materials available for recycling or recovery.

Unfortunately, not all of the activities required by law could be fulfilled by the corporation because of a lack of resources. In the initial year, $177,000 was provided for all of the above activities which included not only staff but overhead, office expenses, as well as the forms of technical assistance. Therfore, decisions were made to concentrate on those areas which would have an immediate and most productive impact on the industrial waste stream. These include waste exchange, passive and active; technical assistance; technology transfer of resource and development information; a handbook on recycling -- technical and economic feasibility; use of the Corporation's pollution control financing authority; and, issuance of a quarterly newsletter.

WASTE EXCHANGE

The exact legislative mandate is for the Corporation to maintain a list of materials that can be recycled or reused. We have interpreted this to mean the development and operation of a waste exchange system, a simple process whereby the waste stream or waste products of a particular industrial application are used or reused by that industry or by another industry. Perhaps the best known example of this is the use of pickling liquors from steel manufacturing operations at municipal sewage treatment plants for nutrient removals.

There are essentially two kinds of waste exchange -- active and passive. In a passive waste exchange, generators list wastes they wish to transfer and potential users list wastes they desire. The information is presented by code in a catalog or brochure in a format where only the quantity, description, availability and general location are identified. Those interested in particular wastes may contact a waste exchange which will forward their inquiry to the lister who in turn responds to the inquiry. Once that has been accomplished, the exchange does not become involved in subsequent negotiations between the two parties to complete the transfer of the material. A graphicl presentation of the passive waste exchange is shown in Figure 1.[1]

In the active exchange, an intermediary serves as liaison between the generators and potential users of wastes. Such exchanges may take possession of the waste and charge fees. This type of an exchange, although time and labor intensive, can result in reductions of difficult waste streams or large quantities of wastes, particularly those produced on a continual basis.

For the past three years, the Central New York Regional Planning and Development Board and the Manufacturers Association of Central New York have been in the process of developing the catalog system for the passive waste exchange. This activity will be described in anothe paper presented to this convention and I shall not elaborate on the details of their operation beyond the general characteristics indicated above and to state that rather than develop a separate waste exchange catalog system, the Corporation elected to financially support the activities of this catalog system by means of a contract. Under this contract, not only is a catalog published but an annual report is required to analyze the results and effectiveness of the passive waste exchange system.

The active waste exchange is patterned after the experience and legislation in the State of California. In California, however, the legislation is stronger in that it requires generators to justify their choice of disposal over recycling for a waste stream. This is not true in New York. To assist us in our initial development, we must give a great deal of credit to Bill Quan from the California Department of Health Services who spent two days with us under an EPA Peer Match Assistance Program.

Under the active waste exchange, the EFC staff has concentrated on making direct contact with industry in the form of surveys of small businesses; an analysis of the New York State Department of Environmental Conservatrion's regulatory program of Manifests; an analysis of the TOSCA listing of chemical users; and contacts by a technical field representative who covers the State offering technical assistance and seeking wastes that could possibly be reused or recycled. In addition, the Corporation not only works with generators and users, but with brokers, resource recovery firms, public interest groups, other State and Federal agencies, and many industrial associations including The Business Council of New York State, Inc.

Although we have not completed the analysis of the results of the second year of operation, a breakdown of the type and number of contacts during the first year is indicated in Table I.[2] Total contacts during the first and second year number 306.

Contacts are made by visitation, telephone or correspondence. The basic information about the concern of the company, its waste or technical problem as well as identifying information is recorded on a standard Contact Information Sheet. This information is then

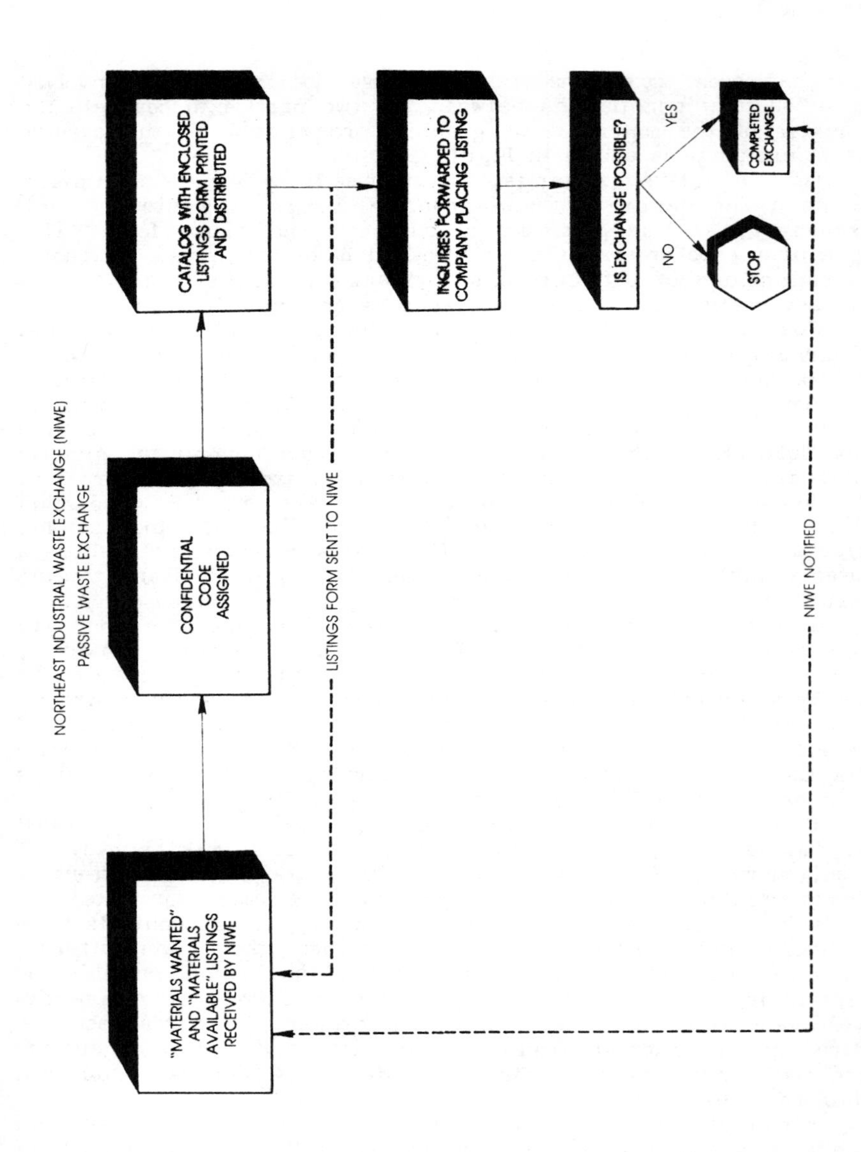
NORTHEAST INDUSTRIAL WASTE EXCHANGE (NIWE)
PASSIVE WASTE EXCHANGE
"MATERIALS WANTED" AND "MATERIALS AVAILABLE" LISTINGS RECEIVED BY NIWE
CONFIDENTIAL CODE ASSIGNED
CATALOG WITH ENCLOSED LISTINGS FORM PRINTED AND DISTRIBUTED
LISTINGS FORM SENT TO NIWE
INQUIRES FORWARDED TO COMPANY PLACING LISTING
IS EXCHANGE POSSIBLE?
NO
STOP
YES
COMPLETED EXCHANGE
NIWE NOTIFIED

FIGURE NO. 1

TABLE I. INDUSTRIAL CONTACTS [2]

Waste Stream Type	Letter and Telephone Contacts	Industry Visits
Acids	26	1
Alkalis	2	1
Other Inorganic Chemicals	2	1
Metals/Sludges Containing Metals	11	1
Organic Chemicals Solvents	35	5
Oils/Fats/Waxes	10	0
Plastics	0	1
Textiles/Leather/Rubber	4	3
Wood/Paper Products	4	3
Miscellaneous	8	4
General Field Contacts		13
Totals	102	33

transcribed into an in-house computer catalog system for access, progress reporting and analytical purposes.

If the contact is for waste exchange purposes, a procedure is set into motion to accomplish same. The graphical presentation of the active waste exchange is shown in Figure 2.[3]

As can be seen from the chart, the analysis work required and the identification of a waste stream is a far more complex, time-consuming and labor intensive effort than that of a merely passive listing. A detailed analysis of a particular waste must be made to determine a specific application. Note that there may be points along the way where it becomes possible to lead to an alternate arrangement or exchange of waste stream. A list of successful and potential active waste exchanges is as follows and includes material from both the first and second year of operation.[4]

TABLE II. ACTIVE WASTE EXCHANGE [4]

Operation	Description	Value
Manufacturing Company	600gal. of ammoniated copper	$1,650 saved in net direct

TABLE II. ACTIVE WASTE EXCHANGE [4] (cont'd)

Operation	Description	Value
	etching, ammonia hydroxide and copper hydroxide	costs
Wastewater Treatment Plant	5,000 gal. of ferrous chloride	$176/ton saved in costs over ferric chloride
Chemical Manufacturer	30,000 lbs/wk of ammonium chloride	Savings in the difference of transportation costs over total price of $.45/lb. for virgin material costs
Sheltered Workshop	385 gal. of trychloro-ethylene were recycled through a recycler	Savings of $1.428 over new stock
Utility Company	220 gal. of waste alcohol recycled	Savings of $860
Manufacturing Company	4 qts. of benzene were donated to a college laboratory	$50 savings to the college
Metal Finisher	100 drums of trychloroethylene were recycled	Savings of $8,000
Tanning Operation	100 drums of various solvents were recycled	Savings of $7,500
Research Laboratory	Recycled oil	$30/barrel savings over crude oil costs
Manufacturing Company	Acetylene calcium hydroxide sludge substituted for lime in sewage treatment	$30/ton savings
Utility Company	640 tons of excess coal sold to a broker	Savings of $6,400
Local Government Unit	40 barrels of highway paint sold to recycler	Savings of $800
Local Government Unit	30 gal. of perchloro ethylene	Savings of
Printing Company	110 gal. of alcohol	Exchange pending
Manufacturing Company	30 tons of glycerine and water sold to broker	Value not reported

TABLE II. ACTIVE WASTE EXCHANGE [4] (cont'd)

Operation	Description	Value
Manufacturing Company	300 tons of glycerine and acid sold to broker	Value not reported
Manufacturing Company	Calcium hydroxide waste substituted for lime in sludge conditioning	$30/ton savings over lime
Cement Company	Calcium hydroxide dust for lime substitute	Action is now pending at various locations
Cement Company	Solvents and other hazardous waste burnables	Sources presently being sought for specific heating value; value yet to be determined

Sample Case Studies

While many reports have been written on the use of pickling liquors as a substitute for various polymers and other nutrient removal chemicals at treatment plants and as a substitution for ferric chloride, it has been our experience that, depending upon the nature of the sludge conditioning process, analysis and match-ups must be carefully arranged. In one instance, a municipal treatment plant ordered a bulk shipment of ferrous chloride (pickling liquor) to be used as a substitute for ferric chloride in an application for sludge dewatering. It was the operator's experience that the ferrous chloride did not work as effectively as the ferric chloride for sludge dewatering because the treatment process did not allow for sufficient oxygenation of the pickle liquor substitute.

Another interesting case involves a brewery and acetylene sludge from another manufacturer. In this particular case, we knew that there was a potential application at the brewery sewage treatment plant for the sludge from the acetylene manufacturer. Next, samples were gathered of the acetylene sludge, chemical analysis made, bench scale tests conducted, and a full compatibility test performed to see if the material could be substituted for the lime used in the treatment process. Also, to guard identities of both the generator and the potential user, we acted as the intermediary as well as the identifier of the potential use until such time as the test results indicated the application was correct.

A third example involves an on-going project we are developing with a cement company. In this case, we are making use of the

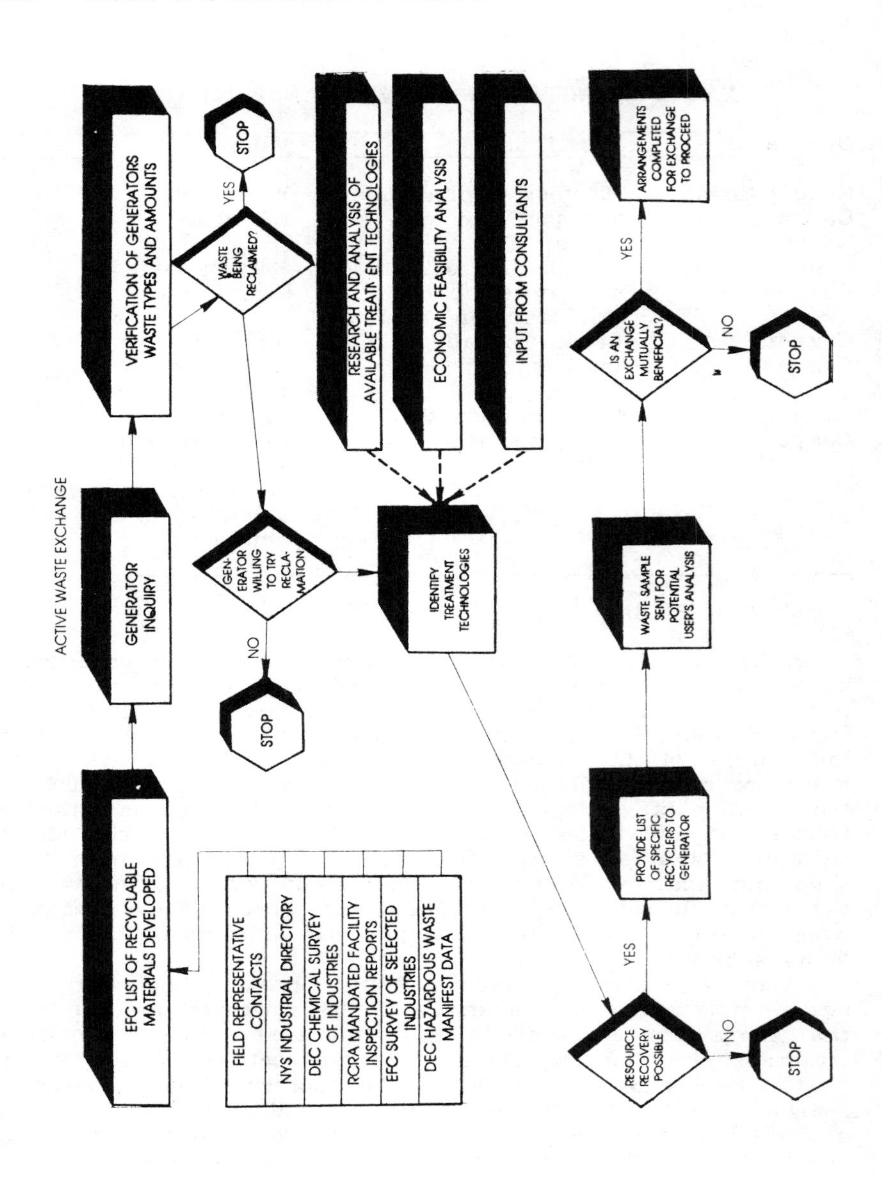
ACTIVE WASTE EXCHANGE
EFC LIST OF RECYCLABLE MATERIALS DEVELOPED
GENERATOR INQUIRY
VERIFICATION OF GENERATORS WASTE TYPES AND AMOUNTS
WASTE BEING RECLAIMED?
YES
STOP
GEN-ERATOR WILLING TO TRY RECLA-MATION
NO
STOP
RESEARCH AND ANALYSIS OF AVAILABLE TREATMENT TECHNOLOGIES
ECONOMIC FEASIBILITY ANALYSIS
INPUT FROM CONSULTANTS
IDENTIFY TREATMENT TECHNOLOGIES
FIELD REPRESENTATIVE CONTACTS
NYS INDUSTRIAL DIRECTORY
DEC CHEMICAL SURVEY OF INDUSTRIES
RCRA MANDATED FACILITY INSPECTION REPORTS
EFC SURVEY OF SELECTED INDUSTRIES
DEC HAZARDOUS WASTE MANIFEST DATA
RESOURCE RECOVERY POSSIBLE
YES
NO
STOP
PROVIDE LIST OF SPECIFIC RECYCLERS TO GENERATOR
WASTE SAMPLE SENT FOR POTENTIAL USER'S ANALYSIS
IS AN EXCHANGE MUTUALLY BENEFICIAL?
YES
NO
STOP
ARRANGEMENTS COMPLETED FOR EXCHANGE TO PROCEED

FIGURE NO. 2

manifest records as well as our own contacts for wastes possessing good heating value. Under a public information request to the Department of Environmental Conservation, we can use their computer research capability to identify several types of waste streams having the caloric value required for the partial use of the solvents as fuel substitutes in the cement kiln operation. Not only is the heating value involved but a dollar price per BTU has been established as an operating parameter. Therfore, this situation requires not only identifying wastes but conducting laboratory tests to determine heating value, quantity, availability and compiling of dollar costs to assure the efficient and effective use of the material at the cement kiln.

TECHNOLOGY TRANSFER

As indicated earlier, the Corporation has been mandated by the Legislature to assist industries in technology and information on the reduction, reuse and recycling of materials. To do this, we have set up our own in-house library to maintain a collection of texts, periodicals, manufacturers' literature, engineering brochures, journals, newspaper articles, and other sources of written information relating to the program.

In providing technical assistance to industries, we must have the in-house capability or gather research from other sources, such as the New York State Library, on the latest, most complete and accurate data on a particular process, chemical or technology. Extensive use is made of data-base searches, including Enviroline, Chem Abstracts, Pollution Abstracts, Pat Search, and several others, to screen for particular items that apply to specific questions raised by the industrial client. This process is displayed in Figure 3.[5]

An illustration of this is a recent request from a chemical company to provide information on co-disposal involving tires and refuse and the incineration of a refuse derived fuel using waste heat boilers. There interest was not only in the technology being employed for both the burning process and air pollution control, but the marketing potential for the refuse derived fuel, as well as the operation and maintenance problems encountered by similar types of facilities elsewhere. A preliminary report was prepared, at no expense to the company, in the areas requested together with an extensive bibliography and a list of locations where excess quantities of tires had been stored awaiting disposal.

In another instance, a paint manufacturer was interested in a heat recovery process whereby their spent solvents could be burned to provide a heat source for the distillation processes at the front end of his manufacturing chain. The search was made for the technology, an evaluation made of the literature available, and the information furnished.

A third example involved analyzing the electroplating

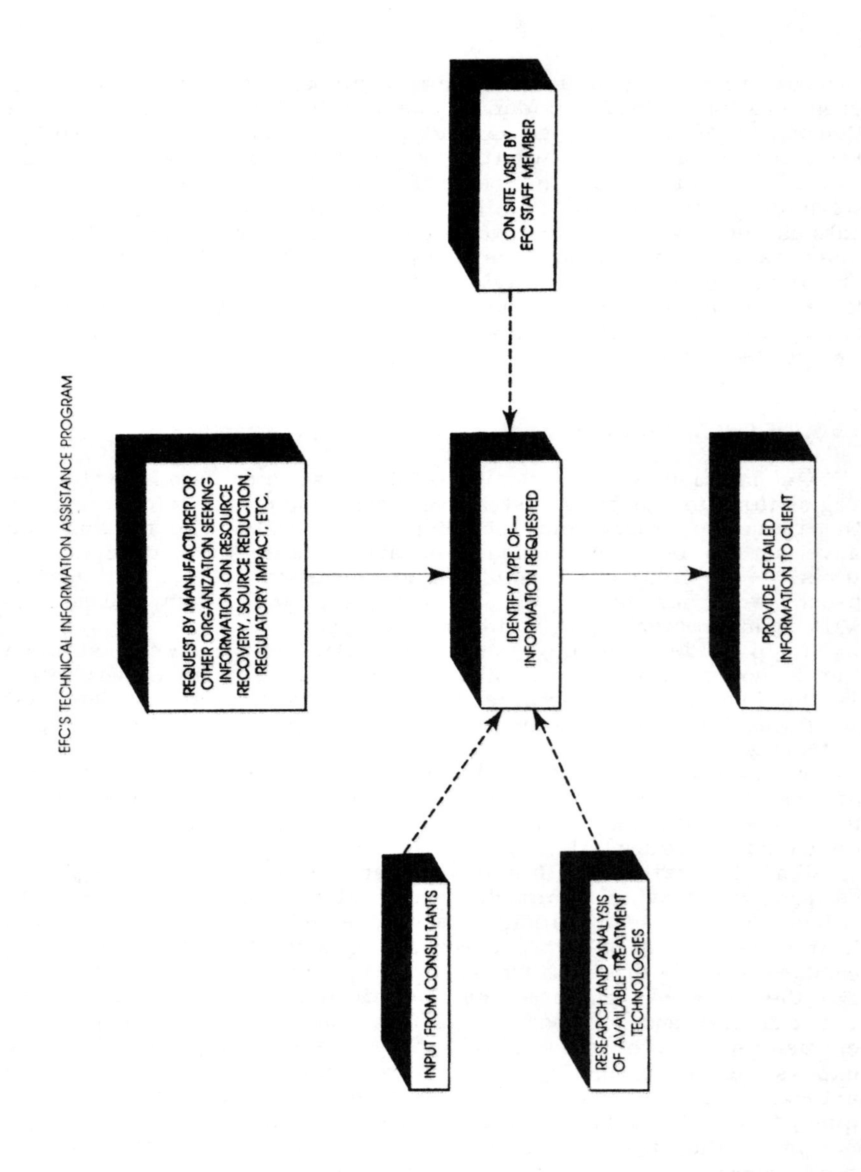
EFC'S TECHNICAL INFORMATION ASSISTANCE PROGRAM
REQUEST BY MANUFACTURER OR OTHER ORGANIZATION SEEKING INFORMATION ON RESOURCE RECOVERY, SOURCE REDUCTION, REGULATORY IMPACT, ETC.
ON SITE VISIT BY EFC STAFF MEMBER
IDENTIFY TYPE OF— INFORMATION REQUESTED
PROVIDE DETAILED INFORMATION TO CLIENT
INPUT FROM CONSULTANTS
RESEARCH AND ANALYSIS OF AVAILABLE TREATMENT TECHNOLOGIES

FIGURE NO. 3

manufacturing process. We are currently, on a region-by-region basis, identifying electroplating waste generators and presenting to them, by means of on-site visits, the alternatives available to reduce the rinse waters and to secure increased potential for the recovery of metals.

In yet another diverse area of service, we are helping companies to prescribe to types of engineering services they may need to design pollution abatement or resource recovery facilities, including the selection of necessary specialists.

INDUSTRIAL FINANCING

The broader powers of the Environmental Facilities Corporation include a provision for the financing of pollution control facilities for private industries. These issues are tax exempt securities under the qualifying provision of the Internal Revenue Code and are backed by the full faith and credit of the private companies seeking the loan and issued in the name of the Corporation as Special Revenue Bonds. The bonds were recently used to finance over $14 million worth of work by the General Electric Company in removal of hazardous materials and treatment of same from former disposal sites located in upstate New York. Because of the difference of definitions of hazardous wastes and solid wastes between the Resource Conservation and Recovery Act and the Internal Revenue Code, a very careful legal analysis must be made as to the eligibility of the facilities dealing strictly with hazardous waste.

A very important feature of the pollution control financing is that several environmental projects at one or more plant sites in New York State can be financed through a single bond issue, an advantage over Local Industrial Development Bonds. If the facility meets the requirements of the Internal Revenue Service and the New York State Department of Environmental Conservation for pollution control, there is no limit on the amount of loan, which can be for a term of up to 40 years. This loan may be used to pay for land, appurtenant buildings, equipment and engineering as well as for design, legal, financing and other related costs.

Since 1976, the Corporation has issued $66 million in bonds under this program thus enabling many industries to meet their environmental responsibilities without high financing costs.

CONFIDENTIALITY

The Industrial Materials Recycling Act includes provisions for the establishment of procedures to protect trade secrets and other proprietary information essential to the operations of private industry. In addition, special provision is made in the law to exempt such information from exposure under the Freedom of

Information Act if information received by the Corporation is identified as confidential or a trade secret. Accordingly, EFC has established procedures to safeguard such material and no material can be released without the express written consent of the generator as to his identity, identifying type information, or specific products, services or trade secret processes.

These procedures include limited access by staff to all mail marked confidential, establishment of a secure area with locked files and limited access, and an office policy and regulations outlining the procedures for safeguarding said material. Failure to keep such information confidential subjects an employee to dismissal or a fine of up to $5,000.

FUTURE DIRECTIONS

The Corporation firmly supports and encourages research in several technical areas that provide information and additional alternatives for treatment and disposal as well as recycling of hazardous waste. These include: development of a transportation risk analysis at Rensselaer Polytechnic Institute which has recently been funded by the Environmental Protection Agency for a generic computer model; the burning of wastes for energy recovery of large volumes of waste at privately owned industries; development of special markets for high volume wastes which appear not to have cost-effective treatment methods (sludges, dilute solutions, etc.); exchange of information and development of similar programs across state lines; the development of incentives for industry to reuse wastes by means of tax incentive; and, special de-regulation of wastes bound for beneficial reuse.

CONCLUSION

Overall, we feel we have developed a comprehensive program to address the issue of source reduction and materials reuse particularly in the hazardous waste field. Program limits are nonexistent except for budgetary limitations typical of any government agency at this time. However, by not being a regulating body we have gained the confidence of the generators and recyclers as to our purpose and sincerity. We now find ourselves in the position of having to deal with increasing demands for the services and information which we offer.

We have just completed our first Annual Report on this program and copies may be obtained by contacting the New York State Environmental Facilities Corporation, IMRA Program, 50 Wolf Road, Albany, New York 12205, or by telephoning (518) 457-4138.

REFERENCES

1. NYS Environmental Facilities Corporation. "Industrial Materials Recycling Act Annual Report," pg. 4 (1982)
2. Ibid, pg. 8
3. Ibid, pg. 5
4. Ibid, Table 2, as amended, pg. 10
5. Ibid, pg. 19

ACKNOWLEDGEMENTS

The author wishes to thank the fine professional staff of the New York State Environmental Facilities Corporation, all of whom make this program work and this paper possible. The are: Diana M. Hinchcliff, Executive Assistant; Peter A. Marini, P.E., Project Manager; Marian J. Mudar, Industrial Waste Program Analyst; Harold Snow, Chemical Engineer; William P. Mirabile, Project Supervisor; William H. Holmes, Technical Fidld Representative; Randal D. Harland, Industrial Waste Intern; and Carol S. Post, Senior Stenographer. A special thanks to Terence P. Curran, P.E., Executive Director of the New York State Environmental Facilities Corporation.

CHAPTER 17

THE ROLE OF A WASTE EXCHANGE IN INDUSTRIAL WASTE MANAGEMENT - DEVELOPMENT OF THE NORTHEAST INDUSTRIAL WASTE EXCHANGE

W. Banning
Northeast Industrial Waste Exchange

NEW INCENTIVES SPARK CHANGES IN WASTE MANAGEMENT PRACTICES

Industrial and hazardous waste management practices in this country are in the midst of a major transition. This slowly evolving transition may be characterized as a change from the relatively easy and inexpensive land disposal practices of the past to the growing emphasis on waste reduction, recycling and resource recovery.

This major change in the concept of waste management is occurring primarily for two basic economic reasons: the first is the dramatic increase in waste disposal costs brought about by stricter waste disposal regulations nationwide and the growing scarcity of suitable waste disposal sites, and the second factor is the rising cost of energy and raw materials which is making it much more economically attractive than in the past for manufacturing companies to investigate recyclying, resource recovery, and raw material substitution opportunities.

In addition, this growing trend is being accelerated by recent prohibitions on the land disposal of certain highly toxic wastes such as PCBs. These bans are usually based on the waste's persistence, toxicity, mobility in groundwater, and on the availability of alternate disposal methods. The uncertainty of the long-term integrity of even state-of-the art land disposal facilities combined with increasingly complex generator liability fears are causing generators to closely reexamine their current waste management strategies.

Therefore, although land disposal still remains the most popular waste management option in 1983, all of these factors taken together are forcing industrial waste managers to explore a variety of options for lowering their waste management costs. Generators are focusing their attention not only on alternate disposal technologies, but also on source reduction, on-site recycling, and on locating off-site recycling opportunities.

Management of Toxic and Hazardous Wastes, H. Bhatt, R. Sykes, T. Sweeney (Editors)

Well-established scrap markets exist for the purchase, collection, and processing of many industrial by-products. However, there is a wide variety of other industrial wastes, especially hazarouds wastes, for which no readily identifiable market exists. In an era where changing economies and technology are shifting in fine definitional lines between a "waste" and a "scrap", the need has developed for a formal, institutionalized transfer agent to help identify and bring together generators of waste with reuse value and those recyclers who can realize its potential value.

ENTER THE CONCEPT OF WASTE EXCHANGE

To meet this need, two types of waste exchanges have developed over time: the materials exchange and the information exchange.

Materials exchanges actively broker the transfer of waste material between generators and users. These exchanges may buy or accept wastes, analyze their properties, reprocess or repackage them as needed, and seek buyers and sellers. Materials exchanges are "for profit" firms, being paid through commissions or from the sale of waste products. They usually operate by acting as consultants for waste generators and are staffed by technical, managerial and marketing personnel. They also may have storage, laboratory, and processing facilities. There are 6 or 7 "for profit" materials exchanges in the United States.

The other type of exchange is the information exchange. The information exchange offers a centralized clearinghouse for collecting, displaying, and disseminating information about wastes available from generators and wastes sought by potential users. The function of the clearinghouse does not normally involve any negotiations that may lead to transfers. The basic clearinghouse service is to receive from companies listings for wastes that a generator has avaliable and listings for wastes that a user would like to acquire, to publish a periodic catalog of these wastes, and to distribute the catalog throughout a specific geographic area. Wastes are usually assigned a confidential code number so that the identity of the listing company is protected. A company interested in a material listed in an information exchange's catalog will send a letter of inquiry to the clearinghouse. The clearinghouse will then forward the letter of inquiry to the listing company. That company will then contact the inquirer and begin the negotiation process, if it so chooses.

There are about a dozen waste information exchanges in the United States. Four of the major exchanges have been designed to serve multi-state areas: the Northeast Industrial Waste Exchange, the Midwest Industrial Waste Exchange, the Industrial Material Exchange Service, and the Southern Waste Information Exchange. Other exchanges have been established primarily to serve single states (such as California, Georgia, and New Jersey) or sub-state areas (for example Louisville, Houston).

DEVELOPMENT OF THE NORTHEAST INDUSTRIAL WASTE EXCHANGE

One such multi-state clearinghouse, the Northeast Industrial Waste Exchange, was established in early 1981 as a non-profit cooperative program of the Central New York Regional Planning and Development board and the Manufacturers Association of Central New York, both based in Syracuse, New York. Developmental funding for the Exchange was provided by a grant to the Board from the U.S. Economic Development Admnistration.

Shortly after this Exchange began operation, the New York State Legislature passed an amendment to the Public Authorities Law specifically requiring the New York State Environmental Facilities Corporation, a public benefit corporation, to establish the reduction, recovery, and recycling of industrial wastes, especially hazardous wastes, throughout New York State. An important aspect of this program included the establishment of an information clearinghouse (waste exchange) to serve industry in the State. Since this service was already being provided by the Northeast Industrial Waste Exchange, the Corporation elected to financially support the activities of the Exchange rather than establish a separate duplicative clearinghouse. This decision reflects the fact that a multi-state exchange provides a better service to industry by exposing listed materials to more potential buyers throughout a larger area. This broader marketing approach is central to the Exchange's basic function of identifying or creating a market for manufacturing residuals.

Although the Exchange currently distributes its quarterly Listing Catalog free of charge to nearly 8,500 firms in 45 states as well as Canada and Puerto Rico, its primary service area is Maryland, New England, New Jersey, New York, and Pennsylvania.

INITIAL RESULTS

To assess the effectiveness of the Northeast Exchange in creating a market for byproducts and in facilitating actual waste transfers among companies, a survey was undertaken in February 1982 of companies that had received at least one letter of inquiry about their listings in the August or November 1981 Listings Catalog. These two Listings Catalogs contained 520 listings and, overall, 75% of the listings generated at least one letter of inquiry. The Northeast Exchange received an average of 55 letters of inquiry per week during its first six months of operation (a total of 1,432 inquiries). Respondents to the survey reported that 22 successful exchanges were completed with an estimated value of $240,000 and a weight of 3,100 tons. According to the survey, more organic chemicals and solvents were sold than any other category (13 exchanges, 267 tons, $90,000). By products that have been exchanged include sulfuric acid, sodium sulfide, bleach, ferric chloride, potassium cyanide, copper sulfate crystals, copper sulfate solution, acetone, trichloroethylene, trichloroethane, ethylene glycol, solvents, ceramic slip, and ceramic sheet. Nearly 75% of the wastes that have been sold were classified as hazardous. The total amount of wastes still being negotiated at the time the survey was completed was nearly 13,000 tons. The major types of materials included: solvents and organic chemicals (15); metal and metal containing sludges (6); rubber (5); plastics (4); other organic chemicals (3); acids, alkalis, oil, paper and miscellaneous (2 each); wax, textiles, leather and wood (1 each).

The Northeast Exchange is currently in the process of conducting a follow-up survey of activities during 1982 to assess its effectiveness in reaching generators and potential recyclers. The results of the 1981 survey are encouraging for they indicate that the Northeast Exchange is beginning to identify trading partners for a variety of industrial residuals.

IMPEDIMENTS TO SUCCESSFUL WASTE EXCHANGE

Despite this positive beginning, the 1981 survey also revealed that successful transactions occur less frequently than companies would hope. Respondents to the survey noted that a growing market exists for waste transfer activities in the Northeast but that current restrictive regulatory and legislative requirements often, by their nature, act as disincentives to waste transfer.

The survey results also indicate that, although the concept of an industrial waste exchange is simple to describe, the actual transfer of wastes between a generator and a user can be complicated. Thus, only a fraction of the benefits that are possible through waste exchanges are actually realized at this time. What, then, are some of the specific impediments to the successful exchange of wastes among companies?

Small Quantities/Contaminated Waste Streams

First of all, many generators have available only a small quantity of material offered on an infrequent basis or only once. Recyclers are generally interested in large volumes of continuously available waste streams. Of course, even small quantities of some waste streams are valuable enough to justify less-than-truckload transportation charges. However, it is generally not economical to treat a small quantity of material. Thus recyclers must face the problem of collecting and aggregating small quantities of wastes dispersed over a broad geographic area. but even similar waste streams are likely to contain contaminants or unique formulations that preclude aggregation. This problem could be lessened to a degree if more generators were made aware of the recycler's requirements and then practice good housekeeping procedures to prevent mixing and contamination of recyclable waste streams.

Permits Needed for Storage of Hazardous Wastes

The problem faced by small generators is aggravated by Reesource Conservation Recovery Act requirements that generators need a permit if they are to store listed hazardous wastes or sludges for more than 90 days (characteristic hazardous wastes that will be recycled are exampted under 40 CFR 261.6). Since many generators are reluctant to undergo the permitting process, they are unable to accumulate enough material to be of interest to a potential recycler. In addition, negotiations for the transfer of wastes typically are not completed within 90 days, making waste exchange transactions very difficult. To overcome these problems, appropriate methods for the collection and recycling of dispersed low-value material are needed and the regulations governing the 90-day storage limitation shoud be relaxed for material that will be recycled.

Regulatory Deterrents to Exchange

Another impediment to waste exchange are the restrictive uncertain, misunderstood, or feared hazardous waste management regulations existing at the state and federal levels. The changes in these regulations through time have confused generators, transporters, and recyclers alike, and this uncertainty deters companies from getting involved. Not enough education has taken place about the regulations as they affect recycling opportunities. For example, some generators are still under the impression that if they purchase a hazardous waste it is no longer a waste and, therefore, not covered by the regulations. Others are not aware of the limited exemptions offered by Part 261.6. In addition, many generators will manifest all their wastes on the assumption that they are "playing safe". However, a manifested waste can only be shipped by a licensed handler and accepted by a TSD facility. Since few manufacturing companies are licensed TSD facilities, manifesting can make finding a recycler more difficult.

Confusion about the regulations and how they affect recycling opportunities is compounded by unique variations each state has incorporated into its own regulations. Since few states possess the full range of recycling opportunities, interstate shipments are necessary but made more difficult by the lack of consisistency among the states. One cannot be optimistic in the short term about standardizing these requirements since the states have not even been able to reach agreement on using a standard manifest form.

Identifying Markets

An additional impediment to waste exchange is that many companies are simply not aware of recycling opportunities for their own wastes, or of what materials are available from other companies and how they could be incorporated into their own manufacturing processes. Even if a company is aware of recycling opportunities, capital expenditures for new equipment or process design changes may be necessary before these opportunities can be realized.

POTENTIAL SOLUTIONS

Despite these obstacles, the fact remains that change is in the air with regard to waste management options. Assuming management's sincere desire to explore recycling opportunities and the availability of affordable capital, the impediments mentioned here could be addressed by the numerous consulting firms that specialize in hazardous waste management and process engineering. "For profit" waste exchanges can be better utilized by waste generators to market unique waste streams or to solve difficult waste management problems. In fact, a recommended strategy for many generators would be to use the services of a regional (multi-state) information exchange and those of a broker/consultant. The former offers the advantage of inexpensively reaching a large and diverse market area while the latter can provide special consulting services and contacts.

The concepts of waste exchange and recycling will continue to be appealing in the years to come because of the high cost of disposal and the intuitive feeling that reuse is preferable to disposal. Maximal recycling will not occur, however, until existing impediments are removed. Attention should be focused, in particular, on the problems faced by small, widely dispersed generators: on disseminating information about recycling opportunities; and on developing legislative incentives to encourage recycling. Success depends on the willingness of generators, waste exchanges, haulers, recyclers, and legislators to develop new skills, procedures, and alliances.

CHAPTER 18

3P: POLLUTION PREVENTION PAYS - A 3M SUCCESS STORY

Michael D. Koenigsberger
3M Company

ENVIRONMENTAL TRUISMS

There are four environmental truisms:

1. Pollution is a visible sign of inefficiency in industrial operations. It is money that is going up the chimney, down the sewer, and out of a plant in waste trucks.
2. Pollution is, quite simply, the discharge of material and energy residues into the environment. Some of those residues are raw materials which are unconverted, some are products which are not fully recovered and some are by-products, but all are waste.
3. Increased corporate effort to reduce pollution can actually help to increase profits.
4. If you make no mess, you have nothing to clean-up.

Pollution prevention is the environmental aspect of conservation-oriented technology, which is based on conservation in all aspects, from raw material supply and production, to consumption and disposal.

The idea is to use a minimum of resources to accomplish objectives and to create a minimum of pollution. It also means learning to create resources from pollution, such as the making of nylon and other materials from the waste by-products of petroleum, as was done some years ago.

The environmetal benefits and economic incentives in this approach to pollution abatement are evident. Financial and natural resources can be saved, and technology innovations can be achieved.

HISTORICAL PERSPECTIVE - TREATING THE SYMPTOMS

To see where we are heading in this direction, however, requires a thoughtful look at where we have been. Government and industries alike traditionally have been preoccupied with

Management of Toxic and Hazardous Wastes, H. Bhatt, R. Sykes, T. Sweeney (Editors)

controlling pollution instead of eliminating its sources.

This occurred at least partially because the original battery of environmental laws and regulations that followed earth day in 1970 was a reaction to problems that already existed, instead of an intent to prevent new ones.

Hindsight has made it clear that we have placed too much emphasis on pollution control. We have been treating the symptoms by relying on complex mechanisms, attached at great expense to the end of a production line.

The tendency is to catalogue these controls as "black boxes" and to avert the eyes from pollution problems, as long as the black boxes work and regulators are satisfied.

These black-box or end-of-pipe controls, however, merely shift the problem from one form of pollution to another and they are contrary to the law of conservation of matter.

The law states that we can change the form of matter, but matter does not disappear. Purifying waste water creates sludge. burning chemical wastes creates particulate matter and fumes. both residues are pollution and create disposal problems of their own.

By using control measures, we also have been responsible for creating offsite pollution - generated by those who produce the energy and materials we use for abatement measures.

That includes such things as the fossil fuel consumed and sulfur dioxide and particulates wafted into the air by a power plant located far away, but which generates electricity for our environmental - control purposes.

COST AND FEASIBILITY OF CONTROL

The end-of-pipe approach also has been a major drain on the U.S. economy in terms of energy and natural resources consumed to control pollution and in the dollars thus diverted away from production.

In one year alone, for example, just 146 chemical companies in North America consumed the energy equivalent of 61 trillion BTUs for pollution control - 7.3% of their total energy requirements and enough energy to supply a year's heat and electricity for 300,000 typical homes in a climate as severe as Chicago's.

These same companies - 130 in the United States and 16 in Canada also estimated that their pollution-control spending for two recent years at 8.1% of total capital investment, which was double their rate of pollution control investment in the decade 1961-1972.

While capital costs remain a significant portion of this spending, operating costs increasingly are becoming a major burden as well, and one which threatens to escalate out of control.

In addition, the cost of pollution control, the resources consumed, and the residue produced increase exponentially as removal percentages rise to the last few points.

In other words, the unit cost becomes four or five time greater when you try to get from a base level of 85% removal to a goal of 95% than it was to achieve the 85%. at the same time, the amount of residue produced per ton of pollution removed is increased by 200% for the advance beyond 85%.

When all of these factors are considered, it is apparent that the end-of-pipe approach, at some point, creates more pollution than it removes and consumes valuable resources out of proportion to its benefits.

A catch-22 situation exists in the use of pollution controls, in that it takes resources to remove pollution. Pollution removal generates residue. It takes more resources to dispose of this residue - whose disposal also produces pollution. Paradox and irony, indeed, abound.

In addition, pollution controls work best with simple pollutants present in large quantities. They are not effective in dealing with those pollutants that are the by-product of sophisticated technology and which exist only in minute quantities - expressed in parts per billion or trillion.

It is only recently that technology has existed with which to measure such tiny quantities of pollution, let alone remove them.

Removal of these tiny amounts of pollution is either technically unattainable or extraordinarily expensive in both money and consumption of energy and other natural resources.

When all of these factors are considered, along with the flagging economy and energy shortages of the mid '70's, it becomes apparent why the emphasis began to shift from pollution controls to pollution prevention. At present, this shift is more evolutionary than revolutionary, but nevertheless, a shift is taking place.

Since the mid-1970's, the use of end-of-pipe pollution controls has become recognized by 3M as a wasteful and ineffective way of improving the environment, which cannot be tolerated if an alternative is available. 3M and most of the world is coming to realize that the natural limits on its resources can form a constricting noose.

INTERNATIONAL AWARENESS

Concern about the environment is shown by a growing international awareness and interest in conservation-oriented technology, with emphasis on pollution prevention.

In 1976, the UN Economic Commission for Europe held a conference in Paris on "Non-Waste Technology and Production", which drew participation from more than 25 countries. Technical papers were presented on all aspects of resource conservation from a model for administering conservation - oriented technology to a case study of how to clean minerals from the residue of coal mining.

A year later, the US EPA and Department of Commerce joined with representatives of industry to hold four regional pollution-prevention conferences.

Also in 1977, Dr. Michael Royston, an internationally recognized environmental specialist from Geneva, published perhaps the first book specifically about resource-conservation-oriented technology. It was titled "Pollution Prevention Pays".

The book had special significance for us at 3M because Dr. Royston selected as the book's title the name of our pollution prevention pays program, which we thought out in 1974 and introduced throughout our company beginning in 1975.

PROGRAM FEATURES AND RESULTS

The 3P Program is an effort that we continue today and will emphasize as far into the future as we can project. It has been so effective that it would be continued for economic and technological reasons alone, even if the environment and conservation incentives were ignored.

In addition to those factors, we were interested in pursuing a pollution prevention effort because of additional requirements imposed by the toxic Substances Control Act and other legislation and regulations that are involved with product use.

3M has been very concerned about the environmental impact in the use of our products. Our scientists, for example, have eliminated a mercury catalyst from an electrical insulating resin, which did away with any mercury-related pollution problem for the user of that resin. This type of activity contributes to one of the measurements we use to gauge the effectiveness of our 3P Program: "Sales retained of products that otherwise might become environmentally unacceptable."

Since 1975, the 3P approach to product design and process development has become standard within the company.

There are four major methods we are using to eliminate pollution at the source:

1. Product Reformulation. For example, we are trying to use solventless coating solutions wherever possible - and we have made some good advances recently.
2. Changing Production Operations. For example, changing from batch feeding to continuous feeding so we do not overload the pollution control system.
3. Modifying Manufacturing Equipment. For example, not long ago we reconstructed the boiler facility at one of our plants so it could burn solvent laden air from the coater oven. This eliminated the need for a million dollar air pollution removal unit. The 1,800 tons of volatile organic compounds now being burned in the boiler has a fuel value of $270,000 a year.
4. Recycling (including recovery for reuse or sale). The best illustration of this is solvent recovery.

While these methods - and the entire idea of preventing pollution at the source - are not new, determined application across-the-board on a continuing basis throughout the entire company is a new twist.

The program has produced considerable dollar savings from pollution-control equipment and operating costs that could be eliminated or delayed. It also has eliminated significant amounts of pollution discharges annually.

From 1975 through 1984, the program has produced total savings of $235 million. This includes $179 million from U.S. operations and $56 million from international.

In the U.S., 3P savings include $62 million for pollution control equipment and facilities not needed; $94.6 million for raw material and operating costs; $6 million for energy savings not included in operating costs; and $16.4 million for sales retained of products that might have been taken off the market as environmentally unacceptable. We have found 60 to 70% of operating cost savings are repeated annually but are not included in the totals.

Each year, the program has eliminated environmental discharges of 98,000 tons of air pollutants, 11,000 tons of water pollutants, 1 billion gallons of wastewater, and 150,000 tons of sludge and solid waste.

In addition, the program's annual energy savings are estimated at the equivalent of 210,000 barrels of oil. These results have occured from 396 projects in the U.S. and another 1132 projects overseas, where the program has been put into use by 20 of our subsidiary companies.

These results have not been achieved by a small band of environmental engineers going through the company and reformulating products, changing operating procedures, modifying equipment and starting recycling programs. That would be impossible - at 3M or probably at any company.

What our environmental engineering group of about 50 professionals did is to start a program that has been aimed at bench level laboratory, engineering, and manufacturing personnel. The idea is to solicit the help of all technical personnel, by asking them to start preventing pollution at the source in the areas for which they are responsible. So instead of putting 50 people to work on the program, we put more than 6,000 technical people to work.

POTENTIAL PROBLEMS

There are some problems associated with such a program. Let us look at the main factors that we have had to consider.

The first and perhaps the most important factor is top management approval. A formal, company-wide effort like this requires the enthusiastic support and commitment of an organization's senior management - starting with the chief executive officer. This is necessary in order to get and maintain the attention of lower management, and because the concept may represent a change in direction for some parts of the company.

Point two. The initial investment for a pollution prevention project may be higher in some cases than the cost of installing conventional pollution removal equipment. However, the annual operating and maintenance cost of the removal equipment will almost always make the total cost of this technology higher than the total cost of preventing pollution at the source.

Point three. Pollution prevention technology may not always be a viable alternative to pollution removal facilities. The steel industry is a good example. Reformulating the steel making process simply is not a practical reality.

Point four. Where pollution prevention is possible, it may be viewed as an untried substitute for a manufacturing process that has a successful add-on pollution removal facility. In such a case, manufacturing management may be reluctant to risk making a change. Unless a dramatic cost reduction can be shown with use of pollution prevention technology, it may take awhile to be implemented. This is a situation where it is good to have the general support of top management.

Point five. Marketing considerations can be a deterrent to a particular pollution prevention project. If, for example, product reformulation is one of the methods, there may be some concern over customer acceptance of the revised product. Will the reformulated product perform exactly like the existing one? If not, will the difference be significant? And keep in mind that a number of government purchases are made on the basis of exact product specifications that would make reformulation difficult without revising the purchase contract - something the sales people may be reluctant to do.

These seem to be the main obstacles to implementing an organized pollution prevention program. But, fortunately, they are largely short term constraints - and they do not detract from the long range significance of the concept, which is particularly well suited for most any industry.

SAMPLE 3P PROJECTS

There have been a number of 3P projects in the area of toxic and hazardous waste management:

- Removing mercury as a stabilizer in microfilm. This eliminates mercury exposure to employees as well as mercury as an air and water pollution problem in manufacture. And it eliminates a product disposal problem for the customer.
- In photoprocessing operations, eliminating chromium, reusing disilvered fix and bench, coupler recovery and recovery of trace amounts of silver.

- Sludge recovery and reuse. Chrome oxide sludge was collected in a holding pond to allow time for reducing hexavalent chromium to trivalent chromium. It was then removed and landfilled. Now it is collected, reprocessed, and sold as an off-grade oxide saving over $40,000/year.

As can be seen, the 3P philosophy requires creative thinking by all 3Mers.

PROGRAM CRITERIA

To be eligible for recognition under the 3P Program, technical activity must meet several criteria. It has to eliminate or reduce actual or potential pollution and have a potential monetary benefit to 3M. It also has to represent a genuine technical accomplishment and personal effort.

Contributors to the program have proven to be amoung 3M's most creative technical employees, and this recognition serves to reinforce their reputations and contribute to career growth. Recognition is the only immediate award, because it is an employee satisfier, and the company avoids prizes or cash awards that smack of contests and promotions.

Because 3M is a new-products-oriented company, products and processes always are being invented or modified, and it has been comparatively easy for us to incorporate pollution prevention into our technical activity.

RECOMENDATIONS

Our recommendation is to use resource-conservation technology wherever and whenever possible and practical. Overall, success in this type of endeavor depends upon individual companies or industries developing their own technologies to prevent pollution, as they develop their own techniques to produce goods and services.

Indications are that society is moving away from the trap of thinking that tomorrow's technology will be the same as today's.

Solutions to our present and future environmental problems will come from new and better technology, based on conservation-oriented systems and the pollution-prevention approach.

Remember - if you make no mess you have nothing to clean up...

CHAPTER 19

EUROPEAN NETWORK OF WASTE EXCHANGES

Thomas E. Crepeau
Ohio Environmental Protection Agency
Philip R. Beltz
Battelle Columbus Laboratories

INTRODUCTION

Throughout most of western Europe there exists an efficiently operated network of waste exchanges which serve as an important element in promoting effective waste management in each country as well as among the countries participating in the system.

A waste exchange can be defined simply as a clearinghouse of coordinating operation between buyers and sellers of industrial production residues which can be used again in the production cycle. The operation of these waste exchanges in Europe is viewed as a valid form of recycling; in addition they serve a number of other purposes as well. Included as secondary benefits are a potential savings of disposal costs, a saving of raw materials and above all a lessening of the amount of production materials which otherwise might end up as waste products requiring incineration or land disposal particularly for materials considered under law to be hazardous.

The first waste exchange in Europe began operations in the Hague, Netherlands in 1969 under the sponsorship of the dutch Chemical Association. Throughout the 1970's many other waste exchanges appeared on the scene some of which were initiated by trade associations, technological or research institutes of chambers of commerce. The latter group, namely the chambers of commerce, serve as the most common operating umbrella in Europe as far as the number of operating waste exchanges is concerned. The national chambers in conjunction with local and regional chambers of commerce participate in waste exchange operations in Austria, France, Germany, Italy and Switzerland.

SCANDINAVIAN COUNTRIES

A unique arrangement exists in Scandinavia in which technological or research institutes maintain waste exchanges in

Management of Toxic and Hazardous Wastes, H. Bhatt, R. Sykes, T. Sweeney (Editors)

Denmark, Norway, Sweden and Finland, and more recently, Iceland. An umbrella-type organization, known as the Nordic Waste Exchange, operates from Stockholm and serves as a coordinating body for the exchange of information in all of the Scandinavian countries. As a result, approximately 10,000 firms in all of Scandinavia are potentially served by its operation. The Nordic Waste Exchange was initiated by various industrial federations in each Scandinavian county because such an operation was viewed as a valid from of recycling. A research and development institute, the Swedish Water and Air Pollution Research Laboratory (IVL), provides for the operation of the waste exchange, including its personnel and the shared use of a computer. THe Institute itself is funded jointly by the Swedish government and Swedish industry. Besides providing waste exchange functions for Sweden, the Nordic Waste Exchange receives lists of available production residues and inquiries of materials needed from Denmark, Norway, Finland and Iceland as well and quarterly, mailing the compiled lists to Scandinavian firms. The result has been the establishment of an efficient, cross-frontier network of waste exchanges with operations far more effective than if each Scandinavian country alone were maintaining an exchange within its own borders exclusively.

GERMANY

Another example of an effective and efficiently operated network of exchanges exists in the Federal Republic of Germany. Operations began in 1974. Most local and regional chambers of commerce collect information on the availability of production residues from industries within their area. This information is then published in their local or regional publications. These regional bodies also forward their lists to the German National Chamber of Commerce office in Bonn. This office serves as a conduit for the dissemination of available production residues in the form of lists in a monthly bulletin from one local or regional office to all others.

FUNDING AND OPERATION

The initial funding and the continued operating costs of the waste exchanges in Scandinavia, Germany, Austria and the Netherlands has been borne exclusively by the chambers of commerce in the respective countries or by technological/research institutes. No fees are levied to any of the users of the exchange. The Danish Exchange, however, seriously considered the establishment of a user fee system in 1983.

The operation of all of these waste exchanges is relatively simple and the number of personnel is kept to the very minimum, generally one quarter or less of a full-time professional and the part-time use of a secretary, or in some instances, a data entry operator. Information on available production residues as well as inquires as to industrial materials which a specific firm may need are received from industry. In most instances, each waste exchange has a specified form which is completed by the inquirer or the one with an industrial material to sell. The publication of lists either in already-existing industrial bulletins or in the form of a bulletin by itself forms the core of the actual operation of the waste exchange. Obviously, the greater the area which the publication reaches, the probable greater effectiveness of the waste exchange itself.

In all but one of the waste exchanges comprising this network, anonymity is maintained, that is, the name and address of the firm either offering to sell or needing a specific material is not published. A coding system is employed instead and the waste exchange puts the interested parties in contact with one another. The only exception exists in Austria in the waste exchanges operated by the regional chambers of commerce and the Austrian Federal Waste Exchange (Bundesabfallborse) in Linz. The actual names and addresses of firms involved are published in the chambers' publications. In their opinion, full disclosure provides for a more rapid exchange of information and apparently is accepted by the industries involved.

INTERNATIONAL NETWORK

The most effective and far-reaching feature of the European waste exchanges lies in the cross-frontier network of exchanges. This exists not only in Scandinavia under the umbrella organization of the Nordic Waste Exchange, but the even larger network maintained by the German National Chamber of Commerce and Industry. The Nordic Waste Exchange as well as exchanges in Austria, Switzerland, Italy, Luxemburg and France periodically forward lists of production residues or materials needed by firms to the Bonn office of the German National Chamber of Commerce and Industry. In 1981 alone, 2631 offers and inquiries were handled by the German office which, in turn, forwards the lists to all countries participating in the network.

In order to resolve any language problems which obviously must arise because of the cross-frontier exchange of information, the German waste exchange office publishes quarterly in its bulletin a dictionary of the more important terminology involved in German, English, French and Italian.

EFFECTIVENESS

Some efforts have been made by the waste exchanges themselves to determine their effectiveness. In-depth studies have not been conducted primarily because of the costs involved in conducting them. The German waste exchange did conduct a limited survey in 1980 by means of a questionnaire. It was determined that more than a quarter of those offering materials for sale were successful and more than a third of those firms desiring certain materials actually received what they needed.

While such figures may appear only moderately positive, all of the exchanges feel that there would be continued operation of these exchanges as long as there appeared to be a need in the industrial community. The continued publication of lists of materials available and materials needed serves as the best form of evaluation of success in their opinion.

COMPUTERIZATION

Entry of data dealing with materials available or wanted into a computer program is conducted only in the Austrian and Swedish waste exchanges. The Danish waste exchange is planning to begin data entry as soon as feasible, possibly in 1983. Both of the waste exchanges employing an ADP system cited the advantages of being able to provide prompt response to telephone ad other inquires. The computer system is provided by the host office and is used on a time-sharing basis. There do not appear to be any plans for a cross-frontier, trans-European use of any computerized services at this time.

CONCLUSION

The unique features of the waste exchanges in Europe include the almost exclusive initial funding and continued maintenance by the industries which the exchanges serve. The chambers of commerce are the primary vehicles for such funding. In addition, the cross-frontier network of waste exchanges coordinated by the German National Chamber of Commerce & Industry (DIHT) has significantly enhanced the effectiveness of the system.

SECTION V

LAND DISPOSAL

CHAPTER 20

HAZARDOUS WASTE LAND DISPOSAL REGULATIONS - AN ENVIRONMENTALIST PERSPECTIVE

Linda E. Greer and David J. Lennett
Environmental Defense Fund

INTRODUCTION

The hazardous waste land disposal regulations, published July 26, 1982, in the Federal Register, are an important piece of the Resource Conservation and Recovery Act (RCRA) regulations, as they cover the predominant mode of hazardous waste disposal in the United States. That they be adequate to protect human health and the environment from harmful exposure to hazardous waste is central to the effective implementation of RCRA. There are, however four major problems with the hazardous waste land disposal regulations: lack of regulation for air emission monitoring and control, lack of requirement to retrofit existing facilities with liners, lack of requirements to show financial capability to perform corrective action, and, finally, lack of requirement to clean up contamination which has migrated beyond the facility property boundary. These will be discussed in detail below.

As a preface to this discussion, however, it is important to emphasize that no matter how strict the regulations, landfills will remain the least desirable method of disposing of hazardous waste. Steps therfore must be taken to shift wastes out of landfills as a priority for sound hazardous wastes out to landfills. After land disposal has been fully minimized, ther will still be a need for stringent and effective regulations to cover existing hazardous waste landfills and new landfills whose necessity has been adequately demonstrated. The July 26 regulations which will be discussed here fall short of providing this strict and adequate protection.

AIR EMISSION MONITORING

EPA has failed to address either monitoring or control requirements for hazardous air emissions at land disposal facilities. The Agency justifies this serious omission with a statement that (1) there is less information and experience on the extent of the air pollution problem at these facilities than on

Management of Toxic and Hazardous Wastes, H. Bhatt, R. Sykes, T. Sweeney (Editors)

surface or groundwater pollution problems, and (2) the problem is dependent upon the nature of the particular waste being placed at the facility.

This position on air emissions from hazardous waste facilities is in complete contradiction to the Agency's previous statements in the February 5, 1981 proposed land disposal regulations. In this proposal, the Agency stated that improperly designed facilities could cause significant air emissions and required ambient air monitoring and appropriate response if contamination was detected.

There is no new data which justifies a change in Agency position on the seriousness of the hazard that these practices can present or the feasibility of monitoring and controlling them. In fact, evidence generated by studies funded largely by EPA itself clearly support the ability and need to monitor and control air emissions from land disposal facilities.

Theoretical studies (using mathematical models) have predicted that high levels of emissions will enter the air from these facilities. Monitoring studies which confirm these predictions have been performed at over 20 operating Superfund sites (abandoned or inactive land disposal facilities).

A study undertaken by EPA and the State of California, for example, showed total carbon loading over 10 times higher in downwind locations than upwind. [1] Levels exceeding those estimated to be safe were found for several chemicals including benzene, toluene, and xylene.[2] Other studies conducted at several vinyl chloride sludge disposal sites and a PCB waste disposal site also showed high levels of air emissions. Vinyl chloride levels up to 1,100 ppb were found at the site, with levels from 70 to 500 quite common. Levels over 30 times the level estimated to be safe (up to 370 ppb for vinyl chloride) were found in residential/public access areas close to the sites.[3] Concentrations of PCBs at the PCB waste site ranged from 5.9 to 540 ug/m^3, greatly exceeding those estimated to be safe.[4]

Most recently, EPA has completed a survey of air emissions from 15 off-site hazardous waste facilities. The study found 180 organic compounds in ambient air of facilities and also found toxic metals associated with the fugitive dust particulate at the site.[5]

It is clear from the literature that there is sufficient information about all relevant aspects of the air emissions problem (extent of problem, sampling methodologies, control strategies) for EPA to take prompt regulatory action. Such action should consist of a requirement that facilities monitor their air emissions at a time of worst-case meteorologic conditions quarterly and respond appropriately to the detection of contamination.

LACK OF REQUIREMENT TO RETROFIT EXISTING FACILITIES WITH LINERS AND LEACHATE COLLECTION SYSTEMS

Without substantial justification, EPA has exempted existing portions of land disposal facilities (landfills, surface impoundments, and waste piles) from the containment design requirements of liners and leachate collection systems. By doing so, EPA abandons the notion of preventive action to control pollution from these potentially dangerous facilities and relies soley upon the groundwater protection and response (Subpart F) requirement to protect human health and the environment. The groundwater protection and response requirements consist of monitoring the groundwater upgradient and downgradient of the facility and taking appropriate corrective action measures when contamination is detected.

There are several significant problems with EPA's approach. First, protection of the public is entirely dependent upon the detection of contamination by groundwater monitoring systems. Although groundwater monitoring is an important part of the regulatory program, it is not fool proof, and our experience in installing groundwater monitoring systems is limited. In fact, by the Agency's own admission, stated in the 1981 Groundwater Research Plan, "major deficiencies [exist]in our capability to perform groundwater investigation." According to this document, such deficiencies are found in nearly every area necessary to effective groundwater monitoring: methods development, contaminant transport and fate, subsurface characterization, specific sources of contamination, and aquifer rehabilitation.[6] The Agency is by no means alone in its opinion that knowledge about groundwater and groundwater monitoring is relatively undeveloped. Similar opinions have been expressed in technical journals[7], conference presentations,[8] and expert testimony before Congress[9] by many prestigious groundwater researchers and consultants.

There are additional indications that there is insufficient groundwater expertise at the state or national level to ensure adequate groundwater monitoring systems. This point is backed up by a recent GAO report to the Congress on Federal/State Environmental Programs which states that they have difficulties in filling positions and retaining experienced personnel.[10] It is a mistake for EPA to design a regulatory program which places complete reliance on skills in technical areas in which the states are not well-prepared.

Second, performing corrective action response is expensive and time-consuming, and it presents a host of technical problems of its own. Again, the corrective action requirement plays a necessary role in protecting the public, but it should be relied upon as a last resort, not as the sole means of protection.

EPA's justification for exempting existing facilities for liners and leachate collection systems is based on two easily refutable grounds. First, the Agency contends that the removal of waste necessary to install containment devices may result in the shutdown of production facilities because some land disposal facilities may be integral to manufacturing processes. This contention is based on no factual grounds and has four important defects:

1. It does not differentiate between facilities associated with manufacturing facilities and those which are not.
2. It completely ignores the fact that production facilities will not be uniformly affected by a retrofitting requirement.

We are not aware of any data to support EPA's claim that manufacturing processes would be affected in many cases. In fact, many manufacturing facilities have back-up impoundments or space to construct new impoundments, or they empty their impoundments to perform routine maintenance or dredging. In these instances, manufacturing processes will not be substantially affected by a retrofit requirement.
Data collected by EDF during a survey of hazardous waste surface impoundments in Region II contradicts the Agency's contention. In this survey, we interviewed 53 facilities (one half of the surface impoundments in Region II, excluding Puerto Rico), and 24 responded. Our findings were:

a. only 1 facility would shut down if its surface impoundment was taken out of service temporarily.
b. 48% of the facilities had backup impoundments.
c. of those without backup impoundments, 82% of the facilities had land to build one.
d. 50% of the facilities without a backup impoundment had nevertheless emptied their impoundments in the past to do routine maintenance/dredging.

We realize that the conclusions which can be drawn from a survey of this size are quite limited. However, the results are extremely useful because they contradict a sweeping allegation made without any supporting data.

3. It fails to take into account the inherent flexibility of the permitting process in avoiding manufacturing shutdowns thrugh the use of compliance schedules.
4. It handles the prospect of temporary closures in a way which is inconsistent with other parts of the regulations. Although temporary closures may be necessary in selected cases, the Agency has recognized this necessity with respect to floodplain location standards[11] and substantial breaches of containment.[12]

Second, the Agency contends that exhuming wastes may create significant hazards for workers. This contention ignores the ready availability and ease of operation of various excavation and dredging techniques.[13] It also ignores the fact that the danger posed by exhuming wastes depends upon a variety of factors, including the quantity and density of the waste to be exhumed.[14] Barely used trenches, for example, are appropriate candidates for retrofitting.

The sole criterion of Section 3004 of RCRA is protection of public health and the environment. EPA's approach to existing facilities fundamentally fails to meet this mandate.

The limitations of the Agency's approach to regulating existing facilities are especially troublesome in light of the tremendous hazards these facilities pose to human health and the environment. In addition, because the siting of new facilities is proceeding so slowly, the population of landfills we will be dealing with over the next 5 to 10 years will be predominated by existing facilities.

EDF proposes that retrofitting should be required for existing portions of facilities unless 1) site-specific considerations render the retrofitting option technologically infeasible using best available technology or 2) the process of retrofitting the facility would pose a greater risk to human health and the environment than not retrofitting (due to the danger of exhuming the waste). The burden would be on the applicant during the permitting process to prove a particular facility qualifies for the exemption. Should a facility qualify for the exemption, the permitting authority should then require containment devices which do not require retrofitting to discourage waste migration. These devices should be required unless the hydrogeology of the site and the existing design of the site provide equivalent protection against waste migration. Examples of such devices include subsurface leachate collection drains, slurry walls, and grout curtains. There are several important advantages to this approach:

1. Exemptions from containment requirements will be granted on the basis of proven site-specific technological and environmental constraints rather than unproven generalizations about the impact of retrofitting nationwide.
2. Retrofitting will be required where it can be done safely, thus incorporating a preventative aspect to the regulation of existing facilities.
3. Where retrofitting is truly infeasible, other containment strategies can be employed where necessary.

THE IMPORTANCE OF FINANCIAL RESPONSIBILITY FOR CORRECTIVE ACTION

There is a significant gap in EPA's regulatory coverage of hazardous waste land disposal facilities. The Agency's regulatory scheme relies on corrective action by the owner/operator to protect human health and the environment, yet requires no financial assurances that the corrective action will be performed. Thus, whether corrective action will actually be performed is out of EPA's control, because it will largely depend upon the financial resources available to the owner/operator. In the present regulatory framework, actual protection of human health and the environment may not be achieved.

The lack of a mechanism to ensure the performance of corrective actions is a serious matter from both a legal and policy perspective. Legally, permitting standards for land disposal facilities must include such qualifications regarding financial responsibility as may be necessary to protect human health and the environment, pursuant to Section 3004(6) of RCRA. If corrective action is the underpinning of the health and environmental protection afforded by the permitting standards, then requiring a facility owner/operator to prove its ability to actually perform a corrective action is necessary to protect human health and the environment, absent an alternative mechanism available to ensure performance.

As a policy matter, requiring evidence of financial responsibility for corrective action is important to discourage abandonment of facilities, assure full coverage of the costs of hazardous waste management, and provide incentives for better hazardous waste management.

The regulatory system should not allow or induce the further abandonment of problem waste sites. Abandonment (in either the voluntary or involuntary form) has already plagued the nation's ability to address the risks posed by facilities operating before the RCRA program took effect. The problem of identifying an owner/operator with sufficient assets to clean up these non-RCRA permitted waste sites is so considerable it prompted Congress to enact the Comprehensive Environmental Response, Compensation and Liability Act of 1980 (most commonly known as Superfund). Yet by not requiring assurance for corrective action, the Agency even may encourage the abandonment of permitted sites.

Moreover, the Agency may be forced to accept substandard corrective actions at some point in the future because a proper corrective action would be so expensive as to result in abandonment, and thus in no corrective action at all.

Corrective action financial assurance requirements would help fill the large void in the financial responsibility area. For example, the Agency has required land disposal facilities to obtain insurance or otherwise provide financial assurances to cover liabilities to third parties for non-sudden releases occurring during the active life of the facilities. Yet, nothing is required from the owner/operator to remedy releases while they are on-site, before the releases become a liability. Assurances that contamination, when discovered, will be addressed on-site will reduce the exposure of any insurer of third-party and should, in the long run, make such insurance easier and cheaper to obtain. Furthermore, to the extent the insurance industry is involved in providing corrective action financial assurances, opportunities would be provided to coordinate premium assessments, site inspections and loss management for both off-site and on-site responsibilities.

Requiring financial assurances for corrective action will also provide a worthwhile non-regulatory incentive to make good siting and design decisions. The requirement encourages an evaluation of the long-term consequences of siting the facility in a particular location. If the assurances are required at the time of permitting and are based on site-specific estimates, the owner/operator will be encouraged to consider the ease of performing and the future costs associated with corrective action in the facility's proposed hydrogeological setting. This is an important consideration since the regulations do not currently consider the ease and/or feasibility of corrective action as a location consideration in siting new facilities.

Financial responsibility for corrective action could be demonstrated by the same financial assurance mechanisms which may be used to demonstrate financial responsibility in other contexts. EDF has submitted extensive comments to EPA on the mechanisms available for making this demonstration in a joint filing of comments on the issue with a major insurance company.[15]

DUTY TO CLEANUP CONTAMINATION PAST THE COMPLIANCE POINT

EPA does not require cleanup beyond the waste boundary of a hazardous waste disposal facility. Their rationale is that "there is no guarantee that the owner or operator could obtain permission to enter neighboring property to conduct corrective action."[16] Instead, EPA seeks to rely upon CERCLA and other authorities to address the off-site contamination.

This position is absurd and unlawful. First, under CERCLA or any other authority, the potential problem of obtaining off-site easements exists. Thus, there is no distinction to be made between the permitting process and enforcement authorities. In the consent decree at Hyde Park, for example, defendant Hooker Chemicals and Plastic Corporation is required to use its best efforts to obtain rights of way, rights of entry, approvals, etc. to perform its off-site cleanup obligations.[17]

Second, it is difficult to imagine the owner of contaminated property not granting the necessary permission to have his property cleaned up by the entity which polluted it. EDF seriously questions whether the problem raised by EPA exists.

The repercussions of the Agency's folly, however, is not hypothetical. Under the current scheme, the Agency would be forced to duplicate its efforts in cleaning up contamination from one source because it would be required to institute two separate actions to address the on-site and off-site contamination. This represents an intolerable waste of resources, assuming both actions are actually undertaken. Significantly, resource constraints may prevent the Agency from addressing the off-site contamination in a timely fashion. In addition, by relying upon other authorities instead of the permitting process, EPA has knowingly shifted the burden of proving whether a risk to human health and the environment posed by the contamination exists from the applicant to the Agency and thus has made its task more difficult and time-consuming. It has also significantly altered the level at which cleanup action is warranted by relying solely upon the imminent and substantial endangerment language of the other authorities. In short, the Agency has crafted an unacceptably difficult methodology protecting the public from off-site contamination.

Releases from permitting sites pose their greatest risk to human health and the environment when they migrate off-site and are thus available to expose the nearby community. EPA has needlessly and irrationally abrogated this authority to require the cleanup of these releases.

CONCLUSION

The issues of air emissions monitoring, standards for existing facilities, financial responsibility for corrective actions and duty to clean up past the contamination point are not the only problems in the landfill regulations. Other problems are outlined in detail in a separate EDF publication.[18] At the core of the issues not discussed here is the following problem: EPA has proceeded as if their engineering designs are fail-proof and is in some areas trying not to find contamination. For example, where a double liner and a leachate collection/detection system has been installed, the Agency exempts a facility from groundwater monitoring requirements. The psychology of this approach is understandable - EPA would like to think that these regulations are based on infallible principles and sound science and will prevent any contamination. However, by their own admission in the preamble to these regulations, these landfills will surely leak at some point in time. Since the Agency is charged with protecting human health and the environment, they are obligated to take maximum measures to ensure that the waste is contained and, just as important, to detect any waste that escapes containment.

REFERENCES

1. Scheible, M., et al, An assessment of the volatile and toxic organic emissions from hazardous waste disposal in California. California Air Resources Board (1982). p.27.

2. The interpretation of the significance of the results of these studies is severely hindered by a lack of ambient air quality standards for hazardous chemicals. We have done our comparisons to a modification of OSHA threshhold limit value levels where the TLV value is divided by 420. The modification we used was suggested for use by industry and cited in an EPA report, Multimedia Environmental Goals for Environmental Assessment. Volume 1. EPA-608/7-77/136a. November 1977. p.61.
3. Markle, R.A., et al. A preliminary examination of vinyl chloride emissions from polymerization sludges during handling and land disposal. Battelle Columbus Labs (1976). p.2.
4. Shen, T.T., "Air quality assessment for land disposal of industrial wastes." Environmental Management, Vol. 6, No. 4 (1982). pp. 297-305.
5. Ase, P.K., Air pollution sampling and monitoring at hazardous waste facilities. ITT Research Institute. Unpublished draft, December 1981. 298 p.
6. Environmental Protection Agency. Groundwater Research Plan (September 1981). p.10.
7. Sgambat, J.P. and J.r. Stedinger. "Confidence in Groundwater Monitoring." Groundwater Monitoring Review (Spring, 1981). pp. 62-69.
8. Parker, F.L. and E. Figueroa. "Hazardous Waste Monitoring in Groundwaters." In: Proceedings of National Conferences on Risk and Decision Analysis for Hazardous Waste Disposal. August 24-27, 1981. Hazardous Waste Disposal. August 24-27, 1981. Hazardous Materials Control Research Institute (1981). p.66-78.
9. Testimony of David W. Miller, Geraghty and Miller, Inc., Syosset, New York before the Subcommittee on Natural Resources, U.S. House of Representatives, November 30, 1982.
10. U.S. General Accounting Office. Report to the Congress of the United States: Federal-State Environmental Programs - the State Perspective. CED-80-106 (August 22, 1980). 95p.
11. Federal Register, July 26, 1982, Volume 47, p. 32290.
12. Federal Register. July 26, 1982, Volume 47, 32320.
13. EPA. Remedial Actions at Hazardous Waste Stes: Survey and Case Studies. EPA 600/2-81-246 (September, 1981). 230 p.
14. EPA. Handbook Remedial Action at Waste Disposal Sites. EPA-615/6-82-006 - (june, 1982). 497 p.
15. Lennett, D.J. and Cheek, L., "Comments of the Environmental Defense Fund and the Crume & Foster Insurance Companies on Financial Assurance Requirements for Corrective Action." EDF Publication No. 774, Environmental Defense Fund, 1525 18th Street, N.W., Washington, D.C. 20036. November 22, 1982.
16. Federal Register, July 26, 1982, Volume 47, p. 32311.

17. Section 8(d) of the Stipulation and Judgment Approving Settlement filed in U.S. v. Hooker Chemical and Plastics Corp., Civ. Act. No. 79-989 (W.D.N.Y.), filed January 19, 1981.
18. Greer, L.E. and Lennett, D.J., "Comments of the Environmental Defense Fund on the Interim Final Hazardous Waste Land Disposal REgulations." EDF Publication NO. 775. Environmental Defense Fund, 1525 18th Street, N.W., Washington, D.C. 20036. November 23, 1982. Price: $5.00.

CHAPTER 21

INFLUENCE OF HAZARDOUS AND TOXIC WASTES ON THE ENGINEERING BEHAVIOR OF SOILS

Jeffrey C. Evans, Ph.D., P.E.
Woodward-Clyde Consultants
Hsai-Yang Fang, Ph.D.
Irwin Jay Kugelman, Ph.D.
Lehigh University

INTRODUCTION

An understanding of the interaction between hazardous and toxic wastes and the engineering behavior of finegrained soils requires the interweaving of three normally distinct technical fields. First, the measurement of the soil properties of interest, particularly permeability, requires a thorough knowledge of geotechnical engineering test equipment and procedures. The influence of the details of the test procedures and equipment upon the test result must be well understood. It requires the knowledge of the testing apparatus, the manifold ways of conducting the tests, the influence of other geotechnical parameters upon the test result, and the natural variability to be expected due to sample inhomogeneity. Secondly, a knowledge of clay mineralogy is required. Thirdly, to understand the interrelationship between the test results and the fundamental clay behavior on a microscopic scale, a knowledge of chemistry is required. This working knowledge of chemistry is essential to the interpretation of a given measurement of the effect of a certain pore fluid on a certain clay property of a specific clay mineralogy. Hence, the study of pore fluid effects on clays requires the interweaving of the technical skills of these three separate disciplines. further, the understanding of pore fluid effects on the engineering behavior of soils is essential for the engineered use of soils for the containment of hazardous and toxic wastes, engineering systems cannot be confidently designed for reliable longterm performance.

It is the purpose of this paper to examine, at a detailed level, the current understanding of clay behavior in terms of colloidal chemistry and clay mineralogy. Next, hazardous and toxic wastes which are the potential source of pore fluids other than

Management of Toxic and Hazardous Wastes, H. Bhatt, R. Sykes, T. Sweeney (Editors)

water will be examined from their basic chemistry standpoint. These separate areas will then be merged together through a review of the available published information which presents data relating the effect on certain clay properties due to their exposure to extraneous pore fluids. An attempt will be made to explain the measured behavior in response to the pore fluid stimuli, based on the developed understanding of clay mineralogy, colloidal chemistry and pore fluid chemistry. Finally, a discussion of an on-going laboratory investigation at Lehigh University will be presented.

CLAY MINERALOGY

Basic Clay Structure

The structure of clay minerals is the results of two basic structural units forming an atomic molecular latice. THe tetrahedral configuration is termed a silicon tetrahedron and consists of one centrally located silica atom equally spaced between four oxygen atoms or hydroxyl ions to form a tetrahedral shape. Silica tetrahedral groups can be arranged to form a hexagonal network which can be repeated indefinitely to form a tetrahedral sheet.

The second basic building block is an octahedron formed of central atoms (such as aluminum or magnesium) surrounded in an octahedral configuration by oxygen atoms or hydroxyl ions. In like manner to the tetrahedron, octahedron units can form in a continuous sheet-like structure. Octahedral sheets with cations of predominantly aluminum are termed gibbsite. When the octahedral sheet cations are predominantly magnesium, these octahedral sheets are termed brucite.

Atomic And Molecular Forces

The forces holding the atoms and molecules within a clay mineral are generally subdivided into primary and secondary forces or bonds. The two types of primary bonds found in solids are ionic and covalent.

Secondary bonds are either dipole, hydrogen or Van der Waal's bonds. When atomic units are bonded with covalent bonds and sharing electrons which are not symmetrically distributed, the resulting molecule is polar. Polar molecules may be electrically neutral as a whole, but there is a distribution of charge along the molecule. The resulting molecular attraction is termed a dipole force, and dipole bonds can result between polar molecules. A hydrogen bond forms between a hydrogen of one molecule and an unshared pair of electrons on another molecule. Hydrogen bonds are stronger than most dipole bonds as they are formed with atoms which have widely different electronegativities. The small size of the hydrogen atom allows the electronegative atom to approach the hydrogen atom more closely. Finally, Van der Waal's forces result in molecular bonding that is generally weaker than either hydrogen or dipole forces. Fluctuating dipole bonds, are also known as dispersion or London forces. These dispersion forces, unlike hydrogen or dipole forces, can result from nonpolar or polar molecules. In summary, secondary intermolecular bonds result from what are generally termed Van der Waal's forces consisting of dipole, hydrogen and dispersion forces.

Structure of Clay Minerals

Clay minerals are formed by the stacking of the basic structural unit sheets in a variety of arrangements. The mineral types are generally classified by their unique combination of octahedral and tetrahedral sheets. If the mineral consists of one octahedral sheet and one tetrahedral sheet, it is termed a one-to-one (1:1) mineral. In like manner, clay minerals which consist of a combination of two tetrahedral sheets and one octahedral sheet to form a single layer are known as two-to-one (2:1) minerals. The third major mineral type is known as a two-to-one-to-one (2:1:1) mineral and consists of alternately a tetrahedral sheet, an octahedral sheet, a tetrahedral sheet and an octahedral sheet. Hence, the three main mineral types a 1:1, 2:1 and 2:1:1 are then subdivided into groups or subgroups depending upon the octahedral sheet configuration.

One-To-One Clay Minerals

The 1:1 minerals are known as kaolinites and actually consist of kaolinite, dickite, nacrite, halloysite (dehydrated) and halloysite (hydrated) minerals. Discussion will be limited to the kaolinite minerals, the most common of the group. This mineral consists of one gibbsite sheet joined with a silica tetrahedron. Within the octahedral layer there are generally aluminum atoms, and within the tetrahedral layer the cations are silicon. Because the interlayer bond is primarily ionic (it is primarily a hydrogen bond), the interlayer bonding is substantial.

Two-To-One Clay Minerals

Within the 2:1 family of clay minerals is a group commonly known as smectites. Montmorillonite, as a 2:1 mineral, consists of two sheets of silica tetrahedron on either side of the gibbsite sheet. Montmorillonite virtually always has some octahedral sheet cations exchanged with magnesium or sodium. Sodium montmorillonite is a common form of the smectitic clay within bentonite. The layers, each consisting of three sheets, are generally stacked with bonding between successive layers by Van der Waal's forces. Due to the charge deficiencies which exist within the lattice, resulting from frequent substitutions, cations may be present between the layers to balance charge deficiencies. These interlayer bonds are therefore relatively weak and easily separated by imposed stresses such as the adsorption of water or other polar liquids. Due to the lattice substitutions within the basic sheets and the cations within the interlayer needed to balance charge deficiencies, the cation exchange capacity is relatively high.

Another of the commonly occurring clay minerals of 2:1 structure encountered in engineering practice are illites. They consist of a biggsite sheet between two silica tetrahedron sheets, and the layers are bound with a potassium cation populated in the interlayer region. The octahedral cations are aluminum, magnesium or iron, and the tetrahedral cations can be aluminum or silicon. Within the structure of these clay minerals, the sheets experience considerable isomorphous substitution. Some of the silicon within the tetrahedral layer are replaced by aluminum ions. This charge imbalance is balanced by the potassium cation populating the interlayer region. The interlayer bonding through the potassium cations is therefore relatively strong. This differs from that of montmorillonite in several ways. The isomorphous substitutions create a larger charge deficiency in illite primarily due to the isomorphous substitution in the silicon sheet which is close to the surface of the unit layer. Hence as attraction or repulsion is a function of distance, the charged deficiency nearer the surface results in a stronger attraction for cations. The cation between unit layers in illite is virtually entirely potassium. Because of these two charge deficiency differences, illite structural layers have a relatively fixed position and polar ions do not easily enter between them and cause lattice expansion. This is not the case for montmorillonite. THe second major behavioral difference is that the interlayer cations which balance the charged deficiencies are not readily exchangeable. Cation exchange capacity of illite is generally from 10 to 40 as compared to 80 to 150 milli-equivalents per 100 grams for montmorillonite.

Two-To-One-To-One Clay Minerals

Chlorites have a 2:1:1 structure. They are basically similar to illite except that an organized octahedron sheet in chlorites replaces the area populated by cations in the case of illite. The basic layer consists of two tetrahedron sheets bounding a brucite or gibbsite sheet. This is then tied to the next layer with a brucite sheet. The structure is further complicated by the often partial replacement of the central ion in the octahedral sheet with magnesium, aluminum or iron cations. Further, the silica sheets are frequently unbalanced by the substitution of aluminum cations for silica cations within the tetrahedron layer. Thus, the various members of the chlorite group differ in the kinds and amounts of cation substitution and the stacking of successive layers.

Isomorphous Substitution

Isomorphous substitution is the replacement or substitution of cations within the basic structure and frequently results in charge deficiencies. The cation exchange capacity is a measurement of the propensity of cations within a clay mineral structure to be replaced or substituted. The replacement of cations within the mineral structure, and thus the cation exchange capacity, can result from (1) broken bonds within the clay mineral structure; (2) substitutions within the lattice structure; and/or (3) replacement of the hydrogens which are part of the exposed hydroxyls (an integral part of the structure). It is the second cause of cation exchange capacity, substitution within the lattice structure, that is the major contributor to the total cation exchange capacity.

To summarize this section of the paper, it is emphasized that the understanding of the clay mineral structure is essential for our understanding of the macroscopic and microscopic behavior of clays in response to various pore fluids. The summary presented herein is based upon the commonly accepted models of clay structure [1].

THE CLAY-WATER SYSTEM

It has been shown that the clay layer, made up of tetrahedral and octahedral sheets, typically carries a net negative charge as a result of substitutions of certain of the cations within the sheet structure. This net negative charge results from the isomorphous substitution of cations by less positive cations. The net negative charge is often compensated by cations located on the layer surfaces. When the clay is in the presence of water, these compensating cations have a tendency to duffuse away from the layer surface. THis tendency for cations to diffuse away from the layer surface in solution and yet be attracted to the layer surface due to the net negative charge gives rise to the term diffuse ion layer. In a static environment, this ion layer has a constant charge which is largely determined by the type and degree of isomorphous substitutions and the resulting net negative charge in the mineral layers. The understanding of the ion distribution relative to the layer surface of the clay mineral is based upon work in colloidal chemistry. Although several models on how this distribution of ions in the diffuse layer can be analyzed, the most common used in geotechnical engineering is the Gouy-Chapman model.

Van Olphen [2] states that the Stern-Gouy Theory may be better suited to more precisely model the clay diffuse ion layer than the Gouy-Chapman model. The ion distribution of other models would typically provide the same trends with respect to pore fluid effects; only the magnitudes would vary. Hence, for the purposes of this paper, only the Gouy-Chapman model will be considered in detail.

Based upon the interaction of the clay minerals and the accompanying diffuse ion layer under various conditions, definitions of different types of soil fabric have been developed. The classification includes seven modes of particle association in clay suspension. These vary in limit from dispersed to flocculated. Edge-to-edge and edge-to-face particle attraction is defined as aggregation. A dispersed sol is one in which the net electrical forces between adjacent particles at the time of disposition produced repulsion. If these interparticle forces were net attractive during deposition, the soil is said to be flocculated. Therefore, flocculated or aggregated structures are those in which the particles tend to join together, whereas dispersed structures are those structures in which the particles tend to move apart from each other.

The Gouy-Chapman model indicates that the tendency toward flocculation or aggregation is usually increased by a decrease in the diffuse ion layer thickness. corresponding to the decrease in the diffuse ion layer thickness is a decrease in the interparticle repulsion force. Conversely, the tendency toward dispersion is increased as the diffuse ion layer thickness increases. Hence, if the diffuse ion layer thickness is increased, the electrical repulsion between particles is increased, which causes the particles to disperse. The tendency towards dispersion is therefore enhanced by an increase in the diffuse ion layer thickness.

In order to more fully understand the behavior of the clay-water system, and therfore the response of clays to hazardous wastes and pore fluids other than water, it is necessary to examine in more detail the variables of the Gouy-Chapman model and how they affect the clay structure and its behavior.

The equation for the double layer thickness as predicted by the gouy-Chapman theory is given as [2]:

$$t = \sqrt{\frac{\varepsilon KT}{8\pi n e^2 v^2}}$$

where t = the diffuse ion layer thickness,
ε = the dielectric constant,
K = Boltzman's constant,
T = temperature,
n = electrolyte concentration,
e = elementary charge, and
v = valence of cations in pore fluid

From this equation, variables in the soil water system which affect colloidal stability are electrolyte concentration, ionic valence, dielectric constant, temperature, size of hydrated ion, pH, and anion adsorption. These items will be subsequently discussed in detail as they affect clay behavior.

Effect of Electrolyte Concentration

An examination of Equation (1) shows that, as the electrolyte concentration in the pore fluid goes up, the thickness of the diffuse ion layer tends to decrease. The tendency towards flocculation is caused by a decrease in the diffuse ion layer thickness. Hence, flocculated structures tend to be drawn together tighter and have thinner diffuse ion layer thicknesses.

Ion Valence

Increasing the ion valence will also cause a decrease in the thickness of the diffuse ion layer. With a decrease in the diffuse ion layer, the tendency towards flocculation is increased.

Effect of Dielectric Constant

Decreasing the dielectric constant causes a decrease in the diffuse ion layer thickness and a tendency towards flocculation. The dielectric constant is a measure of the ease in which molecules can be polarized and oriented in an electric field. Alternately, the higher the dielectric constant of the material, the more the material behaves as an insulator.

Effect of Temperature

According to the Gouy-Chapman model, an increase in temperature causes an increase in the diffuse ion layer thickness, and therefore a corresponding increase in the tendency towards dispersion. It should be pointed out, however, that many parameters are interrelated. For example, if all parameters are held constant while temperatures are allowed to increase, the diffuse ion layer thickness would increase. However, in reality, it is almost impossible to increase temperature without changing pH, and dielectric constant [3]. An example of this interrelationship as given by Mitchell [4] is presented in Table I. Hence, the product of the dielectric constant times temperature is essentially constant with reasonable changes in temperature. This is due to the fact that the dielectric constant decreases as temperature increases. Interrelationships such as this contribute to the difficulty in interpreting findings reported in the literature regarding the effects of temperature on soil properties such as strength, compressibility and permeability.

TABLE I

T(^{o}C)	T(^{o}K)	Dielectric Constant (E)	ET
0	273	88	$2.40x10^4$
20	293	80	$2.34x10^4$
25	298	78.5	$2.34x10^4$
60	333	66	$2.20x10^4$

Size of the Hydrated Ion

The smaller the ion, the closer it can approach the colloidal surface of the clay particle. Thus, the smaller the hydrated ion, the smaller the diffuse ion layer.

The Effect of pH

THe pH can influence the thickness of the diffuse ion layer in several ways. Changing the pH could affect the electrolyte concentraiton as well as the net negative charge on the clay particle by altering the extent of the clay particle by altering the extent of the dissociation of hydroxyl (OH^-) groups at the edges of the clay particle. High pH encourages the dissociation and increases the net charge, thus expanding the diffuse ion layer. Low pH discourages this dissociation and causes a decrease in the net charge, therefore causing a reduction in the diffuse ion layer.

The Effect of Anion Adsorption

If anions are adsorbed on the clay particle, the net negative charge increases. Increasing the net negative charge causes an increase in the diffuse ion layer thickness because of the increased attraction for cations.

Summary of Diffuse Ion Layer Influence

By an examination of an equation of diffuse ion layer thickness for the Gouy-Chapman model, the effect of pore fluid properties upon the diffuse ion layer thickness has been reviewed. Further, it has been shown that a reduction in diffuse ion layer thickness reduces the interparticle repulsion forces and thus increases the tendency for a flocculated or aggregated soil structure. The effect due to diffuse ion layer thickness and soil structures of changes in the value of any of the parameters in the Gouy-Chapman equation are shown in Table II.

Table II. Effect of Pore Fluid Parameters On Diffuse Ion Layer Thickness and Soil Structure Tendency

Pore Fluid Parameter	Change in Pore Fluid Parameter	Change in Diffuse Ion Layer Thickness	Change in Soil Structure Tendency
Electrolyte Concentration	Increase	Decrease	Flocculated
	Decrease	Increase	Dispersed
Ion Valence	Increase	Decrease	Flocculated
	Decrease	Increase	Dispersed
Dielectric Constant	Increase	increase	Dispersed
	Decrease	Decrease	Flocculated
Temperature	Increase	Increase	Dispersed
	Decrease	Decrease	Flocculated
Size of Hydrated Ion	Increase	Increase	Dispersed
	Decrease	Decrease	Flocculated
pH	Increase	Increase	Dispersed
	Decrease	Decrease	Flocculated
Anion Adsorption	Increase	Increase	Dispersed
	Decrease	Decrease	Flocculated

Pore Fluids

In order to fully investigate the effect of pore fluids on clay structure, a background into the various classes of fluids to be considered must be developed. Considering the thrust of this paper toward the application aspects of clay behavior, the pore fluids will be divided following United States Environmental Protection Agency (EPA) guidelines for wastes found in industrial disposal facilities. The EPA divides these industrial wastes into four physical classes. These classes are aqueous inorganic, aqueous organic, organic and sludges. For the purposes of this paper, only aqueous inorganic, aqueous organic and organic liquid wastes will be considered.

Water, a polar molecule, is considered a unique solvent in several ways. It has a strong dipole moment and a high dielectric constant. Hence, the study of clay behavior normally undertaken utilizes water as the pore fluid and is significantly influenced by these two properties of water. The efefect of the dielectric constant has been discussed. Polar molecules can be distinguished from nonpolar molecules by studying their behavior in an electric field [5].

Aqueous inorganic fluids are those in which water is the solvent, and the solute is mostly inorganic. Examples of fluids classified as aqueous inorganic are brines, caustic sodas, inorganic acids and inorganic salts. Wastes in which water is the solvent and the solutes are predominantly organic are classes as aqueous organic. Solutes are organic chemicals such as those from wood preserving wastes, dye wastes, pesticides or ethylene glycol production wastes. It can be inferred by their water solubility that the solvated chemicals are generally polar [6].

Fluids in which an organic fluid is the solvent and the solutes are other organic chemicals dissolved in the organic solvent are termed organic. This class of fluids represents important liquids to be considered in the study of pore fluid effects on clays as they are frequently encountered in engineering practice. Examples of organic wastes include spent cleaning solvents, pesticide manufacturing wastes and petroleum distillates.

Organic fluids can be classified as organic acids, organic bases, neutral nonpolar fluids and neutral polar fluids. Organic acids are those organic fluids which react with bases, and include proton donors. Organic bases are any organic fluids capable of accepting a proton to become an ionized cation. These fluids are positively charged and will strongly be adsorbed to the negative clay platelet surface. By adsorbing to the clay platelet surfaces, these fluids have a potential for causing volume changes in the clays by changing interlayer spacings as well as increasing the thickness of the diffuse ion layer.

Neutral nonpolar organic fluids have essentially no charge and a small, if any, dipole moment. As these fluids have essentially no charge or dipole moment, they have a potential for moving quite rapidly through the clay fabric. They also have the capability of displacing water molecules from within the hydrated clay structure.

Neutral polar compounds have essentially no charge, but exhibit relatively strong dipole moments, as compared to neutral nonpolar organic fluids. Their effect on the clay structure and fabric can be postulated based on their neutral nonpolar properties, combined with effects predicted from the Gouy-Chapman model.

Summary of Clay-Water System

The behavior of clay on macroscopic and microscopic scales is extremely dependent upon the pore fluid. The effect of certain pore fluid properties has been discussed and is summarized on Table II. The variability in the types of pore fluid and their properties has been reviewed.

It should be noted that there certainly exist chemical environments where the electro-chemical phenomena, such as defined by the Gouy-Chapman model, do not govern the clay behavior. For example, under certain conditions of extreme pH, dissolutioning of the soil skeleton may occur. This would result in an increase in void ratio and a corresponding increase in permeability. Other phenomena would include mineral phase changes and physical and chemical desiccation.

In the final section of this paper, basics of clay mineralogy and clay structure, the concepts of the diffuse ion layer theory, and the pore fluid characteristics will be tied together for the interpretation of data involving the interaction of clays and with pore fluids.

PORE FLUID - CLAY INTERACTIONS

The present interest in the interaction of clays with various pore fluids results from concern regarding the mitigation of hazardous and toxic waste migration by soil barriers to horizontal and/or vertical flow [7]. A significant contribution to our body of knowledge has been made by previous investigators considering engineering properties of clay soil and their interaction with water. In order to better understand this interaction, the interaction of clay soils in pure clay mineral systems with water and other liquids was considered. These early studies can be drawn upon to better understand the principals which apply to the present problem under consideration.

Presented in this section of the paper is a review of selected studies from the available literature, where specific investigations involving clay behavior interacting with pore fluids other than water were undertaken. This section is presented on a case study by case study basis. Additional case studies are presented elsewhere [8].

Moum and Rosenquist (1961)

Two artificially sedimented and consolidated clays were prepared in the laboratory to study the effect of the clay properties as a function of cation replacement [9]. Two series of illitic and montmorillononitic clays were sedimented and subsequently consolidated for a period of one year. After consolidation, half of the samples were percolated by electrolytic solutions of potassium chloride.

After this percolation period, the geotechnical properties of the clays including water content and shear strength were studied. The water content was found to be unaffected by the cation replacement. The undisturbed shear strength was increased by the cation exchange.

These results can be examined considering what would be predicted from the Gouy-Chapman model. By the substitution of the sodium ion by a more strongly electropositive potassium ion, the attractive forces between the cation and the negatively charged clay platelet would be increased and hence the diffuse ion layer thickness would decrease. The tendency for a more flocculated structure with this decrease in diffuse ion layer thickness would be increased, and a corresponding increase in strength is expected. The investigators measured an increase in undrained shear strength.

Matsuo (1957)

The effect of sodium chloride, pottasium chloride, hydrochloric acid, calcium chloride and magnesium chloride in solution upon the Atterberg limits of three Japanese soils was studied [10]. The study included various concentrations of each of these electrolytic solutions and their effect on the liquid limit, plastic limit and hence plasticity index. For the Yosaki Bay silt (MH) and the Kitishurakawa soils (ML), no clear change in the plasticity index is observed for the various concentrations of the electrolytes utilized. However, for the Yashidayama soil (CL) the plasticity index as well as the liquid limit show a marked increase for concentrations of all of the salts from 0.01 normal to 0.2 normal solutions. An increase in the liquid limit indicates an increase in the shear strength.

Examining these data in light of the Gouy-Chapman model results in additional evidence that this model is effective in explaining the clay behavior. The Gouy-Chapman theory indicates increasing the electrolytic concentration decreases the diffuse ion layer thickness. With this decrease in diffuse ion layer thickness, one would expect the shear strength to increase. This is the trend shown in Matsuo's data for the Yashidayama soil. The trend is less pronounced, however, in the Yosaka Bay silt and the Kitishirakawa soils. An explanation offered for this result at this time is that, with soils containing a large amount of silt-size particles, the physical-chemical effects on soil behavior are less pronounced than with colloidal soils.

Torrance (1975)

The role of pore fluid chemistry in the development and behavior of sensitive marine clays of Canada and Scandinavia was studied [11]. Liquid limit tests were conducted on Leda clay using its natural pore fluid as well as electrolytic solutions of various salts including aluminum chloride, ($AlCl_3$) potassium chloride (KCl), hydrochloric acid (HCl), sodium chloride (NaCl) calcium chloride (CaCl), Sodium Phosphate (NaP_3O_4), calcium carbonate ($CaCO_3$), hydrogen fluoride (HF), sodium carbonate ($NaCO_3$), and sodium hydroxide (NaOH). The results of the effect of these various electrolytic solutions at various concentrations are shown of Figure 1. It can be seen that, as the concentration of the electrolyte in solution increases, the liquid limit generally increases. As previusly discussed, this is compatible with the predictions using the Gouy-Chapman model, which would indicate a decrease in the diffuse ion layer thickness and an increase in the tendency towards flocculation with an increase in electrolyte concentration. However, there seems to be a saturation type level, as the effect is quite pronounced at low concentrations, but apparently levels off at higher concentrations. This tendency is consistent with the data by Matsuo at higher electrolyte concentrations.

Leonards & Girault (1961)

In a study of the one-dimensional consolidation test, the effect of pore fluid on the compressibility of undisturbed samples of Mexico City clays was examined [12]. In this study, the naturally occurring pore water was replaced with carbon tetrachloride (CCl_4) by circulating the pore fluid through the sample until the effluent was pure carbon tetrachloride.

The investigators found that the alternate pore fluid significantly affected compressibility. First, there was an overall decrease in the compressibility of the Mexico City clay with the nonpolar carbon tetrachloride. Secondly, there was a substantial increase in the observed preconsolidation pressure. An overconsolidation ratio of two was measured on this normally consolidated clay after the introduction of carbon tetrachloride as a pore fluid. It was concluded that a normally consolidated clay can exhibit apparent preconsolidation pressures due to changes in the interparticle forces.

This increase in the overconsolidation ratio can be examined in light of the Gouy-Chapman model. As previously presented, the colloidal model indicates that the total repulsion force between particles decreases as the dielectric constant decreases. Carbon tetrachloride has a very low dielectric constant, and it would therefore be expected that the interparticle forces would be less with this pore fluid than with the naturally occurring pore water. If the total repulsion force between particles decreases, one would expect higher interparticle bonding and hence a higher apparent preconsolidation pressure. The results of this study are therefore consistent with those predicted by the Gouy-Chapman model.

Andrews, Garwarkiewicz & Winterkorn (1967)

A study was undertaken to compare the interaction of three clay minerals with three pore fluids [13]. The pore fluids were water (H_2O), dimethyl sulfoxide (DMSO) and dimethyl formamide (DMF). The clay minerals used were kaolinite, attapulgite and sodium montmorillonite.

In order to study the effect of these various pore fluids on the three clay minerals, an extensive testing program was undertaken. Test included Atterberg limit tests, sedimentation volumes, and cracking patterns of thin clay slurry films. The Atterberg limit test determines the moisture contents at which changes of state from liquid to plastic to semisolid to solid occur. The position of these limits on a moisture scale is based upon the water affinity of the mineral components of the clays, their tendency to form secondary and higher structural units, and their granular composition, which controls packing properties and pore space geometry. The determination of sediment volumes, frequently used in colloidal chemistry, is little used in soil engineering. In general, flocculated particles settle rapidly to high equilibrium values while deflocculated particles settle slowly to low equilibrium values. The test is usually performed by dispersing a known amount of solids in a large excess of liquid to determine the relative rate of settling and the final volume of sediment. Cracking patterns of thin clay slurry films are an indication of the tensile stresses set up by the clay minerals as they lose their liquids by evaporation.

It was found that, when compared to water, the plasticity index for DMSO was higher for the keolinite but lower for the attapulgite and sodium montmorillonite. The most profound effect on plasticity was with the sodium montmorillonite where the plasticity index was lower for DMSO than water by a factor of 7.5.

The sedimentation test yielded the following ratios for H_2O:DMSO:DMF, respectively:

Kaolinite - 0.63 : 1 : 0.64

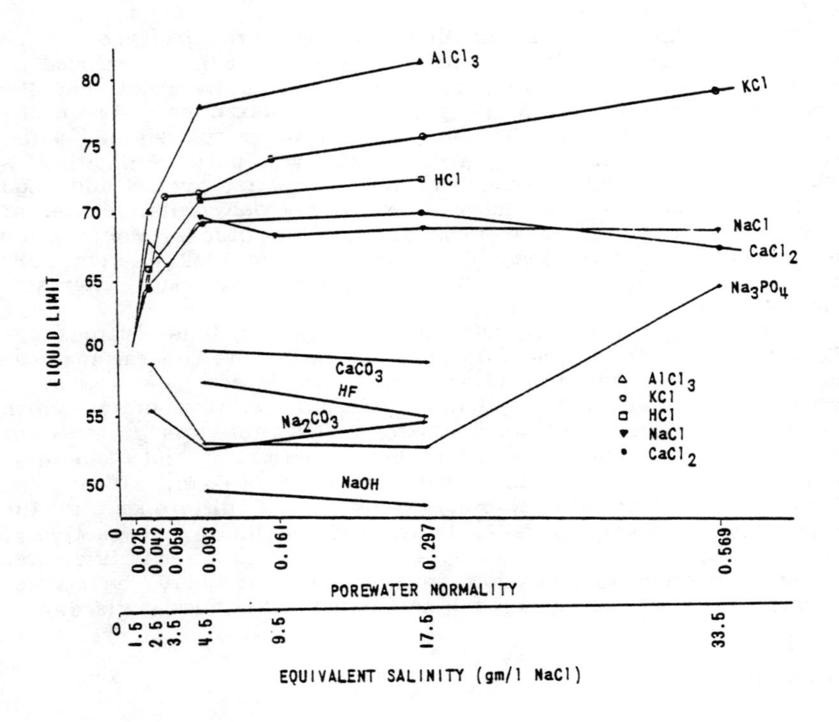

FIGURE 1 INFLUENCE OF SALTS ON THE LIQUID LIMIT

Attapulgite - 2.6 : 1 : 0.65
Bentonite - Infinity : 1 : 1.08

From these data, one can see that kaolinite interacts more strongly with DMSO than either water or DMF. Bentonite interacts most strongly with water, but to about the same degree with both DMSO and DMF. Attapulgite interacts most strongly with water, followed by DMSO and least with DMF.

The cracking patterns of thin slurries were observed for all three clay minerals. THe drying of the kaolinite film produced no cracking pattern. However, the film produced by DMSO has the smoothest appearance, indicating its great interaction between the DMSO and the kaolinite. The aggapulgite water film showed a few cracks, forming relatively large structural units indicating a relatively large tensile strength of the clay-water system and good mobility of the clay particles at relatively low water contents. Finally, the bentonite water films produced no cracks, whereas the bentonite-DMSO system produced a cracking pattern of a larger scale than that with DMF. Additional data regarding cracking patterns is available [14].

The study concludes that, since plasticity index values are taken as a measure of the interaction between the mineral surfaces and the liquids, then this interaction in the water system was, as expected, greatest with bentonite and least with kaolinite. With DMSO, however the absolute plasticity index value was greatest for attapulgite and much less for both bentonite and kaolinite. Further, the plasticity index value for kaolinite with DMSO was about twice that with water. The great difference in the interaction of DMSO with kaolinite and bentonite, respectively, with attapulgite lying in between emphasized the marked differences in the respective surface characteristics and probably the manner in which the water molecules are associated with these surfaces.

Rosenquist (1955)

Considerable investigatory work pertaining to the formation and behavior of Norwegian quick clays was undertaken in the 1940's and early 1950's [15]. The effect of electrolyte concentration upon clay behavior was considered during these studies.

Increasing the electrolyte concentration in the pore fluid has been shown to increase the liquid limit of the soil. The liquid limit is a measurement of a clay's remolded shear strength. It is a measure of the water content at which, under a given energy, a given type of shear failure occurs. Hence, if the liquid limit increased with increasing electrolyte concentration, it meant, for that shear failure to occur in the soil, a higher moisture content was required with a higher electrolyte concentration. Recall that according to the Gouy-Chapman model, as the electrolyte concentration increases, the thickness of the diffuse ion layer tends to decrease. From this, it is concluded that a decrease in diffuse ion layer thickness will cause an increase in the liquid limit of the soil.

Anderson & Brown (1981)

A study was undertaken in which two smectite clay minerals were permeated with organic fluids in what was considered a "standard" permeability test procedure [6]. The permeants were acetic acid, analine, acetone, ethylene glycol, heptane and xylene. The authors note that acetic acid caused decreases in permeability in both soils. However, there was a significant amount of soil piping occurring in the two acid impermeated cores, as evidenced by the presence of soil particles in the leachate. Analine treated cores showed substantial permeability increases with time. The acetone treated cores showed an initial decrease followed by large increases in permeability. In summary, significant increases in permeability were obtained with basic neutral polar and neutral nonpolar organic fluids over those values obtained with water.

From this the authors conclude that the need to test the permeability of clay liners with actual leachates is especially important where organic fluids may be in the waste.

Anderson and Brown describe in detail the test method for determining the effects of waste leachate on the permeability of compacted clay soils. In essence, their tests are conducted in a compaction mold permeameter. Although they use a rather high pressure and a correspondingly high gradient, no backpressure or consolidation pressure is utilized. Clearly, trapped air is a common cause for artificially low permeability values and, in order to provide adequate precaution against the entrapment of air, a backpressure is required[16].

After preparing a sample in the compaction mold permeameter, one pore volume of standard leachate was passed through the clay cores. If the clay has shrunk, they consider it unsuitable as a clay liner, an no further testing is conducted. If it has not changed in volume, they extrude it, weigh it, and remount it in the permeameter. Removing the standard leachate is done by passing at least one pore volume of the various leachates through the sample. Again, if the clay core is shrunk, the authors consider it unsuitable for a clay liner. This test method may not fully assess the actual effect of organic leachates on clay minerals. Firstly, as mentioned, a backpressure is required to insure saturation. Without such a backpressure, varying degrees of saturation within the clay samples can produce misleading results. Further, control of sample stresses is not possible in a compaction mold permeameter. It is therefore suggested that samples be tested in a triaxial cell, where a consolidation pressure can be applied equivalent to that pressure to which the sample will be subsequently subjected in the field. In this way, an accounting of the in-situ stresses may be made. Also, by testing in a fixed-ring permeameter, the swell potential or the volume changes that the material can experience when alternate pore fluids are introduced cannot be measured. Further, if the sample shrinks significantly, there could be piping or bulk transport of fluids. It appears, by an examination of the permeability test results, that certain of the clay minerals are initially consolidating (shrinking) due to the effect of the fluid and hence there is frequently an initial decrease in permeability. At such a point that the sample pulls away, shrinks or potentially cracks (due to a reduction in diffuse ion layer thickness), an open channel is available for bulk transport of permeant resulting in large increases in permeability. From these data, it is therefore very difficult to precisely assess the effect of the pore fluid. For example, the ethylene glycol treated Houston black core showed a steady decrease in permeability following the initial permeability increase. If the swell and/or shrink volume changes were known, one could better assess the meaning of this change in direction and this change in permeability.

An examination of the dielectric constants of the pore fluids reveals that, with the exception of the ethylene glycol, all of the materials have a relatively low dielectric constant. All of the permeants have a lower dielectric constant than that of water. Consideration of the Gouy-Chapman model and the reduction in diffuse ion layer thickness due to the low dielectric constant leads to the conclusion that the materials, upon introduction to these pore fluids, would tend to shrink. Hence piping along the permeameter sidewalls could be expected.

It is noted that, in more recent publications [17,18] the researchers cautioned against misuse of these findings. The researchers reiterate that, based on their studies, compatibility testing is essential to the design of clay liners.

Green, Lee & Jones (1983)

In an effort to evaluate the effects of organic solvents on the shrink/swell characteristics of clays, a laboratory study of the clay-solvent interaciton was undertaken [19]. In these studies, several organic solvents including glycerol, methanol, acetone, trichloriethylene, carbon tetrachloride and xylene were utilized as pore fluids. Three naturally occurring clay solids identified as the Ranger shale, Kosse kaolin and fire clay were utilized. The shrink/swell characteristics were measured in a one-dimensional consolidometer.

The investigators evaluated their data from a physical-chemical and clay mineralogical viewpoint as well as from a macroscopic viewpoint. First, they concluded that the degree of swelling increased as the dielectric constant. Hence, the increase in the degree of swelling with an increasing dielectric constant is consistent with that behavior predicted from the Gouy-Chapman model. It was further obseved by the investigators that, with some solvents with extremely low dielectric constants such as xylene and carbon tetrachloride, the net volume change was negative (i.e., shrinking). This resulted in cracking within the consolidometer. Again, this is consistent with that predicted from the Gouy-Chapman mode. As the investigators approached this research from a clay mineralogical standpoint, several other findings are presented. It has been found that, as the plasticity index increases, the swelling potential increases [20]. This trend was also observed in this study. Review of the clay mineralogy would indicate that the cation exchange capacity should increase with increasing montmorillonitic content. However, this trend was not observed in this study. Also, the montmorillonite content was not found to correlate with the swelling characteristics.

Green Lee & Jones (1983)

In a companion study to the one just discussed, the effect of organic solvents on the permeability of clays was studied [21]. The investigators utilized the same clays and pore fluids as just discussed. The permeability tests were conducted on remolded clays in thick-walled glass columns. No backpressure was utilized nor were the samples consolidated.

The study found that, in general, the permeability of clays was consistently lower for organic solvents than with water. It was found that the permeability decreased with time and attained equilibrium in several weeks. The permeability correlated well with the pore fluid dielectric constant; that is, the lower the dielectric constant, the lower the permeability. Finally, it was found that solvents with extremely low dielectric constants could cause clay shrinkage as previously discussed. Such shrinkage resulted in cracking and rapid breakthrough of pore fluids through the clay columns. This caused the transportation of pore fluids in bulk. These bulk transport data is compatible with those predicted by the Gouy-Chapman model.

The Gouy-Chapman model indicates that, as the dielectric constant goes up, so does the interparticle repulsion, which sould make for a more open or disperse structure. Conversely, as the dielectric constant decreases, the structure becomes more flocculated and tends to shrink.

The breakthrough observed is the consequence of the experiment design, specifically the testing apparatus. That is, the permeameters could not apply a constant confining stress. As the samples began to shrink, the confining stress decreased. Eventually, this stress was reduced to essentially zero, and leakage between the permeameter wall and the sample was observed.

Alther, Evans, Fang and Witmer (1983)

A total of 16 inorganic aqueous solutions were utilized as permeants to determine their effects upon the permeability of both contaminant resitant (polymerized) bentonite and untreated bentonite [22]. It was found that, of the aqueous solutions tested, those with potassium (K^+) cations or (Cl^-) anions or both, induced the largest permeability increases with increasing electrolyte concentration. Conversely, solutions with sodium (Na-) cations or carbonate (CO3) anions had the least impact upon the permeability of bentonite. It was also observed that doubly charged cations (+2) have a greater initial effect on the permeability than do singly charged cations (+1). Furthermore, a "saturation limit" was in evidence for (+2) cations, indicating that beyond certain concentrations the further addition of the soluble salts had only limited additional impact upon the permeability.

For selected bentonite-contaminant combinations, slurry cracking pattern tests were conducted. A correlation between the permeability changes and the results of the cracking pattern tests was demonstrated.

Based upon the findings of this study, it is conlcuded that the character of the solute anions, as well as the primary cations effect the permeability of bentonite clays. Further, the Guoy-Chapman Model of diffuse ion layer was found to be generally consistent with the test data.

ONGOING STUDIES

In an effort to better understand the effects of hazardous and toxic wastes upon the engineering properties of fine-granied soils, a research effort is currently underway. This effort has been divided into two phases, of which only Phase I: Equipment and Protocol Development is completed to date.

Equipment

As indicated in the reviews of the previous studies by others, determination of the influence of hazardous and toxic wastes upon fine-grained soils requires test equipment and procedures beyond those which are presently standardized. Shown in Figure 2 is a schematic of the permeability board system developed to determine the permeability of a sample in a triaxial cell with contaminated pore fluid. This system has several unique features. The ability to measure both inflow and outflow volumes has been incorporated. Further, fluids can be changed and inflow and outflow riser tubes can be filled or emptied without changing the state of stress upon the sample. In addition, the system is designed to accommodate a wide range of stresses up to a maximum of 130 psi. Finally, several systems have been built to allow multiple tests to be run simultaneously. Twelve systems currently being utilized are equipped with all Teflon valves, fittings and tubing for the ultimate in contamination resistance.

As discussed, all permeability tests are run in a triaxial cell permeameter which has been designed and built for these studies. These cells are unique, as the pore fluids are only in contact with Teflon, thus providing equipment resistance to contamination. Although the base is constructed of stainless steel, the end platens, fittings, tubing and parts are of Teflon. in this way, these triaxial cells have all the rigidity and high pressure capability associated with stainless steel, yet all the contamination resistance of Teflon. Triaxial cells with Teflon bases are quite yielding and tend to "creep" under applied stress. A schematic of these cells is shown as Figure 3.

The final major equipment consideration includes the compatibility of membranes with the hazardous and toxic wastes being tested. At present, three different material types have been made available for triaxial sample membranes. These are latex, butyl rubber, and neoprene. It is recommended that waste-membrane compatibility be checked prior to use in testing. This can be done by checking a compatibility chart provided by the material manufactures and/or by submersion testing. Further details are described elsewhere [23].

Protocol

All tests are conducted in a triaxial cell permeameter to simulate an in-situ state of stress. A backpressure is maintained to increase the solubility of air with water. Samples are first consolidated and permeated with sample pore water to equilibrium to determine the "baseline" permeability. The permeation continues with the pore fluid in question. The permeation is continued until equilibrium is re-established, usually within two to four pore volume displacements. The results of a typical long-term permability test are shown in Figure 4.

Upon completion of permeability testing, the sample is tested for strength in isotropically consolidated undrained whear (CIU). Duplicate samples which were not permeated with the waste pore fluid are tested for shear strength to determine the "baseline" strength.

In addition to the above engineering property tests. determination of the Atterberg limits before and after permeation provides insight into the electrochemical changes in grain-size distribution.

Phase II: Pore Fluid Exchange Testing

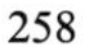

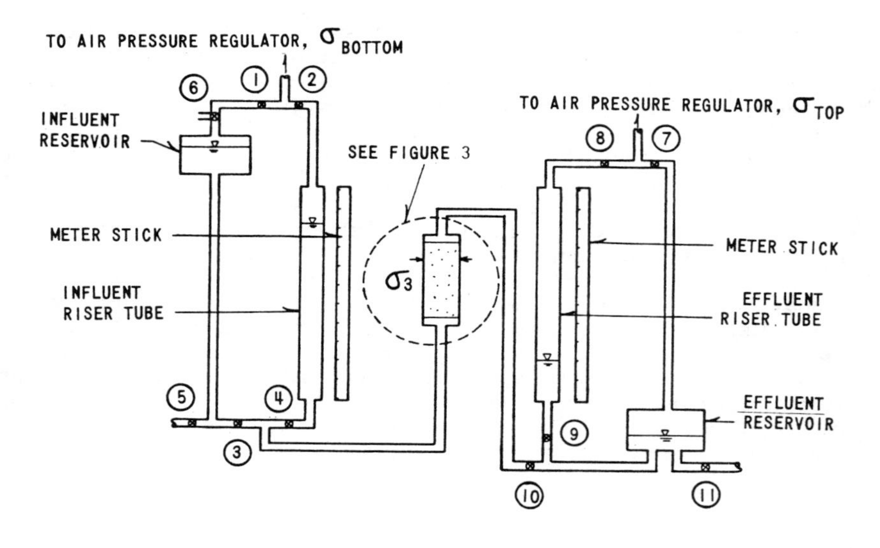

LEGEND:

VALVE	VALVE NAME/FUNCTION - MATERIAL
①	INFLUENT RESERVOIR PRESSURE CONTROL VALVE - BRASS
②	INFLUENT RISER TUBE PRESSURE CONTROL VALVE - BRASS
③	INFLUENT RISER TUBE FILL VALVE - TEFLON®
④	INFLUENT RISER TUBE BY-PASS VALVE - TEFLON®
⑤	INFLUENT RESERVOIR DRAIN VALVE - TEFLON®
⑥	INFLUENT RESERVOIR PRESSURE RELIEF VALVE (THREE-WAY) - BRASS
⑦	EFFLUENT RESERVOIR PRESSURE CONTROL VALVE - BRASS
⑧	EFFLUENT RISER TUBE PRESSURE CONTROL VALVE - BRASS
⑨	EFFLUENT RISER TUBE DRAIN VALVE - TEFLON®
⑩	EFFLUENT RISER TUBE BY-PASS VALVE - TEFLON®
⑪	EFFLUENT RESERVOIR DRAIN VALVE - TEFLON®

FIGURE 2. LEHIGH CONTROL PANEL SCHEMATIC

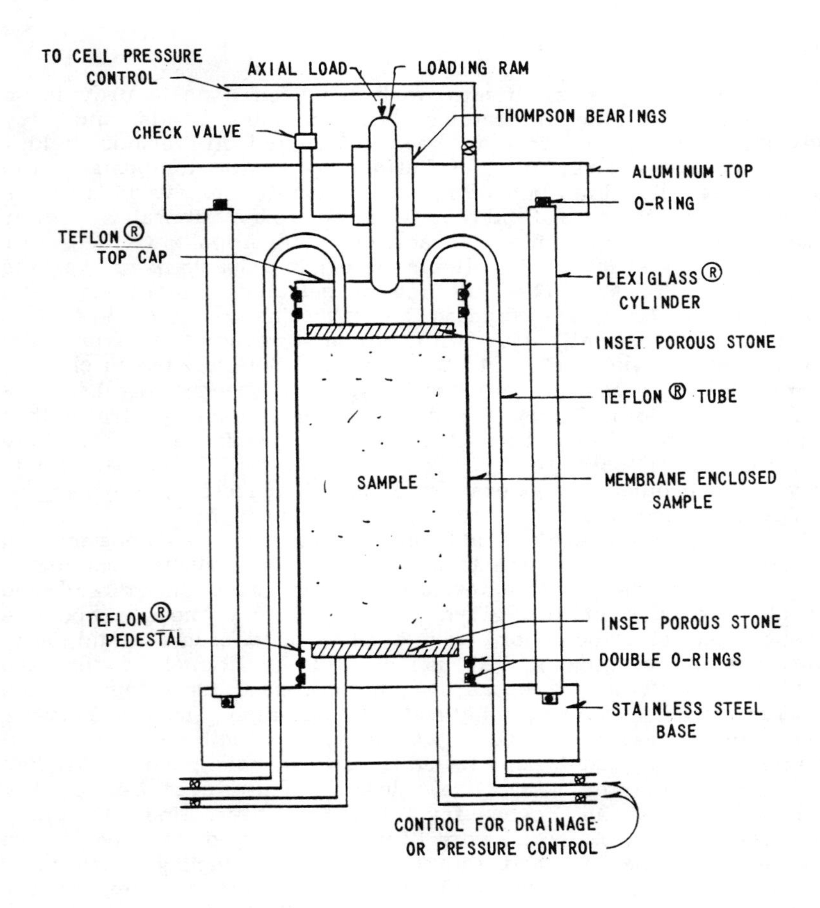

FIGURE 3. LEHIGH TRIAXIAL CELL SCHEMATIC

The research at Lehigh University includes pore fluid exchange testing with pore fluids including water. Testing includes permeability, triaxial shear, Atterberg limits, grain-sized distribution, cation exchange capacity, X-ray diffraction, and scanning electron micrographs. Kaolinitic, and montmorillonitic clays are being used in the testing program. This phase is not completed and presentation of the available is forthcoming [24,25].

SUMMARY AND CONCLUSIONS

It is evident that significatn work has been done to provide an understanding of the interaction between pore fluids and clay behavior. This research has been conducted in various fields. Geotechnical engineers, in their effort to better understand the fundamentals of clay behavior, have conducted various tests utilizing alternate pore fluids and various clay minerals. Other researchers, looking for a practical application to the liner problems, have studied the effects of organic leachate on various oils used as liners. The findings of selected researchers have been reviewed on a case-bycase basis. The results were then examined for compatibility with results predicted from the Gouy-Chapman model. In most cases the clay behavior due to changes in pore fluid composition were consistent with changes predicted by the use of the Gouy-Chapman model. The conclusion is drawn that the Gouy-Chapman theory may be useful as a predictive tool to study the influence of pore fluid on clay behavior. It is cautioned, however, that other phenomena (such as dissolutioning) may govern the clay response under certain chemistry conditions.

To adequately work and understand these phenomena, a characterization of the waste is necessary. In a similar manner to our geotechnical site characterizations, one must understand the general properties of the given waste and how those properties influence the clay behavior from a physical-chemical standpoint. Finally, it is recognized that considerable additional studies are required in virtually all areas of the effects of hazardous wastes on clays from a physical-chemical standpoint. The phenomena investigated herein are extremely complex and all possible influences could not be addressed in this paper. Studies are required on the very basic levels of understanding clay mineralogy, pore fluid chemistry, and the interaction of a clay mineralogical system with pore fluids. Immediate needs are concerned with adequate and safe methods of conducting permeability tests with hazardous pore fluids on clay materials proposed for liners, which adequately reflect field conditions to which these clays will be subjected while in service.

ACKNOWLEDGEMENTS

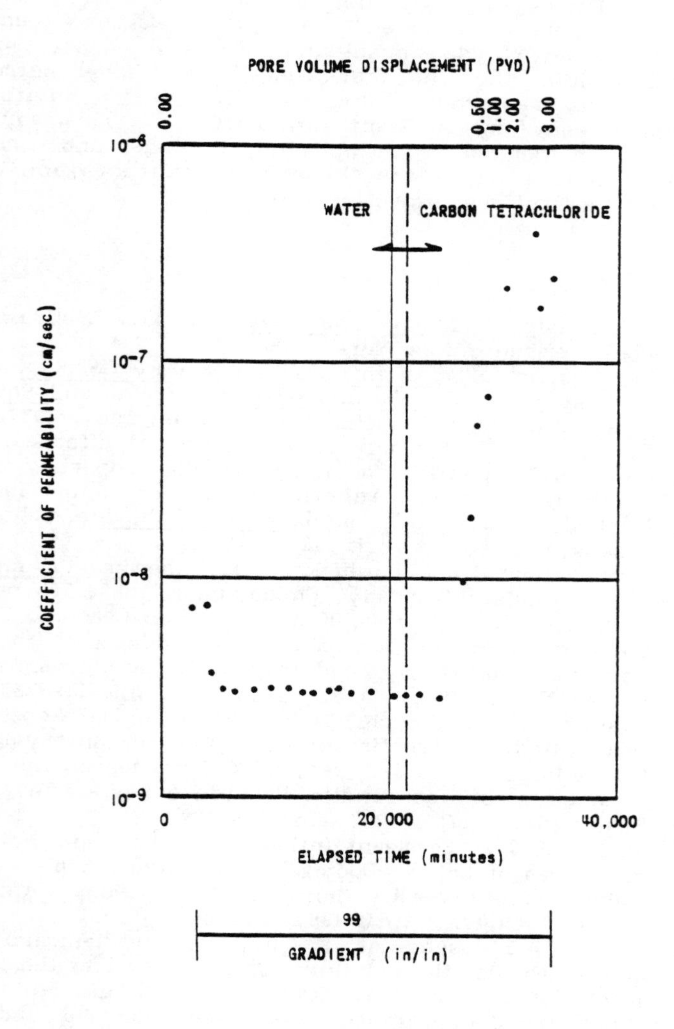

FIGURE 4 PERMEABILITY TEST RESULTS

The financial support for these studies was provided by the Professional Development Committee of Woodward-Clyde Consultants, William S. Gardner, Chairman and the Plymouth Meeting, Pennsylvania office of Woodward-Clyde Consultants, Frank S. Waller, Managing Principal. Funding for the fabrication of additional permeability testing systems was provided by the United States Environmental Protection Agency under Grant No. R810922. This support is gratefully acknowledged. The opinions, findings, and conclusions expressed in this paper are those of the authors and are not necessarily those of the project sponsors.

REFERENCES

1. Grim, Ralph E., Clay Mineralogy, 2nd Ed., McGraw-Hill Book Company, New York, 1968.
2. Van Olphen, H., An Introduction to Clay Colloidal Chemistry for Clay Technologist, Geologists and Soil Scientists, 2nd Ed., John Wiley & Sons, Inc., 1977.
3. Cullen, D.R., "A Study of the Permanent Effects of Elevated Temperature on the Hydraulic Properties of Clay", M.S. Thesis, Auburn University, August 1979.
4. Mitchell, J.K., Fundamentals of Soil Behavior, John Wiley & sons, Inc., New York, 1976, 422 pp.
5. Masterton, W.L. & Slowinski, E.J., Chemical Principles, W.B. Saunders Company, Philadelphia, 1969, 705 pp.
6. Anderson, D. & Brown, K.W., "Organic Leachate Effects on the Permeability of Clay Liners", Proceddings, 7th Annual Research Symposium, Land Disposal: Hazardous Waste. EPA Report No. 600/9-81-002b, March 1981, pp. 119-130.
7. Evans, J.C., and Fang, H.Y., "Geotechnical Aspects of the Design and Construction of Waste Containment Systems", Proceedings of the 3rd National Conference on the Management of Uncontrolled Hazardous Waste Sites, Washington, D.C., 1982, pp. 175-182.
8. Evans, J.C., "Permeant Influence on the Geotechnical Properties of Soils", Doctoral Dissertation, Lehigh University, University Microfilms International, Ann Arbor, Michigan, April 1984, 241 pp.
9. Moum, J. & Rosenquist, I. Th., "The Mechanical Properties of Montmorillonitic and Illitic Clays Related to the Electrolytes of the Pore Water", Proceedings, 5th International Conference on Soil Mechanics & Foundation Engineering, Vol. 1, 1961, p. 263.
10. Matsuo, S., "A Study of the Effect of Cation Exchange on the Stability of Slopes", Proceedings, 4th International Conference on Soil Mechanics & Foundation Engineering, Vol. 2, 1957, pp. 330-333.

11. Torrance, J. Kenneth, "On the Role of Chenmistry in the Development and Behavior of the Sensitive Marine Clays of Canada and Scandinavia", Canadian Geotechnical Journal, No. 3, August 1975.
12. Leonards. G.A. and Giroult, P., "A Study of the One Dimensional Consolidation Test", Proceedings of the 5th International Conference on Soil Mechanics and Foundation Engineering, Vol. 1, Paris, 1961, p. 213.
13. Andrews, R.E., J.J. Gawarkiewicz & Winterkorn, H.F., Comparison of Three Clay Minerals with Water, Dimethyl Sulfoxide, and Dimethyl Formalmide:, Highway Researcy Record No. 209. 1967.
14. Fang. H.Y. Chaney. R.C., Failmezger, R.A., and Evans, J.C., "Mechanics of Soil Cracking", Proceedings of the 20th Meeting of the Society of Engineering Science, University of Delaware, Newark, Delaware, 1983.
15. Rosenquist, I. Th., "Investigations in the Clay Electrolyte-Water System", Norwegian Geotechnical Institute, Publication No. 9, Oslo, 1955, 126 pp.
16. Olson, R.D. and Daniel, D.E., "Field and Laboratory Measurement of the Permeability of Saturated and Partially Saturated Fine-Grained Soils", Geotechnical Engineering Report GR 80-5, University of Texas, June, 1979.
17. Anderson, D., "Does Landfill Leachate Make Clay Liners More Permeable?", Civil Engineering, ASCE, September, 1982, pp. 66-69.
18. Morrison, Allen, "EPA's New Land Disposal Rules - A Closer Look", Civil Engineering, ASCE, January, 1983, pp. 44-49.
19. Green, W.J., G.F. Lee & Jones, R.A., "Introduction of Clay Soils With Water and Organic Solvents: Implications for the Disposal of Hazardous Wastes", Environmental Science Technology, Vol. 17, No. 5, May, 1983, pp. 278-282.
20. Seed, H.B., Woodward, R.J. & Lundgren, J. "Fundamental Aspects of the Atterberg Limits:, ASCE Journal, Soil Mechanics & Foundation Engineering Division, Vol. 90, No. SM6, November 1964, pp. 75-106.
21. Green, W.J. Lee, G.F. & Jones, R.A., "Clay-Soil Permeability and Hazardous Waste Storage", Journal of the Water Pollution Control Federation, Vol. 54, No. 8, 1983, pp. 1347-1354.
22. Alther, G.R., Evans, J.C., Witmer, K.A., and Fang, H.Y., "Influence of Inorganic Permeant Upon the Permeability of Bentonite", Proceedings of the ASTM Symposium on Impermeable Barriers for Soil and Rock, Denver, Colorado, June, 1984.

23. Evans, J.C. and Fang, H.Y. "Equipment for Permeability Testing with Hazardous and Toxic Permeants:, Fritz Engineering Laboratory, Report No. 720.2, Lehigh University, August, 1984, 24 pp.
24. Evans, J.C., Fang, H.Y. and Kugelman, I.J., "Organic Fluid Effects on the Permeability of Soil-Bentonite Slurry Walls", Accepted for Publication, Proceedings of the National Conference and Exhibition of Hazardous Wastes and Environmental Emergencies, Cincinnati, Ohio, May, 1985.
25. Evans, J.C., Kugelman, I.J., Fang, H.Y., "Organic Fluid Effects on the Stregth, Compressibility and Permeability of Soil-Bentonite Slurry Walls", Accepted for Publication, Proceedings of the 17th Mid-Atlantic Industrial Waste Conference, Lehigh Unviersity, June, 1985.

CHAPTER 22

SITE SELECTION AND DESIGN CONSIDERATIONS FOR HAZARDOUS WASTE LAND DISPOSAL FACILITIES

Dr. Paul A. Hustad and Dr. John A. Ruf, P.E.
Burns & McDonnell Engineering Co.

INTRODUCTION

The current federal and state solid waste regulations have revealed the complexity of hazardous waste land disposal. To meet and surpass these regulations as designers and operators of disposal facilities, it is necessary that individuals from the disciplines of geotechnical, geology, covil, chemical, and environmental engineering closely coordinate their activities to achieve the goal of a secure landfill. This goal not only requires that they have a throrough knowledge of their particular discipline, but also an appreciation for the interrelated aspects of the overall disposal system. This paper deals with the process of site selection and design considerations that are necessary for the successful selection and design of a hazardous waste landfill.

SITE SELECTION

There are many factors which should be evaluated prior to the selection of a site for a disposal facility. Examples of some of the factors which play a role in site selection are (1) a demonstrated need for such a facility, (2) nearness to waste generators, (3) the size of the required facility, (4) availability of large parcels of land, (5) nearness to neighbors, (6) potential for obtaining regulatory approval, (7) zoning, and (8) the client's overall company growth plan. Each of these items is normally evaluated by the client, especially if he has had previous experience owning and operating similar facilities. While all of these factors are important, in our opinion, geologic features have the most pronounced effect upon the development and total cost of the facility. Based upon this premise, it is logical that during the initial stages of evaluating potential sites, the client involve the engineer and geologist to obtain input concerning the influence of geologic conditions.

Management of Toxic and Hazardous Wastes, H. Bhatt, R. Sykes, T. Sweeney (Editors)

From a geologic standpoint, the selection of an appropriate site requires extensive office studies, field investigations, and laboratory investigations. These studies and investigations can require as much as 18 months to perform and, thus, represent a considerable expense. In most cases, the nature and character of the subsurface conditions are not initially well defined. Therefore, as the studies and investigative efforts proceed, the disclosure of hidden detrimental subsurface conditions could reveal a site to be unfavorable. Because of this inherent risk, the required studies and investigations are divided into three major phases. As will be seen, the chronological arrangement of these phases produces a maximum amount of information for the required monetary investment. Our experience indictates that by following a phased program of study and investigation, the client can easily be made aware of potential engineering, construction, and operational costs required at a particular site. The client's timely awareness of these potential costs provides him or her the opportunity to economically evaluate a potential site as information is developed. Normally, undesirable sites are eliminated during Phase I, however, in some instances investigations do proceed to Phase II before elimination is confirmed.

Phase I involves office studies of existing geologic information, field inspection of the potential site, a preliminary field subsurface investigation, and an evaluation of the accumulated data. Phase II is composed of formal recommendations for future studies and the performance of the primary subsurface investigation. In Phase III, additional field tests and observations are made to positively confirm trends exhibited by the subsurface data. Also, final design of the facility, is accomplished and the permit application is prepared for submission to appropriate agencies during this phase.

PHASE I

The primary purpose of Phase I is to deterimine whether the potential disposal site warrants additional study and investigations which are required in Phase II. Phase I is divided into a sequence of four steps, these being (1) office studies, (2) field observations, (3) field investigations, and (4) data evaluation. It is emphasized that potential sites may be found to be unacceptable after the completion of any one of these steps. Furthermore, these steps are chronologically performed to minimize costs to the client.

Office Studies

This initial phase of the investigation consists of gathering all available information that pertains to the site or sites in question. It most often involves a review of the latest U.S. Geological Survey topographic maps, state and federal publications, and regulations which pertain to the area. Publications that often yield the most significant information are those which deal with geologic characteristics and groundwater evaluations.

In Missouri, for example, the Division of Geology and Land Survey has prepared a series of maps which divide the state into five general areas [1]. Each area is classified according to its suitable geologic characteristics from the standpoint of development of a hazardous waste landfill. While these maps are generalized and do not provide a specific means for an evaluation of a potential site, they do form guidelines for formulating the direction of future studies. When used in conjunction with publications on the geology and groundwater resources of local areas, these maps are a powerful tool in evaluating aspects of hazardous waste sites before substantial amounts of the client's money are spent for drilling.

Often municipalities have subsurface information that was derived in connection with water supply wells that can provide an indication of the soil overburden thickness. Likewise, private well drilling firms often have a good general knowledge of aquifer locations. This information can be extremely beneficial in regard to determining the general course of the groundwater studies.

At this time, the totally unacceptable sites are discarded and the groundwork is developed for future selection of geologically sound hazardous waste facility locations. It is also during this time that extremely important and beneficial meetings should be held with the client, attorneys, realtors, and regulatory agencies which are charged with the duties and responsibilities of reviewing hazardous waste management facility designs. These meetings set the stage for developing channels of communications which must remain open for several years, if the design and operation of the facility are to be successful.

Field Observations

The basic purpose of this phase of the investigation is to determine whether the site warrants future investigation. this involves a visit to the site primarily for the purpose of examining the surface features. Often, rock outcrops or exposures of the soil overburden in stream valleys or gullies and the occurrence of springs can indicate the potential for development. Also, the visit can reveal recent utilities, structures, and road locations which may not be shown on the most recent maps. These visits are most useful if they are attended by the project geologist and/or geotechnical engineer, the project manager responsible for the overall design of the facility, and the client. In some cases, a local realtor and representatives from regulatory agencies provide extremely useful information.

Field Investigations

This portion of the preliminary investigation has two purposes. First, it should confirm the positive decisions that were made concerning the site during the office and field study. Secondly, it should provide information to efficiently plan a more detailed subsurface investigation, which should be tailored to fully characterize the geological and engineering properties of the site. The major portion of the preliminary field investigation involves subsurface drilling and sampling. Drilling should be performed at strategic locations an to depths which encompass the area to be developed and possibly affected by the proposed facility. Typically, investigations range from 10 to 25 borings. While the basic purpose of the drilling and sampling is to define the thickness and character of the overburden, several of the borings should be extended into bedrock to confirm the location of the site within the stratigraphic profile.

The major portion of the soil sampling is accomplished with the standard split-spoon sampler and a limited number of undisturbed samples are recovered with thin-wall sampling tubes. All of this work should be conducted under the direction of an experienced geologist or geotechnical engineer and boring logs should be prepared by the same. Soil and bedrock sampling should be often enough to provide accurate descriptions of material types and stratification.

It is extremely important that the technical person responsible for conducting the field portion of the investigation is well acquainted with the requirements of a hazardous waste landfill site as this portion of the investigation must be flexible. This flexibility is required, as the information being developed with each foot of subsurface material penetrated is the first factual information on what actually lies beneath the site. This person should be equiped with the knowledge to know what materials are and are not expected, and to handle whatever circumstances arise. In particular, it is beneficial if boring logs are correlated in the field so that stratification and depositional sequence becomes apparent as soon as possible. This allows the drilling and sampling program to be modified without having to remobilize to the site for the purpose of obtaining additional information.

In addition to the performance of test borings, it may be desirable to perform electric logging, surface electrical resistivity, and seismic surveys for correlation of subsurface conditions between bore hole locations and sample depth intervals. Occasionally, test pit excavations are made, especially where secondary soil structure or continuity of shallow sand or gravel lenses are in question. These test pit excavations may be in the form of isolated pits or trenches dug for a considerable distance to observe continuity of soil profiles between boring locations.

Since the occurrence and protection of groundwater is of paramount importance to the hazardous waste landfill facility design, more than casual observations should be made as to the occurrence, quantity, and quality of the groundwater. If at all possible, test pits should be left open for several days or weeks and the quantity and location of groundwater infiltration should be observed and documented. In regard to test borings, if they are rotary wash bored and an artificial drilling fluid is added to the hole during advancement, all circumstances surrounding the addition and level of this fluid should be documented. In those bore holes where piezometers are not to be installed, grouting should be performed throughout the hole depth. This process reduces the risk of having poorly backfilled bore holes provide future conduits for leachate.

Selected bore holes should be designed to receive piezometers. These piezometers should be installed to various depths and designed such that in fine-grained soils inflow does not create clogging or sluggish response. Also, it is important that the piezometers be well sealed at the surface so that infiltration does not occur and distort the groundwater readings. Two-inch-diameter PVC pipe that is either slotted or equipped with a stainless-steel wellpoint and surrounded by a well-graded granular material usually provides an acceptable water level monitoring installation. In areas where perched water or isolated zones of groundwater exist, it is necessary to seal intermediate portions of the piezometer with bentonite pellets. It is also important that once the piezometer is installed it be bailed or pumped for several hours to ensure that it is properly functioning. Should malfunctioning result, it is important to document and provide specific reasons for the improper performance.

Enough water and soil samples should be collected for laboratory tests to leave little doubt as to whether the subsurface materials encountered at the site have favorable engineering and chemical properties. Normally, the laboratory testing performed on the soil overburden consists of numerous classification tests. These classification tests are composed of grain-size analyses and Atterberg limit determinations. Select undisturbed samples are chosen for density and permeability tests. The collection and testing of water samples is of vital importance for establishing the quality of groundwater. More importantly, the piezometers should be examined and readings recorded periodically during and following this phase of the investigation to gain an insight into the occurrence of groundwater and the hydrologic setting.

Data Evaluation

At this stage of the investigation, hopefully, a great deal is known about the subsurface characteristics. Occasionally, the previously discussed phases yield more questions than answers. If this should be the situation, there is a strong possibility that the site may not be acceptable because it may be too difficult and costly to satisfactorily show that the site is environmentally acceptable. However, careful evaluation of all existing data should be undertaken in an objective manner. At this point, it is important that the engineering firm performing the required studies have at least on person that assumes the "devil's advocate role." Using this methodology, worstcase situations are considered and the client adequately informed as to the probable economic risk he may be undertaking with further investigations of the site. The results of these efforts are to clearly define critical areas and provide formal recommendations for future studies.

PHASE II

The primary objective of this phase is to develop and implement a detailed plan. Since the cost involved in implementing this plan is substantial, it is highly desirable that all parties involved arrive at a consensus of opinion as to what investigative procedures will be employed. Similar to Phase I, the steps outlined in Phase II may be terminated at any point should undesirable subsurface features be encountered.

Recommendations for Future Investigations

At this point, the client should be in a position to commit himself to the performance of a detailed subsurface investigation and design development of the site under consideration. It is important that all parties concerned understand the risks involved and the potential requirements for development of the site as a hazardous waste landfill facility. The interest of the client, engineer, and regulatory agencies can be best served by formally outlining a detailed program for all future studies based upon the information gathered to date. As consultants, we find this step is valuable in that both clients and state agencies have highly qualified personnel that provide valuable input for future studies. Also, it becomes apparent whether future studies are warranted, or the difficulties are such that success of the project is extremely doubtful. In any event, at this time the consultant should be in a position to outline to the client what future studies will be required and their approximate cost so the client can evaluate his potential profit-to-risk ratio. The consultant's recommendations should describe a specific scope of subsurface investigation and any contingencies which would result in a primary modification of the same.

Subsurface Investigation

The purpose of this step is to substantiate the interpretations that were made during Phase I, and also to resolve questions that were raised by client and state technical representatives. This investigation involves much more detail when compared to the preliminary field investigation. In comparison, the primary investigation should involve the recovery of considerably more undisturbed samples with sampling of subsurface materials at much closer intervals. All of this work should be supervised by an experienced engineering geologist or geotechnical engineer and preferably, the same individual that was involved with the field portion of the preliminary investigation. While expensive, some borings should be chosen for continuous sampling to provide a continuous profile of materials. This procedure aids in answering the question of whether the site contains strata of highly permeable materials that will transmit pollutants to off-site areas.

As stated earlier, soil sampling during the preliminary investigation (Phase I) generally involved the use of the standard split-spoon sampler or thin wall tubes. However, materials such as clean sands, coarse gravels, weak shale and sandstone bedrock, and hard gravelly clays, are often difficult to sample with standard samplers. Attempts at sampling these materials often result in little or no sample recovery and extreme disturbance. Thus, special sampling devices may often be necessary. Some of the more common samplers which can penetrate and recover samples in hard materials are the Pitcher Sampler and the Denison Barrel. Piston samplers are also employed to obtain relatively undisturbed samples of clean sands and fine gravels. Since drilling firms normally do not equip their drill rigs with these samplers, prior arrangements are often necessary to attain the flexibility of using these devices when necessary. Ideally, the investigations in Phase I will have indicated if these types of sampling devices are necessary.

At locations where bedrock exists within 100 feet of the ground surface and the bedrock geology of the area is not well established, it is necessary that an adequate number of borings be extended well into the bedrock using diamond bit coring equipment. Generally, NWX-size coring equipment will suffice. NWX-size cores are approximately 2 inches in diameter and can rovide a relatively continuous sample of the bedrock materials. Coring to depths of 50 to 100 feet beneath the bedrock surface may be necessary and in some situations, where deep aquifers exist, deeper penetrations will be required to establish the complete geologic profile. Field packer tests in core holes provide a means of determining the water tightness of the bedrock formations. These tests are performed in much the same manner as would be accomplished in an investigation for a major reservoir.

Test pits and slit trenches can be extremely beneficial in that they provide exposures that can answer questions concerning water infiltration, joint patterns within the soil, and depositional continuity of soil strata. Often, in glacial areas where five- to ten-foot thicknesses of sand are encountered, these pits and trenches indicate that the granular deposits are not continuous, but isolated pockets or lenses within a silty clay matrix. Normally these exposures are dug with a backhoe and common depths range between 14 and 19 feet; however, larger backhoes are available which can reach depths approaching 30 feet. Deeper exposures can be obtained, if necessary, by using truck-mounted auger rigs that are utilized in the drilled shaft industry. These rigs vary in capability, but it is not uncommon to have ready access to equipment that can drill 5-foot-diameter holes, 100 feet deep. Visual exposures should be documented with the same accuracy as that applied to bore hole logging. In many cases these exposures provide for more accuracy as a continuous profile is exposed for direct observation. Photographs of these exposures can also provide well-documented information. It is suggested that for those individuals not fully versed in photography, a professional photographer be retained to perform this task.

It is vitally important the investigation establish the location and number of all water tables that will have bearing on the landfill design. In addition, enough information should be developed to estimate piezometric gradients and recharge of all water-bearing strata which may be affected by the disposal facilities. These determinations require that a considerable number of piezometers be set at various levels and properly isolated to measure the piezometric levels at specific elevations. That is to say, a piezometer which interconnects several levels of water within the subsurface profile will yield misleading information. For example, it is typical in glacial areas that given enough time, most piezometers will indicate shallow water tables when proper isolation techniques are not utilized. This situation primarily develops because the majority of the shallow glacial soils contain joints and fissures which are filled with water. At deeper depths, the glacial clays exhibit fewer joints and fissures and are often very impermeable. Thus, the shallow water levels indicated in the piezometers are a result of movement of water through the shallow fissures and joints. When placed in open contact with the piezometer, they will drain into the opening and fill the piezometer to a level within several feet of ground surface. In situations where piezometers are installed at greater depths and the unscreened upper portions of the piezometers are backfilled with bentonite and clay, and sealed in the upper 10 to 15 feet with a grout encasement, the piezometer often remains dry. This reflects the low moisture content conditions which often exist in the underlying dense glacial clays which form the relatively impermeable barrier to leachate migration from beneath the landfill.

Another situation which has been found to occur in glacial areas is where bedrock may be buried 100 to 200 feet beneath deposits consisting primarily of glacial clay. Often, it is found that granular deposits consisting of sand and gravel lie immediately above the bedrock and beneath the glacial clay. The drilling of a bore hole and setting a piezometer in the transition zone between the bedrock and glacial clay may indicate that artesian flow will occur. In the event that such a situation occurs and it can be established that an upward gradient exists throughout the area, the downward migration of leachate or contaminatns to thes deeper water-bearing zones is unlikely to occur. This upward gradient can be established by installing piezometers which are isolated at various depth levels. Also, it is important that water samples from the various levels be collected and tested. In some cases, the determination of the total dissolved solids in the collected samples will indicate whether the source of groundwater in the glacial material was sourced from the bedrock or from above infiltration.

Although properly constructed piezometers may also serve as monitoring wells, several larger diameter holes and casings should be installed as permanent long-term monitoring wells. These wells should be placed at locations which will indicate the quality ofthe groundwater entering and leaving the site. It is important that all wells be constructed with clean water and materials so they do not become artificially contaminated. The samples recovered from these wells, when compared to off-site well samples, can indicate probable sources of groundwater and contaminants.

In some cases, it may be necessary to install wells and perform pumping tests to determine aquifer properties and the influence of lower and higher levels of water. In these situations, the well casing is generally six inches or larger in diameter. To provide for sufficient granular packing around the screened zone, the bore hole size may approach 12 to 18 inches in diameter.

The cost of installation of piezometers and small diameter wells is substantial, and careful consideration should be given to constructing these installations at locations which can remain in-place as permanent groundwater monitoring points. The information derived during the preliminary (Phase I) and the primary (Phase **) subsurface investigations provide background data on the quantity and quality of water prior to constructing the hazardous waste landfill disposal facilities. If the piezometers and well installations remain permanently in-place, a direct comparison of preconstruction and post construction groundwater conditions can be affected without increased cost to the client. If this is to be accomplished, consideration must be given to locating groundwater monitoring installations in relation to the physical location of the planned disposal facilities. This objective requires that all field investigation and monitoring installation points be accurately surveyed so that their horizontal and vertical locations are known.

The number of test borings and piezometers depends upon the complexity of the geology and the size of the proposed landfill. However, it is anticipated that for new sites the number of borings will probably range between 25 and 75, in addition to those performed in Phase I. In most cases, it seems reasonable to estimate that approximately 25 percent of the bore holes will contain either temporary or permanent piezometers. Several may be enlarged and constructed as monitoring wells.

Laboratory Testing

The required laboratory testing may be conveniently divided into two groups, those tests performed on the collected water samples and those tests performed on the recovered soil and rock samples. The required water sample tests include those, but may not be limited to, tests which are normally required to establish the quality of water for potable use. Since several of these tests require performance within a certain period, or the sample refrigerated immediately after collection, advanced planning is necessary to ensure that collected samples are handled in a proper manner. Collection bottles may also require washing with distiled and/or deionized water, and sterilization with chemical additives for preservation of samples. The collection and handling of undisturbed soil and rock samples is also of vital importance if one expects to obtain realistic laboratory-measured soil properties.

A wide variety of laboratory soil and rock tests are required to adequately characterize the geomechanical properties of both natural subsurface materials and those materials which will be utilized in the construction of the disposal system. These tests may be conveniently divided into classification tests and physical property tests. Classification tests are those which are utilized to determine index properties which provide the geologist and geotechnical engineer with numerical values that allow positive identification of subsurface materials. Many of these tests can be performed on samples which are collected in a disturbed manner.

The laboratory equipment and methods for the performance of tests on soil and rock for engineering purposes are found in publications by the American Society for Testing and Materials [2,3]. It should be noted that these publications contain both standard and suggested methods. Generally, the majority of the required classification tests involve standard or adopted methods.

On the other hand, more sophisticated tests which measure strength, hydraulic, and volume change properties are often referred to as state-of-the-art tests. In these situations, standard methods may not have been adopted and testing procedures are labeled as "Tentative" or "Suggested."

PHASE III

Prior to initiating this phase, all of the data accumulated in Phase I and II should indicate that the potential site is acceptable for development of a secure landfill. However, minor questions concerning subsurface features may still exist and must be necessarily answered before completion of the final design drawings, operation manuals and permit application. During this phase, these questions are answered through additional laboratory and field investigative efforts. The final design is formulated and all of the documents prepared which are necessary to submit a complete permit application to state and federal agencies.

Final Field Tests and Observations

This step may, may not be necessary depending upon the complexity of the geologic conditions and the closeness of the working relationship between the engineer, client, and state personnel. If the geologic setting is complex, several interpretations of the data generated from Phase II maybe plausible. The various interpretations of the data should be examined and analyzed to determine whether additional subsurface tests or observations are required so that all parties arrive at a consensus of how the data relates to the proposed hazardous waste landfill. Quite often, this consensus cannot be obtained without additional field investigation or extended monitoring of observation test wells or piezometers.

In arriving at a consensus of opinion concerning the interpretation of the subsurface data or the need for additional field tests and observations, the most expedient method is to schedule a meeting with all parties concerned in attendance. Prior to this meeing, the engineer should have his interpretation of the subsurface data finalized. In addition, design drawings for the facilities should be approximately 50 percent complete so anticipated behavior of the landfill can be discussed in a definitive manner. Also, it is helpful if an agenda for the meeting is agreed upon prior to the same. At the completion of this meeting, the person assigned as notetaker should provide a verbal summary to stress the points of agreement and disagreement. In the event of disagreements, decisions should be made as to how they can be rectified prior to finishing the final design and submitting the required permit application. It is suggested that within one week of the meeting a "rough draft" summarizing the content and decisions made in the meeting be circulated to all participants for comments. It is important these minutes be rough draft and in an unofficial form as the minutes may not totally agree with each person's perception of what transpired. Rectification of differences can usually be remedied through phone conservations. As soon as practical, the official documentation of the meeting should be issued. This procedure permits open and direct communication and more importantly, provides a public record which more clearly reflects the outcome of the meeting.

Since this meeting involves the establishment of a specific interpretation of the subsurface conditions in relation to the planned facilities, it is important that the project geotechnical engineer and geologist attend. Also, it is important that state or federal agencies charged with the responsibilities and duties of administering the applicable landfill laws and regulations are present. These meetings are most often attended by a dozen or more people and as a result, a tape recorder often provides a convenient unofficial record that can be replayed during preparation of the meeting minutes. Most importantly, the outcome should be to establish whether or not additional field tests and obsevations are necessary to clarify the characteristics of the geologic setting under consideration.

Final Design and Permit Application

The documentation of the geologic setting is a critical portion of the overall permit application process. Thus, considerable thought should be given to provide clear and concise information. It is also advisable to present all data, void of interpretations, in the appendix portion of the report or a separately bound volume. The binding of the data in a separate volume provides the advantage of being able to distribute this information to either an independent consultant or an internal technical person. The information can be reviewed in the raw form without the person being biased by previous interpretations made by those technical persons closely associated with the project.

The presentation of the interpretation of the subsurface data can be conveniently divided into three logical segments. First, the data should be interpretated and discussed in relation to the conditions which exist in the geologic setting prior to construction of the proposed landfill. Second, the interrelationships between the geologic setting and the proposed construction should be defined. THird, the anticipated post construction performance should be discussed. It is important that the second segment relate directly to design drawings and that specific references are made to design plans and specifications to illustrate the interrelationship between the geologic setting and the overall design concepts for the facility. It is also important to discuss the planned sequence of construction and expansion of the site so that future requirements for additional subsurface information, or the installation of additional monitoring wells, are well understood by the client, engineer, and regulatory personnel.

It is vitally important that the second and third segments of the report relate to the final plans and specifications that are actually submitted with the permit application. Since the preparation of plans and specifications can easily require four to eight months of concentrated efforts by many design team members, the drawings and specifications will generally undergo several revisions. Following each revision, Segments 2 and 3 of the report should be reviewed for consistency and completeness in regard to site development.

The geologist and geotechnical engineer have the important role of interpreting the subsurface data for the civil or environmental engineer. This communication should provide a design that takes the fullest advantage of the subsurface conditions. In many instances, the geotechnical engineer or geohydrologist designs the detailed aspects of the leachate collection system. For example, filter designs should be in accordance with accepted practice so that plugging or clogging does not occur. Another example would be the identification and quantity estimates of the more plastic soils for use as compacted soil liners beneath hazardous waste, treatment ponds, and final cover.

Another example of required input is in evaluating the in-place moisture content of the existing soils that are to be utilized as liner, cover, bedding, or fill materials. In some instances, the natural moisture content of the soil may be well below or above the optimum water content. These situations require that the soil be brought to a moisture content which is near the optimum moisture content. THe addition or removal of moisture from the soil requires that processing be performed prior to attempting compaction. Often, treatment ponds are not of the size or configuration which permit efficient processing whithin the pond area and thus, nearby areas should be designated for processing.

In some situations, it may be beneficial to add bentonite to the natural soil during processing so lower coefficients of permeability are achieved. Here again, the geologist or geotechnical engineer's knowledge of the behavior of bentonite-soil mixtures and the blending and processing procedures is often found to be useful by the design engineer in preparing specifications. The selection of manufactured membrane liners and proper installation procedures often require input from the geotechnical engineer. The location of permanent monitoring wells and their design are another aspect of the overall project design which requires the assistance of the geohydrologist and geotechnical engineer. We have also found it extremely helpful to provide contour drawings depicting the upper surface of the bedrock which lies beneath the site. A comparison of this contour drawing with other drawings denoting the base of the leachate collection system illustrates the thickness of soil that underlies the facility at various locations. Isopach drawings are also helpful to illustrate overburden thicknesses.

Engineering drawings are developed for three primary purposes, i.e., permit application, construction phase, and operation phase. These drawings present the following information.

A. Plan View
   1. General site location
   2. Boring locations
   3. Facilities location
   4. Floor plan
   5. Construction phases

B. Topography
   1. Initial
   2. Excavation
   3. Construction phases
   4. Final
   5. Top of bedrock
   6. Isopachs

C. Building/Facility Designs
   1. Gatehouse
   2. Office, showers, lunchroom, toilets
   3. Laboratory
   4. Maintenance
   5. Storage
   6. Processing
   7. Truck wash
   8. Water treatment facilities

D. Roads, parking lots, drainage structure designs

E. Cross sections

F. Three-dimensional boring profiles

G. Miscellaneous details

1. Security fencing
2. Piezometer, monitoring wells
3. Utilities
4. Signs
5. Sample, tarp platforms
6. Ponds, tanks
7. Liners
8. Leachate collection
9. Landscaping
10. Cell construction
11. Final Cover

Coordinates should be provided on the drawings and horizontal and vertical controls established in the field so that operating personnel can readily determine locations from the drawings in the field. Surface runoff water should be diverted from the facilities, cells, and trenches to prevent this water from being contaminated. Contaminated waters should be diverted to storage ponds or tanks for treatment. Air pollution control equipment should be designed for storage, mixing, and treatment facilities to prevent the emission of dangerous pollutants. The thickness of soil required under the trenches and cells should be established utilizing the geotechnical data and the quantity of soil available for cover material determined. The soil quantities should be balanced within certain stations to ensure adequate soil and reduced haul costs. We use a computer program we developed at Burns & McDonnell to do this which also establishes the final elevations, provides soil balance tables, and develops the cross sections by computerized graphics. The landfill liners and soil cover details should also be designed.

The permit application, reports, specifications, and operations/maintenance manuals should then be prepared. This includes waste characteristics, process information, surface and groundwater monitoring, security, inspections, contingency plan, personnel training manual, and closure and post-closure plans. It is also necessary to evaluate routing of hazardous waste vehicles the site and obtain concurrence from regulatory agencies. All the necessary regulatory forms should be completed and insurance documents presented.

FACILITY CONSTRUCTION

Once construction has begun, large excavation exposures will provide excellent opportunities to verify subsurface design assumptions. Also, it is not uncommon to uncover subsurface features that were not disclosed by the previous subsurface investigations. These features may require that modification of the original design be accomplished to remain consistent with the overall design constraints. In addition to examining subsurface conditions, these site visits offer an excellent opportunity to observe the actual construction. It has been our experience that often landfill operators utilize equipment and employ personnel that may not be thoroughly familiar with processing and compaction of earth materials. Advice and simple explanations of the behavior of soils at varying moisture contents can be extremely helpful to the landfill manager and his operators for planning and conducting their earth-moving operations. Also, at this time the field testing procedures, which generally involve the determination of moisture content and percent of compaction, can be observed and modified as necessary to establish an acceptable quality control program.

Since the construction of the facilities will occur over a number of years, it is most desirable that communication between the operator, engineer, state, and federal agency personnel is maintained throughout the life of the facility. Scheduled and unscheduled visits by the engineer and regulatory personnel to the site can provide assurance that the facilities are being constructed in a manner that is consistent with the permit application.

CONCLUSION

The intent of this paper has been to outline the phases required for the siting, design, and construction of hazardous waste landfill facilities. The important role of the geologist, geotechnical and environmental engineer is self-evident from the nature of the facilities, which involve the handling of large quantities of earth on a daily basis. By comparison to other construction projects, hazardous waste landfill developments are most likened to the construction of earth dams, as both require attention to detail from beginning to end. While most construction projects require performance to be satisfactory over the economic life of the project, which often is a 30- to possibly 100-year period, secure landfill facilities are expected to perform properly for hundreds of years. Unfortunately, the systematic disposal of wastes has only developed on a large-scale basis in the last 30 years and long-term performance records are lacking. The monitoring of the long-term performance of these facilities promises to be the most advantageous means of increasing our knowledge for future and better landfill designs.

REFERENCES

1. Stohr, C.J., St. Ivany, G., and Williams, J.H., "Geological Aspects of Hazardous-Waste Isolation in Missouri," Engineering Geology Report No. 6, 1981, MDNR, Division of Geology and Land Survey, Rolla, Missouri.

2. ASTM Committee D18 on Soil and Rock for Engineering purposes, "Special Procedures for Testing Soil and Rock for Engineering Purposes," ASTM SPT 479, 1970.
3. 1982 Annual Book of ASTM Standards, "Part 19 Soil and Rock; Building Stones," ASTM, Philadelphia, Pennsylvania.

CHAPTER 23

EPA'S LAND DISPOSAL REGULATIONS - WASTE DISPOSAL INDUSTRY'S PERSPECTIVE

Reva Rubenstein, Ph.D
U.S. Environmental Protection Agency

INTRODUCTION

The Institute of Chemical Waste Management includes those member firms of the National Solid Wastes Management Association that are active in storage, treatment and disposal of hazardous and other industrial wastes. The member companies are active in all aspects of hazarouds waste management. These methods include treatment (e.g., solidification), recovery, incineration, and deep well injection. Like everyone else committed to recovery and treatment, all of our members are dependent on land disposal for management of those wastes that are not economically or technically feasible to treat. Treatment processes themselves often result in residues, often hazardous if to a lesser degree than the original waste itself, that must be disposed of in the land. Even the most committed advocates of treatment rely on land disposal and several treatment oriented companies actually own and operate land disposal facilities as part of their waste management system.

The Institute has campaigned vigorously for strict federal regulation of land disposal practices. As history has shown, this is sorely needed to protect the environment and the public health. Moreover, only through strict regulation will it be possible to build public confidence that disposal will be done safely and thus provide the opportunity for public acceptance of the siting of needed disposal facilities.

For the regulations to be effective they must be enforced uniformly without regard to the location or the ownership of the facility. It is an indisputable fact that environmental damage can be caused by facilities located either on or off the site of generation. The interim list of priority Superfund sites leaves no doubt on this score. Moreover, facilities owned by a unit of government should be regulated under exactly the same rules and to exactly the same degree of enforcement as facilities owned and operated by private industry.

Additionally the promptness with which the U.S. EPA expects to put the new regulations into effect, is a major concern of the

Management of Toxic and Hazardous Wastes, H. Bhatt, R. Sykes, T. Sweeney (Editors)

Institute. This is especially sensitive because of the regulatory definitions of "existing portion". Major components of the new regulatory program will not apply to a facility until the permit is actually issued. Only the ground monitoring requirements pursuant to Subpart F, apply to all existing units which received hazardous waste after January 26, 1983. This means that continued construction of land based units which do not meet the design standards of the new regulation is probable unless an absolute cutoff date is established. The Institute has sued the U.S. EPA on this issue. We propose that the design exemption be limited to interim status units for which construction had been completed and operational use commenced by January 26, 1984.

Why choose these dates instead of the date the groundwater monitoring regulations went into effect, January 26, 1983? We favor January 26, 1983 but because EPA states it cannot possibly permit interim status facilities for at least a year, we have suggested January 26, 1984. But uncontrolled expansion of the current operating area must be curtailed. We suggest that a maximum of 10% of the original Part A area be available for operation without application of the new design standards. If the active life of a landfill is anticipated to be 20 years, 10% of the original Part A would represent a maximum of 2 years of active life. This should give EPA sufficient time to examine RCRA Part B applications without causing unnecessary capacity shortfall.

LIQUID WASTE DISPOSAL

Concerning the regulations: the Institute strongly supports EPA's stated liquid management policy as the means of protecting groundwater. Clearly this is the major environmental thrust of these regulations and, at long last, the Agency seems to have provided the framework for a coherent strategy. The EPA program calls for minimizing the amount of liquids allowed to enter the landfill during the operating life, removing and treating any liquid that does enter and closing the facility so as to prevent entry of liquid after closure. We have some disagreements however with the Agency on specifics of that program.

First, we disagree with the continued disposal of bulk liquids in a landfill if the facility has a liner and leachate collection system. The Agency's reasoning that bulk liquids will pass rapidly through the fill and be withdrawn in the leachate collection system and hence is not of environmental concern is misguided. As bulk liquids pass through the fill, they come into contact with various wastes within the fill and potentially mobilize (i.e., leach) constituents already properly managed. The treatment of the resultant leachate is likely to be more difficult, and costly. Bulk liquids should be treated prior to disposal and only the residues or sludges should be landfilled.

Second, we disagree with EPA's emphasis on synthetic liners.

We believe such exclusive reliance on synthetic liners is foolhardly and unfortunate. Synthetic liners are thin films of plastic materials that in concept are impermeable to liquid. Plastic liners are relatively new in application for landfills. The quality of the barrier they provide depends completely on the quality of the installation, far more than the liner material itself. A tiny puncture in a critical location can completely void the protective element. An imperfect seam can leave a hole of major proportion. The question of chemical compatibility of the liner with contaminated leachate has been examined only superficially. EPA's rules would accept a landfill with a plastic liner placed directly over a highly permeable base such as sand. We think synthetic liners should be backed up with clays to provide a second measure of protection from leachate permeation. Moreover we believe clay liners by themselves should be allowed when the wastes to be accepted are compatible with clay liners.

Clay liners, either natural or installed, provide a much greater margin of safety than plastic liners and are not subject to castrophic failures. Clays have the ability to retain a volume of liquid without migration of that liquid, i.e., field capacity. Both the field capacity and permeability of a clay can be measure. Therefore, the amount of penetration during the active life of a facility and the depth of clay required to distribute the leachate so as to reach field capacity can be calculated. For example, if there will be one foot of penetration during the active life and one-third of the saturated volume equals field capacity, then three feet of clay are required to distribute the leachate to a point where no further migration would occur. In such a situation, a five foot liner (which includes two feet of buffer) should suffice. Such a liner would accomplish equivalent theoretical protection to that of the required synthetic liner and it would not be subject to catastrophic failure as a result of tiny holes.

No single design can satisfy all situations. Both synthetic and natural liners and combinations thereof should be allowed. The liner should be engineered for the specific site conditions.

"AQUIFER" AND "GROUNDWATER"

Third there is a failure to adequately define the terms "aquifer" and "groundwater". While all aquifers contain groundwater, not all groundwater is in an aquifer. The key issue, however, is that not all groundwater in the uppermost aquifer should be afforded regulatory protection. Rather, such protection should be extended only to underground waters which reasonably may serve as media for the transmission of hazardous wate constituents from land storage, treatment or disposal units to significant off-site sources of water for human consumption. The definition of such waters would then be consistent with the definition of an "underground source of drinking water" pursuant to the Safe

Drinking Water Act. Indeed, EPA is under a statutory directive (Section 1006(b) of RCRA) to integrate RCRA with the Safe Drinking Water Act and other environmental laws.

Furthermore, an aquifer exempted from the protections of the Safe Drinking Water Act also should be exempted from the groundwater monitoring requirement under RCRA. An aquifer may be exempted under the Safe Drinking Water Act only if it is not now used as a "source of drinking water" and it "cannot water" or it has high dissolved solids content and "it is not reasonably expected to supply a public water system" (40 CFR 146.4). Exemption may occur only after public hearings, and is subject to approval of EPA (40 CFR 122.35). In light of these qualifications and approval processes, there is no ratinal basis to apply the RCRA groundwater program to an exempted aquifer.

The Institute has also pointed out that the uppermost aquifer subject to regulatory protection must also originate beneath or upgradient of the waste management area and terminate at a location beyond the facility property boundary downgradient of the area, as well as be part of a continuous, hydraulically connected groundwater formation. With regard to the former, the purpose of the entire groundwater strategy is to prevent offsite migration of contaminants originating in a regulated unit, thus it must be determined which of various groundwater-bearing zones has the capability of delivering such contaminants off-site. The requirement that the uppermost protected aquifer be of a continuous, hydraulic connection is based on the fact that the regulations focus on a upgradient/downgradient monitoring comparison, so it is essential that the same underground water be the medium monitored by all wells. Without specifying that the uppermost aquifer be of a continuous, hydraulic connection, it is entirely possible that the uppermost aquifer be of a continuous, hydraulic connection, it is entirely possible that the uppermost aquifer upgradient of the regulated unit is different from the uppermost aquifer downgradient of the regulated unit. All groundwater-gearing strata are not necessarily interconnected. It would be capricious for EPA to allow monitoring to occur in different aquifers, since this would defeat the logic of the entire monitoring program.

Fourth, in the United States there are a few unique locations where there is no usable groundwater at those locations is extremely remote, if not nil and the siting of land disposal facilities in those locations should be encouraged. Unfortunately, several sections of EPA's regulations actually discourage use of these superior locations. If the owner/operator wishes to site a facility at a location with natural protective features we must show in advance that the groundwater must be protected "for all future time". Will it be possible to provide absolutely that there will be no contamination of groundwater forever? EPA concedes that making such a showing may be difficult at best. Unless the operator can make this existential showing, he cannot take advantage of the unique natural site. The owner/operator of a land disposal facility that uses synthetic liners and leachate collection to contain and remove leachate is responsible for assuring protection of groundwater only for the operating life of the facility plus a post-closure period of thirty years. This is none of several ways in which the regulation in effect discourages use of what we consider to be the best natural locations for land disposal facilities.

ENVIRONMENTAL STANDARDS

Finally, the greatest deficiency in the program, in our opinion, is the absence of environmental standards for all but the 14 substances for which there are interim primary drinking water standards. Only for eight metals and six pesticides are there quantitative numerical limits on the concentrations allowed in groundwater. For the hundreds of other hazardous constituents, EPA's standard is no increase over background.

Many of the hazardous constituents are used extensively and are routinely discharged to other environmental media--surface water, air of the oceans-by persons using them. An example is the class of materials known as halogenated organics--specifically chlorinated methanes, ethanes and ethylenes. These are widely used as paint strippers, degreasers, dry cleaning fluids and industrial solvents. They are used to unplug septic systems clogged with grease. Not surprisingly these compounds are commonly detected in public and private water supplies. There should be numerical limits on the concentration of these and many other hazardous constituents sometimes found in drinking water. The zero increase over background limits applicable to this one activity only-disposal of wastes--is unjustifiable.

The job of standard setting has been assigned to EPA by the Congress and the Agency has filed to complete the essential task.

CONCLUSION

In conclusion, the waste service industry is committed to a program of strict environmental regulation for the land disposal of hazardous wastes. We feel that the EPA program aimed at limiting the amounts of liquids destined for land disposal is right on target and deserves public and congressional support. Although we have some technical problems with the regulations as written and we encourage EPA to make the necessary changes, we are ready to work with any group or agency to assure that hazardous waste is well managed now and in the future.

SECTION VI

DISPOSAL SITE CLEANUP

CHAPTER 24

WATERWAY CONTAMINATION - AN ASSESSMENT OF CLEANUP PRIORITIES

John C. Henningson
Malcolm Pirnie, Inc.

PROBLEM DEFINITION

Many of the nations waterways are contaminated by significant levels of toxic and/or hazardous materials. The nature of contaminants, amount, and areal extent varies greatly, as indicated in Table I.

Table I. Examples of Contaminated Waterways

Location	Name and Type of Waterway	Nature and Extent of Contamination
Illinois	Waukegan Harbor	PCB; source to L. Michigan
Indiana	Grand Calumet River and Indiana Harbor Canal	PCB; source to L. Michigan
Massachusetts, Conneticut	Housatonic River	PCB; over 100 mi. of river
Massachusetts	Acushnet Estuary	PCB: -
New York	Hudson R. & Estuary	PCB; over 150 mi. of river and NY Harbor
New York	Oneida Lake	Mirex; -
New York	Niegara River	Dioxin; -
Virginia	James R./Chesapeake Bay	Kepone; -
Virginia	Shenandoah River	Mercury; -
Washington	Commencement Bay	Various Contaminants; -
Wisconsin	Fox River	PCB; source to L. Michigan
Wisconsin	Sheboygan Harbor	PCB; source to L. Michigan
Wisconsin	Milwaukee Harbor	PCB; source to L. Michigan

In addition, many ports and harbors have substantial contamination by heavy metals.

These conditions often result in significant risks, impacts and

Management of Toxic and Hazardous Wastes, H. Bhatt, R. Sykes, T. Sweeney (Editors)

economic effects. For example, in the Hudson River System the contamination has resulted in the closing of the commercial fishing for most species and advisories against consumption of sport fish caught by recreational fisherman.[1] In addition, the contamination of sediments has made the maintenance of navigable waters difficult and expensive because highly contaminated dredged material must be placed in upland sites suitable for such materials.[2] Furthermore, the continued utilization of the river as a public water supply has been the subject of considerably controversy.[1] Similar widespread impacts on the fishery and other resources have occurred in Chesapeake Bay, Housatonic River, Acushnet Estuary and numerous other locations.

This extent of exposure and impacts is often much greater than that associated with upland uncontrolled abandoned sites. The currents, tidal movements and storms tend to exacerbate the original contamination of sediments by industrial outfalls. Storm surges are difficult to predict and may result in significant redistribution, greatly increasing the costs of cleanup and decreasing the feasibility of complete cleanup. For example, the contamination of sediments in the upper Hudson River was discovered in 1973. Before action could be taken, a 100-year flood occurred in 1976. It moved more than 300,000 cubic yards of highly contaminated material (>1000 ppm PCB) downstream, forming over 40 "hot-spots" exceeding 50 ppm. It is fortunate that the upper river is composed of a series of dams and pools that trapped most of this material before it reached the estuary.[3] However, another storm of similar magnitude may result in even greater redistribution and make river cleanup infeasible.

In Chesapeake Bay, the conditions for sediment distribution have already resulted in extensive movement of Kepone prior to cleanup. It is probable that effective cleanup has now been precluded by the cost of removing such an extensive volume of contaminated sediments.[4]

A hotspot of almost 100,000 cubic yards of material exceeding 500 ppm PCB has been identified in the Acushnet Estuary. Unless prompt action is taken, the opportunity for an effective cleanup may be lost.[5]

A portion of Waukegan Harbor has PCB levels that exceed 33 percent in the sediment. This sleeved distribution results in a condition where approximately 90% of the PCBs are in 5% of the contaminated sediments. The annual steady state loss of PCB to Lake Michigan has been estimated at up to 20 pounds. However, it is estimated that the loss during a single storm event may equal or exceed this level.[6]

The above examples clarify the unique problems associated with waterways contamination. In particular, the possibility for catastrophe loss or redistribution of contaminated sediment would seem to warrant special consideration and a high priority for cleanup.

CLEANUP METHODS

The determination of the most effective methods of remediating the extensive contamination of sediments has been the subject of several comprehensive evaluations.[1,3,4,5,6] A particularly good summary was prepared by Dawson (1979) and is summarized in Tables II and III. [7]

In general, current feasible techniques are limited to dredging and upland disposal or in some cases stabilization in place. The cost of such measures are expensive. However, redistribution will result in even higher costs or loss of the opportunity to clean up at all.

One of the most difficult aspects of dredging contaminated materials is the identification and development of a suitable disposal site. Table IV summarizes criteria for selecting a site for a disposal facility. These limitations make the siting process costly and time consuming. In addition, considerable effort is often required to convince the public that such projects are acceptable.

CLEANUP FINANCING

The five most probable mechanisms for financing the cleanup of contamintated waterways are listed below:

- Direct action by industry or other "Responsible Parties"
- Clean Water Act, Section 115. In Place Toxic Pollutants
- Clean Water Act, Title II, Grants for Construction of Treatment Works
- Clean Water Act, Section 311, Emergency Spill Cleanup
- Superfund
- State Funds
- Extension of U.S. Army Corps of Engineers Responsibilities for Waterway Maintenance

The most desirable is direct action by "responsible parties". However, in most cases lengthy litigation occurs before a settlement is achieved. In the interim, storms or other mechanisms may cause substantial redistribution of contaminants. The litigation with responsible parties with respect to Waukegan Harbor has gone on for over six years and still has not been resolved.

Section 115 of the Clean Water Act directs the USEPA to identify the location of in-place toxic pollutants and authorizes removal and disposal. Fifteen million dollars was authorized, but has never been appropriated. This work would be done in conjunction with the U.S. Army and, therefore, is related to an extension of the Corps of Engineers responsibilities as noted below.

In New York, the U.S. Congrss provided Title II funds under Section 116 of the Clean Water Act to clean up the Hudson River.

TABLE II. CLEANUP OF OIL WASTES

Alternative	Advantages	Disadvantages
OILY WASTES		
Containment and Removal		
Booms and Skimmers	Effective in calm waters, easy to stockpile, low associated impacts, removes oil	Costly, not effective in high seas or rough weather
Gelling Agents	Can facilitate clean-up under adverse conditions	Difficult to apply and harvest
Physical Sorbents	Low cost	Difficult to apply and harvest, create associated disposal problem
Destruction		
Combustion	Amenable to large spills; low cost	Effective only on unweathered crudes, light oils or spills in arctic conditions
Biodegradation	Low cost	Not shown to be effective on large spills
Dispersal		
Sinking Agents	Low cost, can apply in rough weather	May re-release oil, may cause impacts on benthos
Chemical Dispersants	Applicable in rough weather, competitive costs	May be more toxic to aquatic life

CLEANUP OF SOLUBLE WASTES

Alternative	Advantages	Disadvantages
SOLUBLE WASTES		
Ex-Situ Treatment	Only proven, available technique	Must handle large volumes of water, required materials may not be readily available, may be costly
In-Situ Treatment		
Chemical Addition	Inexpensive, simplified logistics; no disposal requirements	Hazards associated with proper dosing and by-products
Biochemical Degradation	No hazardous reactants involved	Requires extensive inventory of specific cultures; may create significant DO problems
Tea bags	Easily deployed and collected	Low effectiveness due to poor contact
Buoyant Media	Effective and adaptable to a variety of water bodies and conditions	No current source of floating activated carbon; buoyant exchangers are expensive, specialty products

TABLE III. SEDIMENT DREDGING TECHNIQUES

Mode of operation	Availability	Advantages/Disadvantages	Estimated Costs
Mechanical and Wireline Dredges			
Clamshell Dredge	Common sealed bucket type not available in U.S.	Difficult to provide even coverage, highly turbid	$\$2.50/yd^3$
Dragline dredge	Common	Difficult to control out, highly turbid	$\$2.94/yd^3$
Dipper dredge	Common	Turbidity augmented by violent digging action, difficult to control out	$\$2.50/yd^3$
Bucket Ladder Dredge	Only a few private units in U.S. used for mining	Disperses sediments widely	
Hydraulic Dredges			
Hopper Dredge	Most units owned by U.S. COE, some private units now available	Elutriate and fines discharged overboard, prohibitions to discharge have resulted in high water-to-solids ratio and hence lower productivity, hard to manipulate in confined areas	
Cutterhead Pipeline Suction Dredge	Common	Reduced turbidity, releases can be minimized if cutterhead is not required. Required disposal area of hopper dredge within piping distance	$\$1.50/yd^3$ mobilization-demobilization could raise to $\$4.00/yd^3$
Dustpan Dredge	All units owned by U.S. COE, not available in Great Lakes areas	Reduced turbidity	
Mud Cat Dredge	Available on rental basis from National Car Rental System, Inc.	Small units, mobile, and tailored to site, cut depths up to 10-15 feet, can't work in water over 15 feet deep	$\$2.40/yd^3$ depends on length of rental period
Sidecaster Dredge	Three units in U.S., all are owned by COE	Required hopper dredge in tandem or produces highly turbid discharge, use is highly restricted and not well suited to in-place toxics	
Pneumatic Dredges			
Airlift Dredge	Constructed as needed	Susceptible to cratering in one location, would require support of hydraulic dredge-type equipment	
Pneuma Dredge	One unit in U.S.	High noise levels, low solids ratio in discharge, poor cut control, limited data available	$\$0.70/yd^3$
Oozer Dredge	Units in Japan only	Law prevents use in U.S. at this time, good turbidity control, high solids ratio in spoils	

TABLE IV. PRIMARY SITE SCREENING CRITERIA HAZARDOUS DREDGED MATERIAL

Parameter	Unacceptable	Reference
Soil	Permeability greater than 1×10^{-5} cm/sec.	Part 360 NYCRR
	Less than 3 ft thick in situ.	Pirnie
	Class I or II agricultural soils.	Part 360 NYCRR
	Unified Soil Classification System Group Symbols: GW, GP, GM, GC, SW, SM, SP, ML, OL, MH, Pt.	40 CFR Part 250
Slope	Deep gullies, slope over 15%.	Pirnie
Surface Water	Closer than 300 ft to any pond or lake used for recreational or livestock purposes, or any surface water body officially classified under state law.	Pirnie
	In 100-year floodplains.	40 CFR Part 257 & Part 761
	In official state-designated wetlands	Pirnie
Bedrock	Closer than 30 ft to highly fractured rock or carbonates. Closer than 10 ft to all other rock.	Pirnie
Ground Water	Closer than 10 ft to ground-water table.	Part 360 NYCRR
	Closer than 500 ft to any water supply well or recharge area.	40 CFR Part 250
Committed Land	Located in designated agricultural districts or closer than 1000 ft to parks, residential areas, historic sites, reservoirs, etc.	Pirnie
Biologically Sensitive Areas	Endangered plant or animal habitats, unique or regionally significant habitats.	Pirnie

The USEPA opposes this approach and recommends that a more appropriate mechanism would be the use of the Superfund. The net result has been a lack of the approvals necessary to take decisive action and an increased risk of catastrophic loss of PCB to the estuary.[9]

Some funds have been allocated from Section 311 to study the problem in Waukegan Harbor.[6] However, the extent of such funds is limited and cannot facilitate cleanup on a broad nationwide scale. This mechanism appears to be intended for emergency response to spills rather than contamination resulting from long term industrial discharges.

In most cases contaminated waterways are clearly abandoned uncontrolled hazardous waste disposal sites. Waukegan Harbor, the Acushnet Estuary and the Niagara River sediment deposits have been earmarked for the receipt of monies from Superfund. Hopefully this program will accelerate from its present pace so that these problems may be dealt with in a timely fashion. The most difficult problem is the allocation of priorities between upland sites such as the Woburn Landfill in Massachusetts and the Acushnet Estuary.

Some states, such as New Jersey and New York, have established state superfunds. However, it is unlikely that the available funds can remedy extensive waterway contamination. For example, the cost of cleaning up the major hotspots in the Hudson River will exceed 20 million dollars.

On the James River and Chesapeake Bay, the corps of Engineers has financed limited studies to evaluate certain cleanup effort related to maintenance dredging.[10,11] An expansion of Corps authorization beyond normal maintenance may be an effective mechanism for accelerating waterways cleanup.

CONCLUSION

The contamination of sediments in the nation's waterways is a major problem. The redistribution of contamination by currents, tides and storm events often make the potential risks and impact on natural and economic rsources greater than for upland sites. The potential for catastrophic releases of contaminants probably warrants that greater priority be given to waterways contamination than to many upland disposal sites. The potential for castrophic releases of contaminants probably warrants that greater priority be given to waterways contamination than to many upland disposal sites. The remedial methods for contaminated sediments have been evaluated extensively and the most feasible action is usually to dredge concentrated areas before redistribution can occur and place the dredged material in secure upland disposal sites. Unfortunately, such actions are relatively costly and compound the difficulties in assessing priorities. The cleanup of an extensively contaminated waterbody such as the upper Hudson River may be several times more costly than stabilizing an upland abandoned uncontrolled waste disposal site. Several possible mechanisms are available for financing cleanups. However, the delays associated with many proposed programs have greatly incrased the risk of irretrievable loss of contaminated materials due to storms or other catastrophic events less common to upland sites. It is recommended that a higher priority be considered for allocating resources to the cleanup of contaminated waterway in recognition of the special risks and potential impacts associated with such problems.

REFERENCES

1. Malcolm Pirnie, Inc., "Draft Environmental Impact Statement PCB Hot Spot Dredging Program, Upper Hudson River, NY", for NYS Department of Environmental Conservation, White Plains, NY, 1980.
2. Malcolm Pirnie, Inc., "Draft Environmental Impact Statement and 10-Year Management Plan, Hudson River Federal Channel Maintenance Dredging", US Corps of Engineers NY District, White Plains, NY, 1981.
3. Henningson, John C. and Richard F. Thomas, "Hudson River Cleanup", Proceedings of the Workshop on Environmental Decontamination, Oak Ridge National Laboratory, Oak Ridge, TN, 1979.
4. USEPA, "Report on Mitigatin Feasibility for the Kepone Contaminated Hopewell/James River Areas", Washington, DC, 1978.
5. Malcolm Pirnie, Inc., "Final Report: The Commonwealth of Massachusetts Acushnet River Estuary PCB Study", Massachusetts Division of Water Pollution Control, White Plains, NY, 1982.
6. USEPA. "The PCB Contamination Problem in Waukegan Harbor", Chicago, IL, 1981.
7. Dawson, Gaynor W., "Decontamination of Water Bodies", Proceedings of Workshop on Environmental Decontamination, Oak Ridge National Laboratory, Oak Ridge, TN, 1979.
8. Henningson, John C. and J.S. Reed. "Social and Environmental Concerns Associated with the Hudson River PCB Decontamination Projects", ASCE National Convention, New Orleans, October 1982.
9. Knickerbocker News, "New Twists may block dredging of PCBs" Albany, NY, August 28, 1982.
10. Haller, D.L. "Demonstration of Advanced Dredging Technology- Dredging Contaminated Material 'Kepone' James River, Virginia," Norfolk District, U.S. Army Crops of Engineers, undated.
11. Vann, R.G., "James River, Virginia Dredging Demonstration in Contaminated Material (Kepone) Dustpan versus Cutterhead," Norfolk District, U.S. Army Corps of Engineers, undated.

CHAPTER 25

CLEANUP OF A VINYLIDENE CHLORIDE AND PHENOL SPILL

Albert R. Posthuma, John G. Kraus, and Julie A. Rutherford
Williams & Works/Environmental Data Inc.

BACKGROUND

In February 1978, a freight train derailed at Woodland Park, Michigan, a wooded, semi-rural resort area approximately 80 miles north of Grand Rapids. Four tank cars were damaged, and approximately 30,000 gallons (300,000 lbs) of vinylidene chloride, 40,000 gallons (330,000 lbs) of phenol, and 15,000 gallons (112,000 lbs) of ethylene oxide were lost. Because of the cold temperatures at the time of the derailment, virtually all ofthe vinylidene remained as a liquid and percolated into the sandy soils of the site, while the phenol solidified and remained as a solid on the surface. The ethylene oxide vaporized into the air.

The initial cleanup included the excavation and removal of approximately 5,000 cubic yards of contaminated soil. This removed most of the phenol.

By the time Williams & Works was called in and groundwater cleanup operations began, the vinylidene chloride had spread over 17 acres and was approaching several private wells and East Lake. concentrations as high as 300 mg/l of vinylidene chloride were found in monitoring wells. Later investigations showed that approximately 70,000 pounds of residual phenol had been left in the soil and was flushed into the groundwater table during the cleanup program. This also had to be removed and placed additional limitations upon the type of treatment system that could be utilized to clean up the groundwater.

PHASE I CLEANUP

The initial treatment system was placed in operation in September, 1979. It consisted of seven purge wells to capture and control the contaminated groundwater, two aerated ponds, and six 20,000 pound activated carbon adsorbers (see Figure 1). The aerated ponds removed approximately 90% of the vinylidene chloride by air stripping. The residual vinylidee chloride and phenol was removed by the activated carbon. The system had a design capacity

Management of Toxic and Hazardous Wastes, H. Bhatt, R. Sykes, T. Sweeney (Editors)

of 700 gallons per minute, which was the flow rate needed to contain and capture the plume of contaminated groundwater.

The effluent from the treatment system was discharged to East Lake. The effluent limitations set by the Michigan Department of Natural Resources (MDNR) were very stringent, 0.02 mg/l for vinylidene chloride, the detectable limit, and 0.60 mg/l for phenol. Each pair of activated carbon adsorbers was operated in series to insure that a breakthrough of the first unit would not violate the effluent limitations. The combination of aerated ponds and activated carbon did an effective job of removing the chemicals down to the levels required for discharge during the first two years of the cleanup program.

Figure 1. Aeirial View of Treatment Site (looking west), Spring 1980. Phase I Treatment included aeration ponds and activated carbon asorption. Tracks, lake system, and nearby residences are also visible.

The operation of the activated carbon units did present some problems. Even though the units were treating relatively clean groundwater, a bacterial slime soon built up on the surface of the carbon, partially plugging the units. A chlorination system was installed to kill the bacterial slime. The units had to be backwashed frequently to remove accumulated organics and sediment.

A second problem with the activated carbon units was discovered when the discharge caused an algae bloom in East Lake. It was found that phosphorus was leaching out of the carbon into the discharge water. The added phosphorus discharged in the effluent was enough to upset the ecological balance in East Lake. As soon as the source of the problem was discoveed, the carbon in the units was replaced with a different type that did not leach phosphorus.

PHASE II CLEANUP

By the end of 1980, over 95% of the chemicals had been removed from the groundwater. As the concentrations decreased in the groundwater, the cost of treatment per pound of chemical removed increased dramatically because of the high cost of leasing the carbon units. Alternatives were then evaluated to find a more efficient method of stripping the vinylidene chloride and for biological treatment to remove phenol.

Pilot studies demonstrated that spray irrigatin of aerated lagoon effluent would reduce the residual vinylidene chloride to non-detectable evels provided the initial concentration was less than 1.0 mg/l. The tests also showed that the phenol could be biologically removed given the proper environment. Based on these studies, the MDNR agreed to removal of the carbon adsorption units.

In July 1981, a spray irrigation system was installed in the marsh area (West Springs) adjoining East Lake (see Figure 2). As predicted, continuous irrigatin of aerated pond effluent at rates of 800 to 1,000 gallons per minute reduced the vinylidene chloride to nondetectable concetrations. The marsh also provided a suitable microbiological environment for reducing phenol to normal background concentrations before the water entered East Lake.

Figure 2. Spray Irrigation System. Phase II treatment involved spray irrigation to reduce vinylidene chloride t a non-detectable concentration. This system also provided for microbiological breakdown of residual phenol in the adjacent marsh.

The amounts of chemicals removed from the groundwater during the cleanup program versus the original spill are shown in Figure 3. By November 1981, more than 99% of the chemicals were removed, and irrigation ceased for the winter months. Limited pumping and irrigation were resumed during teh summer of 1982. The entire system was shut down and partially dismantled at the end of October 1982.

Hydrogeological studies showed that the natural groundwater flow from the site of the original spill is into the marsh area used as the discharge point for the irrigation system. Any residual chemical left in the groundwater will be removed by natural processes in the marsh. Monitoring of the groundwater in the area and the discharge from the marsh into East Lake will continue over the next 2-3 years to insure that no traces of the chemical in excess of the discharge limitatins enter East Lake.

FLUSHING OF SOILS AT SPILL SITE

Hydrogeological investigations showed that some residual vinylidene chloride and phenol remained in the soil at the site of the original spill. This residual chemical continued to bleed into the groundwater due to the flushing action of normal rainfall and seasonal fluctuations in the groundwater table.

A series of soil samples were collected at various depths around the site of the original spill to determine the extent of the soil contamination. Most of the residual vinylidene chloride was found in the zone immediately above the water table, which is approximately 50 feet below the ground surface. Most of the residual phenol was found near the ground surface.

A spray irrigation system was installed along the tracks in the ara of the original spill to flush the residual chemical down into the groundwater. Two additional purge wells were also installed in the source area to remove and treat the additional water from the flushing. The flushing system was operated during the summer months of 1980 and 1981, and was successful in removing most of the residual chemicals from the soil.

HYDROGEOLOGICAL STUDIES

Extensive hydrogeological investigations were performed at the cleanup site to determine the extent of the groundwater contamination and the geological and hydrogeological conditions. More than 200 soil borings and monitoring wells were installed on site to determine soil conditions and to sample the groundwater. Most monitoring wells were installed in clusters, with the screens set at 10-15 foot vertical intervals to monitor the vertical as well as horizontal extent of the chemical contamination. The monitoring wells were sampled quarterly throughout the cleanup

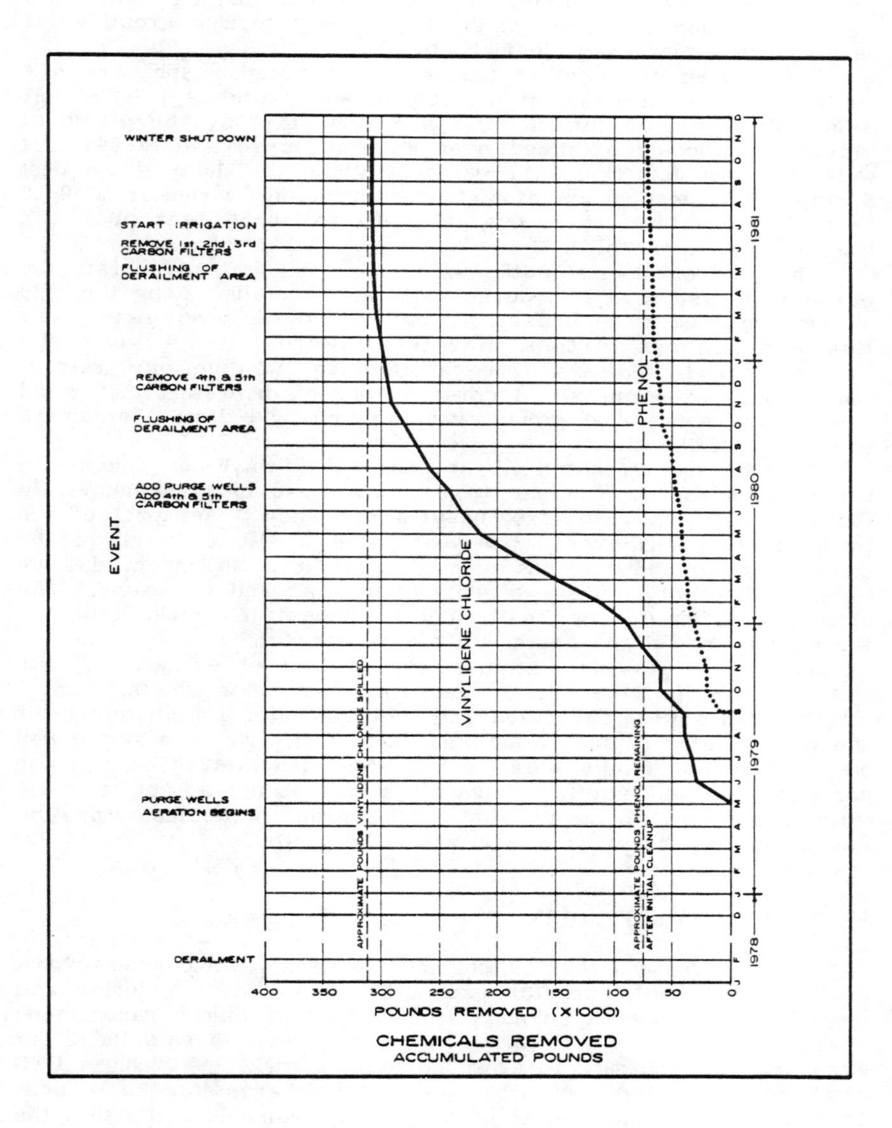

Figure 3. Chemicals Removed.

program. Water levels were also measured quarterly to evaluate groundwater movement under purging and non-purging conditions. This information was used to develop a computerized groundwater model for the complex geological site.

Groundwater movement in the area of the original spill was in a south-southwest direction at the rate of approximately 1.3 feet per day. By the time the cleanup program started, the plume of vinylidene chloride extended over an area approximately 600 feet wide by 1,000 feet long, and was approaching the shore of the West Spring. The vertical extent of the contamination varied from 10-20 feet at the site of the original spill to 40-50 feet near the leading edge of the plume.

The hydrogeological data was used to design and operate the purge well systems. The purge wells were installed along the axis of the contaminated groundwater plume to intercept the groundwater flow and prevent further migration towards East Lake. The hydrogeological studies showed that a purge well pumping rate of 700 gallons per minute would create a cone of depression that would control the contaminated groundwater movement and draw it in toward the center of the site.

The average concentrations of phenol and vinylidene chloride in the groundwater in October 1980 and August 1981 are shown in Figures 4 and 5. The isoconcentrations show the impact of the purge well system over a one-year period. Virtually all of the contaminated groundwater was removed from the south end of the site by the fall of 1981 after two years of purging and treatment. The cleanup program during the third year concentrated on cleaning up the source area at the north end of the site.

The hydrogeological studies were also used to locate a safe water supply for the three residents adjacent to the spill site. The residents were furnished with bottled water immediately after the groundwater contamination was discovered. Later a single well was installed to supply water to all three residents. The well was screened in an aquifer below the contaminated aquifer and upgradient from the spill site. The groundwater contamination never reached the private wells of the homeowners.

ENVIRONMENTAL STUDIES

A major question faced throughout the cleanup program was "what are the safe limitations for exposure to vinylidene chloride and phenol?". Vinylidene chloride is a suspected human carcinogen; however, at the start of the cleanup program, no safe limits for exposure had been set. To avoid any possible risk, the Michigan DNR established a limitatin of 0.02 mg/l for discharge into East Lake. This was the detection limit of vinylidene chloride at the time the program was started.

Phenol occurs naturally in the environment, but studies have shown that phenol is toxic to fresh water aquatic life at average

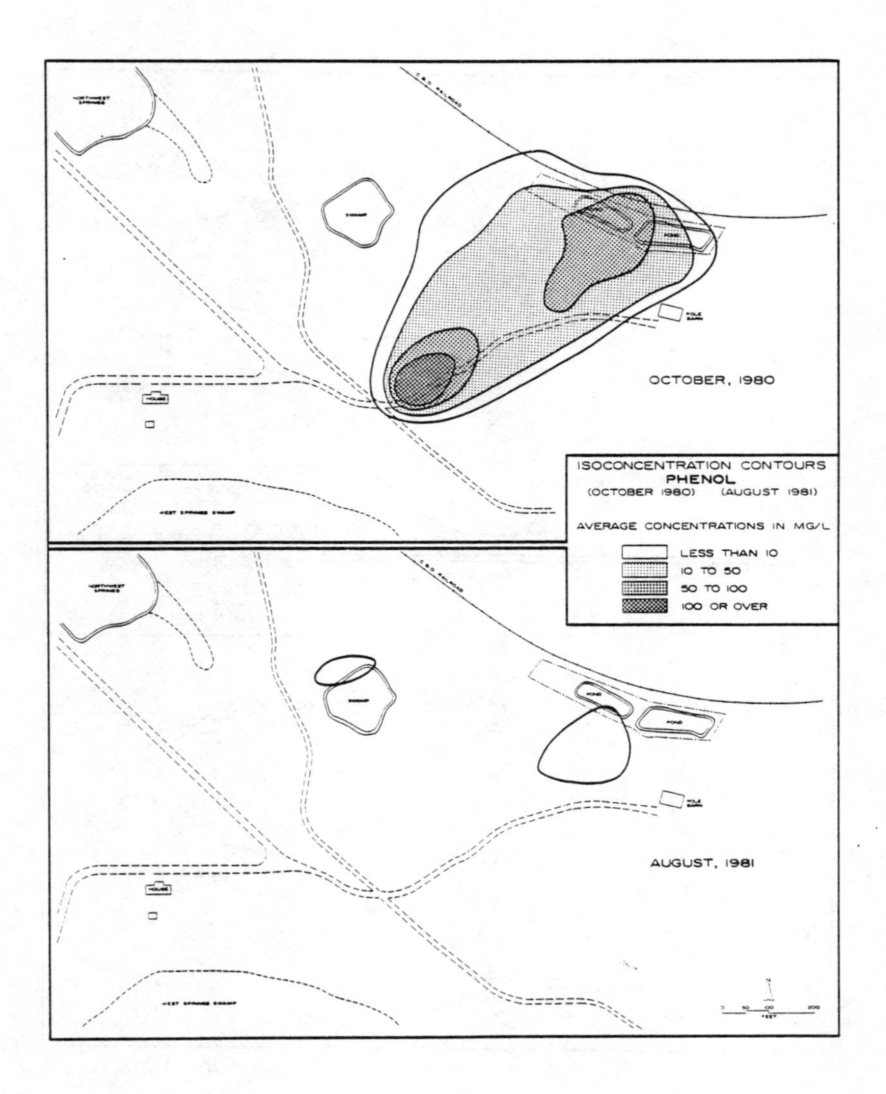

Figure 4. Average Phenol Concentrations.

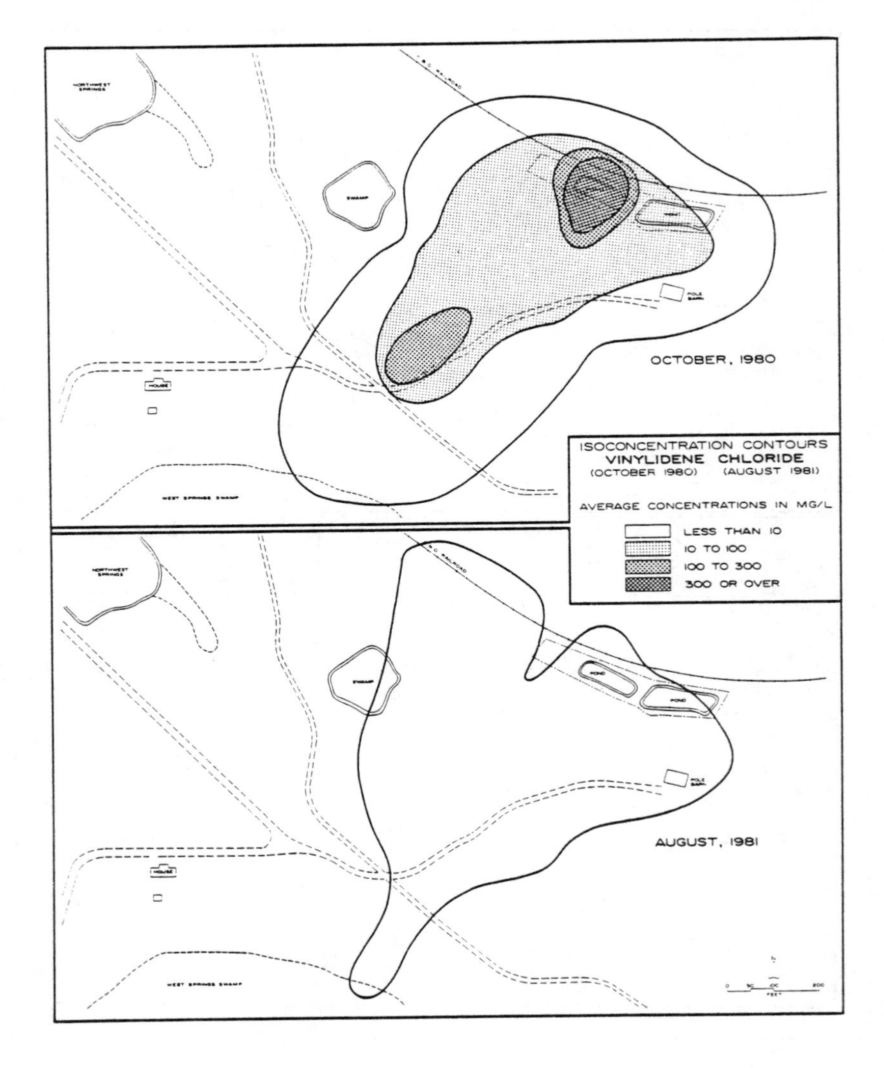

Figure 5. Average Vinylidene Chloride Concentrations.

concentrations greater than 0.6 mg/l. Therefore, the Michigan DNR established an effluent limitation of 0.6 mg/l for phenol.

Another concern was the general health of teh employees working at the cleanup site. Most of the vinylidene chloride was being stripped into the air above the aerated ponds, providing a constant source of exposure to the staff. Vinylidene chloride vapor also acumulated in some of the monitoring wells and exposed employees to high concentrations in teh air when the caps were removed from the wells for routine monitoring.

A toxicologist was retained to evaluate the potential exposure of employees on site during the cleanup program and establish a safe limitation for vinylidene chloride in the air. Based on this evaluation, it was established that the air quality around the lagoons was safe. The air quality escaping some of the monitoring wells did exceed the safe limitation, however, and breathing apparatus was furnished to employees for wearing whenever the concetrations exceeded the safe limit. A procedure was established for testing the air in each well prior to sampling.

Frequent safety seminars were held with the employees to insure that they understood the potential dangers of the chemicals and practiced safe working habits. Blood samples from each employee working on site were analyzed before and during the cleanup program ot insure that no one was being exposed to excessive amounts of the chemicals accidentally.

The safety of the water and fish in East Lake was a frequent concern of local residents. To alleviate their concerns, frequent samples were collected from various areas of the lake to verify that chemicals were not present in the lake. Also, representative species of the fish in the lake were netted and analyzed to verify that they were not contaminated. None of the samples collected from the lake ever showed any traces of vinylidene chloride, or concentrations of phenol greater than the normal background levels.

Hydrogeological studies showed that traces of vinylidene chloride and phenol left in the groundwater will eventually discharge into the West Spring. Pilot studies were done by our own staff biologist to determine the volitalization rate of the binylidene in the marsh environment and the potential impact of the vinylidene chloride on the vegetation. The studies showed that concentrations of vinylidene chloride less than 1.0 mg/l could be assimilated by the West Spring. This data was then used to establish what would be an acceptable level of vinylidene chloride to leave in the groundwater.

PUBLIC RELATIONS

An important part of the cleanup program was maintaining good public relations with the local residents and government officials. At the beginning of the cleanup program, the local residents were convinced that their own private wells and East Lake were contaminated with the chemicals. Frequent public meetings were held for the residents to show them the extent of the contaminated groundwater and bring them up to date on the status of the cleanup program. Representatives from the Michigan Department of Natural Resources and Department of Public Health were also invited to attend these meetings.

The various types of media in the area were also utilized. The results of the lake and fish studies were published in the local newspapers. Two on-site interviews were conducted with the local television station to publicize the positive aspects of the cleanup program.

The success of the cleanup program and the associated public relations can best be illustrated by the fact that the program never received any bad publicity. Much of the credit for this can be given to the railroad for their commitment to cleaning up the chemicals before they contaminated nearby water supply wells and East Lake.

This project won the ACEC Grand Conceptor Award for Engineering Excellence in 1982. The Chessie System Railroad, Williams & Works/Environmental Data, and the State of Michigan are all very proud of the good public relations that this success story has achieved.

CHAPTER 26

CASE HISTORY - REMEDIAL INVESTIGATION RE-SOLVE, INC. HAZARDOUS WASTE SITE

Jeffrey A. Cassis and Dana Pedersen
Camp, Dresser & McKee Inc.

INTRODUCTION

The U.S.E.P.A. selected Camp Dresser & McKee Inc. (CDM) to lead engineering studies for immediate remedial action in Superfund Zone 1 (U.S.E.P.A. Regions I, II, IV), which covers 15 states east of the Mississippi River. Two additional zones cover the midwestern and western states. As a result, CDM was tasked to conduct the remedial investigation and source control feasibility study for the Re-Solve, Inc. hazardous waste site in Dartmouth, MA.

The purpose of the Remedial Investigation at the Re-Solve, Inc. site was to define the extent and degree of contamination on and off the site. The Re-Solve, Inc. site will require a long-term cleanup action to provide adequate protection of the public health, welfare and the environment. The specifics of the Remedial Action Program are described in Phase VI, Section 300.68 of the National Contingency Plan (NCP), published in the Federal Register, Vol. 47, No. 137, July 16, 1982, pursuant to the comprehensive Environmental Response, Compensation, and Liability Act of 1980 (CERCLA). The remedial investigation and alternative actions recommended are consistent with the permanent site remedy and the actions taken to prevent and mitigate the release of hazardous substances into the environment.

STATEMENT OF THE PROBLEM

The Re-Solve, Inc. hazardous waste site was listed on the U.S. Environmental Protection Agency's Interim List of 115 top-priority disposal sites and is on the proposed national priority list of 419 hazardous waste sites published December 30, 1982. The site was ranked 156. As such, it is eligible for funds under the comprehensive Environmental Response, Compensation, and Liability Act of 1980 (CERCLA), known as "Superfund". Remedial actions are subject to the requirements of State participation pursuant to Section 104(c)(3) of CERCLA.

The Commonwealth of Massachusetts has negotiated with the owner

Management of Toxic and Hazardous Wastes, H. Bhatt, R. Sykes, T. Sweeney (Editors)

of Re-Solve, Inc., Mr. William Jackson, a compromise for settlement on past disposal practices. Operation and reclamation of waste chemicals at the Re-Solve, Inc. site began in 1956 for reusable chemicals, primarily solvents. Mr. Jackson has been the owner since May 1976.

The major known sources of contamination at the site are the four unlined hazardous waste lagoons (with PCB concentrations greater than 500 ppm), oil spreading areas and contaminated soils, a cooling pond contaminated with organic and inorganic chemicals, and the structural remnants. Figure 1 shows the location of these areas. USEPA, the Commonwealth's Department of Environmental Quality Engineering (DEQE), and CDM have sampled the known sources of contamination on-site and off-site, surface water, groundwater stream/river sediments, and nearby private drinking wells. The major area of off-site contamination is in the wetlands to the north, and in sediments to the southeast, where high levels of organics, notably PCBs and heavy metals have been identified. The major pathways for the off-site migration are via the surface water runoff and groundwater.

The public health is not immediately threatened by the existing contamination on or off the site. The private drinking water wells downgradient of the site do not have contaminants that violate the National Primary Drinking Water Standards or pose a threat to public health. The hazardous waste lagoons and other contaminated areas on-site are stable in their present condition but pose a public health threat from the potential for direct contact with the contaminants and remain a continuous source of contaminants. The levels of volatile organics in the air at the boundary of the site are not above background and do not present a public health threat.

BACKGROUND

Site Location and Description

The Re-Solve, Inc. site is located in a rural area of the town of Artmouuth, Bristol County, Massachusetts, as shown on Figure 2. The site is situated on the east side of North Hixville Road on approximately 6 acres of land. The site is surrounded by wetlands except for a pine and mixed hardwood forested area to the south and west. The Copicut River is located about 500 feet directly east of the site, draining into Cornell Pond, approximately one quarter of a mile down river. Cornell Pond drains about 2 miles into Noquachoke Lake which is designated as a secondary water supply for the City of Fall River.

Remedial Actin Master Plan (RAMP)

A RAMP, prepared by Camp Dresser & McKee Inc., under contract

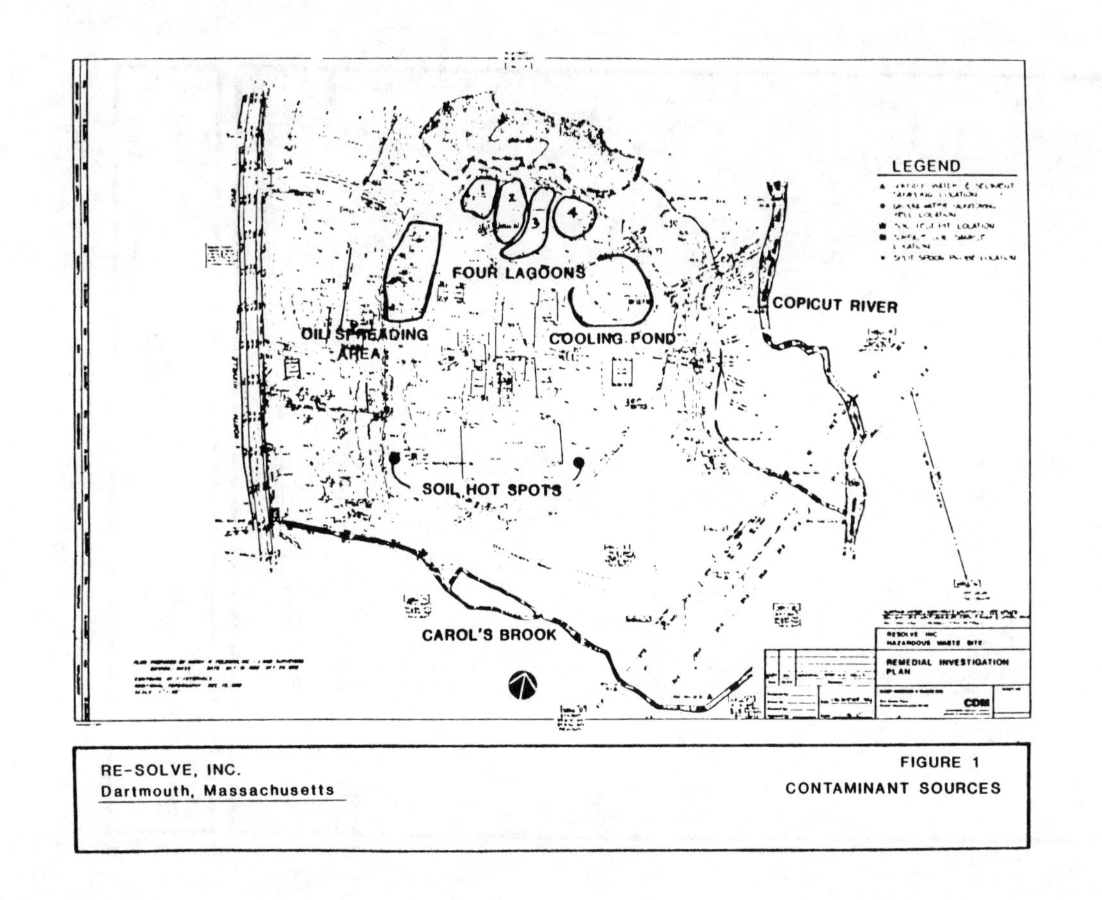

RE-SOLVE, INC.
Dartmouth, Massachusetts

FIGURE 1
CONTAMINANT SOURCES

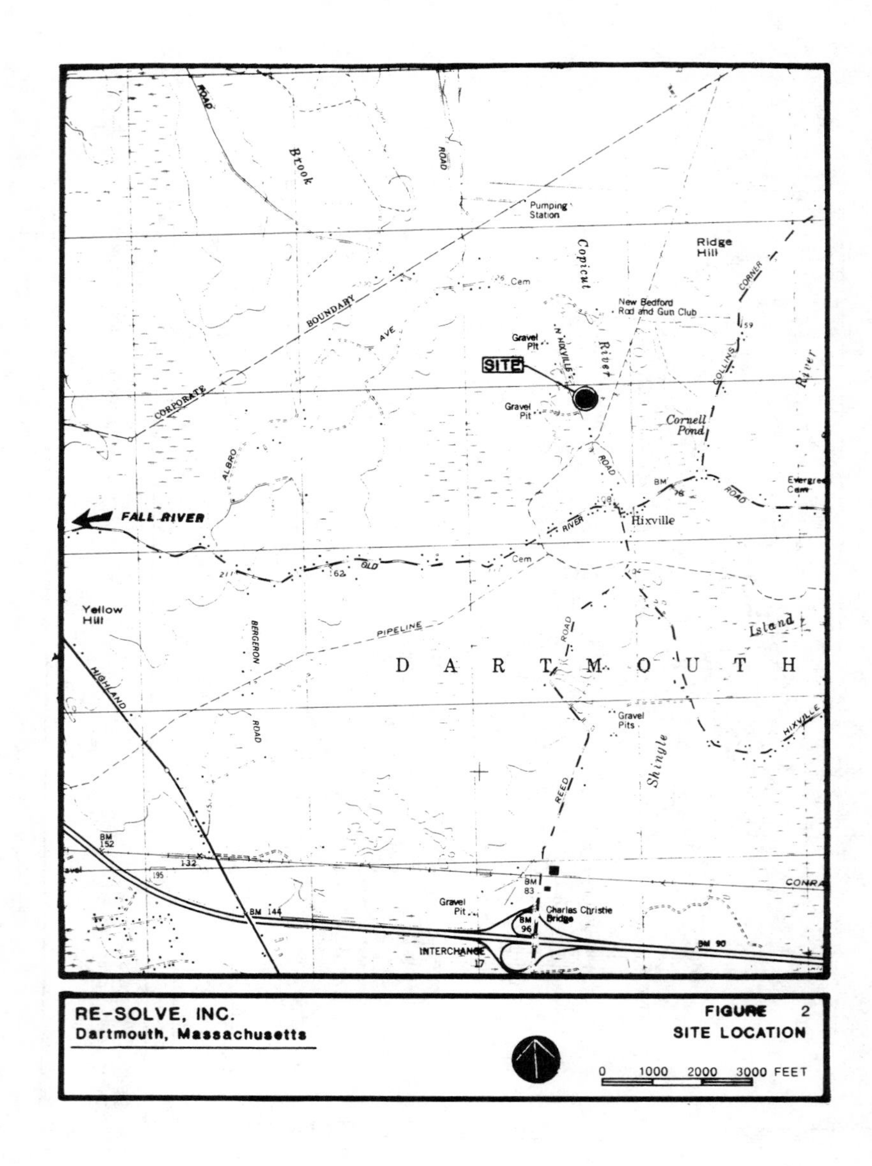
Brook
ROAD
Pumping Station
Ridge Hill
Copicut
Cem
New Bedford Rod and Gun Club
BOUNDARY
CORPORATE
AVE
Gravel Pit
N HIXVILLE
River
SITE
COLLINS
CORNER
River
Gravel Pit
Cornell Pond
ALBRO
ROAD
BM
Evergreen Cem
FALL RIVER
RIVER
Hixville
OLD
Cem
Yellow Hill
BERGERON
PIPELINE
ROAD
Island
D A R T M O U T H
HIGHLAND
Gravel Pits
ROAD
Shingle
HIXVILLE
REED
BM 152
BM 83
CONRAIL
Gravel Pit
BM 144
Charles Christie Bridge
BM 96
INTERCHANGE 17
BM 90
RE-SOLVE, INC.
Dartmouth, Massachusetts
FIGURE 2
SITE LOCATION
0 1000 2000 3000 FEET

with the USEPA as the Zone I Remedial Action Contractor for the Re-Solve, Inc. site was published on July 26, 1982. The primary purpose of the RAMP was to assess the available data on the site and identify the type, scope, sequence and schedule of remedial projects which may be appropriate.

Budget level cost estimates were prepared for the first projected phase of activity, along with a detailed statement of work. Brief project descriptions and order of magnitude cost estimates for future projects were also included. Other key components of the RAMP were the evaluation of existing data and any limitations associated with the data such as sampling/analysis protocols and chain-of-custody requirements; assessment of the need for additional data to evaluate remedial alternatives; community relations requirement; and a discussion of administrative and procedural requirements, including any special problems that may be encountered during project implementation.

CONTAMINANT SOURCES

Lagoons

The four unlined lagoons leached contaminants into the local groundwater system. The bottoms of all four lagoons were found to be in the seasonal high groundwater table. The lagoons were excavated into groundwater. There were volatile organics found in the four lagoons at concentrations greater than 65,000 ppm, heavy metals greater than 1,000 ppm, and PCBs at a concentration greater than 1,000 ppm, and PCBs at a concentration greater than 500 ppm.

For example, toluene was found at a concentraion in the lagoons ranging from 5,000 ppm to 250,000 ppm and in the groundwater immediately down-gradient from the lagoons (Location B), toluene was found at a concentration of 91,000 ppb. Toluene can be found off-site on the east wide of the natural gas pipeline at a concentration of 6,200 ppb. Other volatile organics were also found at high concentrations and exhibit a similar pattern. The most problematic waste compound are the PCBs because of the disposal limitations for PCBs >50 ppm and because of their inclusion in the site soil in the groundwater, PCBs are associated with sediments in the sample. At Well B, PCBs wree found in the groundwater at a concentraion of 660 ppb. PCBs wre found in the down-gradient off-site monitoring wells at concentrations less than 10 ppb.

Based on our remedial investigation, we have determined that the four lagoons have overtopped on a number of occasions in the past, contaminating the wetlands and unnamed tributary to the north and east of the site. PCBs were found in the sediments in the wetlands and in the unnamed tributary that drains the wetlands to the Copicut river. Concentrations in the sediments have recently been measured at up to 100 ppm. Volatile organics and heavy metals were also found in the sediments. The surface water in the wetlands and tributary were also contaminated, primarily with volatile organics. The highest concentration detected in recent sampling (January 1984) was 7,400 ppb of methylene chloride which was found in the wetlands.

Filled Cooling Pond

The cooling pond in the eastern portion of the site has been filled with clean coarse sandy loam. The DEQE (October 1980) sampled the active cooling water pond and found methylene chloride at 1.45 ppm, acetone at 1.5 ppm, trichloroethyene at 860 ppb, methyl ethyl ketone at 780 ppb, and other organics less than 100 ppb.

The cooling pond has added contaminants to the contaminant plume. Contaminated groundwater was pumped from an on-site well downgradient of the lagoons and the oil spreading area and was used as the non-contact cooling water for the solvent distillation towers. This practice created a situation that allowed contaminatns to spread out in all directions, re-introducing contaminants to the groundwater. The bottom of the cooling pond was in the seasonally high groundwater table which would allow easy entry of contaminants into the local groundwater system. This practice was terminated and based on CDM's test pits excavation in the cooling pond, the majority of organic contaminants have been flushed through the pond, except for a 4 inch to 6 inch band of residues at the bottom. The residues are composed of PCBs at a concentration less than 10 ppm and the only other organic contaminant detected was bis (2-ethylhexylphthalate) at a concentration of 60ppm. The coarse sand used to fill the cooling pond was not found to be contaminated.

Oil Spreading Area

This area is the western portion of the site and occupies about 10,000 sq ft. Based on the high levels of PCBs found at depth in this area via test pits and soil borings, it is theorized that this area was used for disposal of waste oils, unsalable solvents and for PCB contaminated oils. The groundwater fluctuates between about 3 and 6 feet beneath the surface in this area. Just to the east of this area, the groundwater is at the surface and ponds during periods of heavy precipitation. The surface soils in this area had PCB concentrations ranging from 15,000 to 52,000 ppm. Volatiles were detected at low concentrations on the surface, generally less than 1 ppm. PCBs were found at concentrations greater than 50 ppm at 6 feet beneath the surface. Highest concentration detected for PCBs were 200,000 ppm at the 4 feet to 6 feet level. Based on the test pit (TP4) in this area, there are volatile organic detected at over 1,000 ppm - the same contaminants seen in the groundwater at Well B and Well K.

Contaminated Soils

Another source of significant contamination is the surface soils ("Hot Spots") throughout the entire site where chemical wastes were spilled during loading and unloading activities. These areas include the soils adjacent to the unloading platforms and the on-ground drum and vehicle storage areas. Based on the test pit (TP-03) in one hot spot area, it was determined that the contamination was confined primarily to the upper 2 feet of soil. PCBs were found at a cncentration of 17,360 ppm at the surface, was less than 4 ppm at 2 feet and was not detected at 4 feet. At 5 feet, PCBs were found on to of the groundwater table at a concentration of about 200 ppm. At well location K, the soil boring at 2-4 feet showed PCBs at 4,274 ppm but it was not detected at 5-6.5 feet. Tetrachloroethene was detected at 40 ppm. The K location is near an unloading platform where oil or solvents contaminated with PCBs were spilled.

REMEDIAL ACTIVITIES

The open waste lagoons and contaminated soils at the site pose a public health problem from individuals coming into accidental or intentional contact with the waste. There are three types of remedial activities identified in the NCP that are appropriate for the Re-Solve, Inc. site. These are:

- Initial Remedial Measures
- Source Control Remedial Action
- Off-Site Remedial Actions

Initial Remedial Measures

A number of Initial Remedial Measures were conducted at the Re-Solve, Inc. site. In accordance with the NCP, Initial Remedial Measures are only used when a significant threat to public health, welfare or the environment are found to exist and when the problems causing the threat has a straight-forward solution available. The Initial Measures for this site were initiated in September 1982 and before the selection of a final remedy. The Initial Remedial Measures impemented for the Re-Solve, Inc. site was funded by the Comonwealth's Department of Environmental Quality Engineering and supervised by CDM. A 6 foot high chain link fence and 4 gates were installed around the entire work area (about 1,500 feet), with two strands of barbed wire. Hazardous waste warning signs were posted on the fence and gates. The fence achieved its objective to restrict unauthroized access to the site and the accidental or intentional contact with the PCB contaminated soils and 4 lagoons.

Source Control Remedial Actions

Source Control Remedial Actions are recommended when substantial concentrations of hazardous substances remain on site, barriers to retard migration of hazardous substances are inadequate, and/or there is a serious threat to public health, welfare, or the environment. Source control remedial actions are typically not appropriate if all hazardous substances have migrated from the area where originally located or if the lead agency (federal/state) determines that the material cannot be adequately contained.

Source Control Actions are necessary at the Re-Solve, Inc. site because hazardous substances have migrated away from the original disposal area and the sources are not contained. Hazardous substances include PCBs in sediment in the unnamed tributary off-site, Carols Brook and wetlands. In addition, volatile organics have been found in high concentrations in the local groundwater. The Source Control Action(s) evaluated for the site were:

1. No Action
2. Blacktop Entire Site
3. Microbiological Treatment
4. Source Removal and Stockpile
5. Containment/Isolation
6. Source Removal, Treatment and Encapsulation On-Site
7. Source Removal, Treatment (On/Off-Site) and Off-Site Disposal
8. Combinations of the above actions

No Action

The No Action alternative would do nothing to alter the site. The contaminants would continue to migrate off the site and the lagoons, oil contaminated soils and structural remnants would remain on-site in their present condition. The No Action alternative would not have any technological uncertainties associated with its implementation. It would require no time to implement and there would be no costs associated with the implementation. The only costs would be associated with a long-term monitoring program and the intangible potential impacts associated with the continued environmental and potential public health impacts. Re-Solve, Inc. has filed for bankruptcy leaving the financial burden for any remedial action with the USEPA and DEQE. The components of the No Action Strategy are:

- Potential for lagoon dike failure placing creating a major environmental and public health impact;
- Leave the four unlined lagoons, oil spreading area, and "hot spots" in place to continue to contaminate ground and surface water and pose a public health hazard from direct contact;
- No collection or detention of surface runoff, thus allowing sediments containing PCBs, metals, and organics, to be transported off-site which would ultimately degrade the water quality of the Copicut River and Carol's Brook, the two notably off-site drainage receptors;
- No leachate collection.

As described in the July 16, 1982 Federal Register concerning the National Oil and Hazardous Substances Contingency Plan (NCP), "no action alternatives are appropriate, for example, when response action may cause a greater environmental or health danger than no action." The Re-Solve, Inc. site has been a source of contaminants for over 24 years. The results of the remedial investigation indicate that the contaminants have moved off-site and that they will continue to move off-site unless remedial actins are implemented. The No Action stragegy is presented as a control to assess the environmental impacts associated with the Re-Solve, Inc. site in its present state and to compare it with other alternative source control remedial actions. The No Action alternative is evaluated based on the major pathways for human exposure that presently exist at the site.

Based on the present environmental and public health impacts associated with No Action, this alternative was eliminated from further consideration. The remaining alternatives were screened based on cost and non-cost criteria. The cost analysis addressed capital, transporation, disposal, and operation and maintenance costs. Non-cost criteria are reliability, ability to implement, technical effectiveness, environmental/health concerns, safety, regulatory requirements, public acceptance and timeliness.

Preferred Alternatives

As a result, the preferred alternative recommends removing the material from the four unlined lagoons, a significant amount of the soil in the oil spreading area and soil contaminated areas or "hot spots." In addition, CDM recommends placing a cover over the entire site to eliminate the potential for direct contact with the soils and to control the surface water runoff from the site.

The two primary objectives to be met by source control are to reduce and/or eliminate the off-site migration of contaminants and to eliminate the existing direct contact hazards at the site, primarily from the four unlined lagoons and oil spreading area.

Based on CDM's evaluation, two feasible remedial alternatives exist to handle the excavated and highly contaminated material with a total PCB concentration greater than 20,000 ug/g (ppm) dry weight. These alternatives are to transport the wastes off-site to a USEPA approved treatment and disposal facility or to chemically solidify the contaminated material and encapsulate it on-site. The present worth cost for on-site solidification and encapsulation would range from $2,270,000 to $3,020,000 and the present worth cost for off-site disposal is $3,360,000 assuming land burial. The non-cost criteria used to evaluate the feasible alternatives was more favorable than the other alternatives evaluated. Both alternatives would achieve the same objective, namely to reduce and eliminate the threat to public health and the environment.

Off-Site Remedial Actions

Off-Site Remedial Actions are taken when Source Control Remedial Actions are inappropriate or not effective in mitigating a significant threat posed by the migration of hazardous substances from the site. Off-site remedial action(s) were not part of CDM's original scope of work. The off-site remedial action alternative is currently being evaluated by CDM, under contract to NUS Corporatin.

Off-site actions undertaken by CDM in October 1984 consist of a remedial investigation (RI) and feasibility study (FS). The focus of the RI is to determine the level of contaminants in the sediments and waters of the wetlands, unnamed tributary, the Copicut River and Carol's Brook, and in groundwater. The FS will then identify and evaluate appropriate remedial alternatives. In addition to the no action alternative, pumping and treating of contaminated groundwater is being considered, as is removal of contaminated sediments.

SUMMARY

Re-Solve, Inc. operated a solvent reclamation facility for over 24 years on Hixville Road, terminating activities in 1980. Hazardous wastes (e.g., PCBs, inorganic and organic contaminants) have migrated off the site contaminating the nearby Copicut River, wetlands and the local groundwater. Based on CDM's remedial investigation, the contamination sources at the site are:

1. Four unlined lagoons in the norther part of the site,
2. A former cooling pond area filled with sand,
3. An oil spreading area in the western portion of the site.
4. Other cotaminated soil areas ("Hot Spots").

These four sources were confirmed by the analysis obtained from the installation of 35 groundwater monitoring wells, surface water/sediment sampling, soil borings, test pit excavations and lagoon depth probing and anlalyses.

However, the entire site remains a continuous source of contamination with the two major sources, the four unlined lagoons and an oil spreading area, contributing the majority of the contaminants to the environment. The Copicut River and Carol's Brook are presently acting as hydraulic barriers, confining the movement of the contaminant plume to within these two surface water bodies. The contaminant plume is moving in a southeastern direction towards Cornell Pond. Contamination has been found in one well to the east of the Copicut River. However, groundwater movement at this well appears to be back towards the river, so that further migration of the contaminant plume in an easterly direction is not a concern. In addition, the private well water in the immediate vicinity of the site does not present a threat to public health at the present time.

EPILOGUE

As of December, 1984, materials in the lagoons have been excavated and mixed with quicklime in preparation for transport to a secure landfill. Once removal is completed and the excavated areas have been refilled with clean fill, the entire site will be covered with a clay cap, loamed and seeded. Access to the site will continue to be restricted by the chain-link fence currently in place.

CHAPTER 27

WASTE STABILIZATION BASIN DISCHARGE ELIMINATION AND REMEDIATION - A CASE STUDY

William H. Bouck, P.E., Andrew N. Johnson, P.E., and Stephen J. Fleischacker
O'Brien and Gere Engineers, Inc.

INTRODUCTION

For a period of approximately 40 years, process wastewater and non-contact cooling waters had been discharged to an on-site waste stabilization basin from a plastics manufacturing operation that has been involved ina wide range of research and developement activities, and the manufacture of power transformer-related products, monomers, polymers and industrial resins. Elimination of effluent discharge from this two acre earthen basin, and remediation of the remaining aqueous and sediment/sludge contents was accomplished through the implementation of a four-year, $4.2 million hazardous waste management program.
Phases I and II of this four-year program are discussed briefly to indicate the strategies and procedures that are required in achieving an effective hazardous waste management solution while simultaneously phasing out an active system from production activities. The major emphasis of this paper centers on the Phase III remediation that resulted in an effective closure of the waste stabilization basin.

WASTE STABILIZATION BASIN DISCHARGE ELIMINATION

As a first step in achieving the goal of eliminating the discharge of waste stabilization basin effluent to a nearby brook, the client initiated Phase I by constructing an oil/water separator to provide treatment for an oil-laden wastewater and storm water stream that had previously been discharging to the basin. This Phase I also included in-line process modifications resulting in the further reduction of waste stabilization basin influent strength, by the construction and start-up of organic solvent extraction units designed to separate organic constituents from the aqueous wastewater stream via multistage steam distillation.

Management of Toxic and Hazardous Wastes, H. Bhatt, R. Sykes, T. Sweeney (Editors)

Concentrated organics removed by distillation are currently treated as a waste product in accordance with Federal and State requirements.

In order to identify treatment needs in pursuit of eliminating the usage of the waste stabilization basin, all process wastewater, storm water and non-contact cooling water flows were identified by extensive mapping of the on-site sewer system and characterized with respect to discharge frequencies, flow rates and wastewater composition under Phase II.

Wastewater Characterization and Treatability

Wastewater composition was shown to vary considerably. Approximately 95 percent of the organics loading was found to be incorporated into five (5) batch loads which were discharged at almost any time during weekly plant operation. From the information collected during field investigations, approximately 98 percent of the wastewater flow was found to be non-contact cooling water.

Based on the high percentage of non-contact cooling water present in the total wastewater flows, the most practical approach in treating process wastes and eliminating waste stabilizatin basin usage, was to segregate the process waste streams from the non-contact cooling waters, and discharge these wastes to the local municipal sanitary system if treatability studies proved favorable. Consequently, wastewater treatability studies wre conducted under batch reactor conditions to evaluate toxicity and biodegradability of the individualized industrial waste streams.

Treatability results indicated that following 24 hour equalization, the process wastes could enter the municipal facilities and not be expected to upset treatment plant operations. In fact, the local wastewater treatment facility which utilizes a two-stage, fixed film and suspended growth system (trickling filter and activated sludge processes) was found to provide certain advantages for the biological degradation of these industrial-organic process wastes. These advantages included: long mean cell residence time (particularly amenable to the development of microbial populations capable of oxidizing industrial process-related wastes), and inherent system resistance to shock loadings. Based on these treatability studies, a source control and waste segregation plan was adopted and implemented to separate various process waste and cooling water streams in order to minimize flow requiring on-site pretreatment and eventual discharge to the municipal treatment facilities.

Upon implementation of the waste segragation plan, the process sewers that had been discharging combined process, cooling and storm waters to the waste stabilization basin were rehabilitated by cleaning and relining with polyethylene inserts. Upon completion of this procedure, these former process lines were used to deliver non-contact cooling water flow, thereby eliminating the need to construct new cooling water lines. To insure that this proposed plan could be implemented, the existing sewers leading to the waste stabilization basin were first internally inspected with closed circuit television equipment.

To allow for the completion of construction activities associated with the above designs, the Phase I and II project work was divided into three contracts exclusive of internal plant piping modifications and pretreatment facility modifications. Contract No. 1 included the construction of a new cooling water interceptor tying together the former process wastewater lines (to be decontaminated under Contract No. 3 and used for cooling water); the installation of additional piping required to deliver the segregated process flows from the plant buildings to the appropriate process interceptor lines and the construction of a separate process wastewater trunk line leading to the pretreatment facilities which were constructed under Contract No. 2. Contract No. 3 included piping rehabilitation, which involved: decontamination of existing lines as described above; connection of the cooling water and process wastewater lines to their respective interceptor lines; and the installation of a cooling water monitoring station.

WASTE STABILIZATION BASIN REMEDIATION

Coincident with the design of the pretreatment facilities, an investigation was undertaken to characterize the waste stabilization basin contents and determine whether it was a source of groundwater contamination. The first step of this investigation was to analyze the basin contents and groundwater conditions in the immediate vicinity. Observation wells were drilled around the waste stabilization basin perimeter. Samples from both shallow and deep deposits were collected from the cluster wells in an effort to map potential vertical contaminant distribution. Once identifying the presence of groundwater contaminants at the basin perimeter, the next step of this investigation was to characterize the remaining basin aqueous layer and sludge contents for constituent similarities.

CHARACTERIZATION

Characterization of the aqueous layer was accomplished by obtaining samples from various locations in the basin and conducting priority pollutant analyses. These analyses indicated the presence of a two layer aqueous system consisting of organic contaminants, similar to those found in the preliminary groundwater investigations with the lower aqueous layer (adjacent to the basin sediments) exhibiting higher concentrations of each constituent than the upper aqueous layer.

Characterization of basin sediments was initiated by obtaining core samples from the waste stabilization basin sludge blanket. Visual inspection of the core samples revealed two to four feet of stratified layers of reddishbrown, resin-impregnated sludge overlying natural deposits of fine grained, wet, dark grey sand and silt; an easily identifiable line of demaracation defining the physical extent of the basin sediment/sludge layer.

In addition to the visual distinction of the two layers, the concentrations of lower molecular weight organic solvents and priority pollutant metals found in the sediments were consistently greater than those present in the underlying natural material. PCBs were detected in the top eight to twelve inches of two to four foot sediment/sludge layer that had accumulated at the bottom of the waste stabilization basin, but in no instances were PCBs found in the underlying natural material.

Upon completion of the basin characterization study, the source of the environmental contamination problem was defined to be the continued existence of the contaminated sludge layer resulting from years of accumulation due to settling of solids from storm water and process wastewater discharges. To eliminate the continued transmission of organic contaminants to the natural underlying strata below the basin, a remedial program was developed and implemented.

Review of Remedial Alternatives

Based upon the above-described characterization, a number of remedial alternatives addressing corrective actions required to prevent the release of contaminants from the basin's accumulated sludge layer were identified:

1. Solidification/Fixation of Basin Contents and On-Site Landfill.
2. Incineration of Basin Contents.
3. In-Place Containment of Basin Contents with Isolation from Groundwater.
4. Excavation and Removal of Basin Contents for Off-Site Disposal in a Secure Landfill.

Program Selection and Implementation

After considering the available alternatives in terms of economics, compliance with current and anticipated regulations, short and longer term liabilities, ultimate site use, safety during construction, ease of implementation, erosion, surface water run-off, infiltration and site management, a remedial program involving the excavation and removal of the waste stabilization basin contents to a secure landfill was selected.

The three (3) major components of this program included the installation of a cement/bentonite cap, disposal of the aqueous layer to the municipal treatment facilities following pretreatment, and excavation and removal of the basin sediment/sludge layer to an off-site, secure landfill.

Cement/Bentonite Cap Installation

Bench-scale investigations indicated that due to the nature of the chemical constituents present in the basin sediments, a potential volatilization problem would exist if the aqueous layer was removed without effectively covering the sediments. Installation of an interim cap was considered necessary as a control mechanism in limiting volatilization potential upon removal of the aqueous layer.

Installation of the cement/bentonite cap was accomplished by first installing a Dupont TYPAR stabilization fabric, followed by a 6-inch, cement/bentonite slurry layer. The stabilization fabric was weighted down through the aqueous layer of the basin in such a manner as to minimize disturbance to the existing sediment layer and stirring of bottom sediments into the aqueous layer solution, as it was unrolled across the full width of the basin.

Bentonite slurry was prepared by first mixing water and clay until the bentonite particles were fully hydrated and appeared homogeneous. Storage basins were installed and used on-site to hydrate the bentonite slurry for a minimum of 24 hours. Cement was added and thoroughly blended just before pumping the freshly mixed cement/bentonite slurry directly into the stabilization basin. The supply lines feeding the slurry into the basin were secured onto floating devices such that the lines would not drag along the basin bottom, and disturb the stabilization fabric.

Air monitoring conducted during the completion of the aqueous and sediment/sludge layer removal indicated that the safeguards employed were effective in limiting the volatilization of organics.

Aqueous Layer Pretreatment and Disposal

Based on the results of aqueous layer toxicity and biodegradability studies, discharge of basin supernatant to the municipal secondary treatment plant was selected as the most favorabe disposal option. To comply with General Pretreatment Regulations, pH adjustment was required. Furthermore, in accordance with aqueous layer treatability studies, a pretreatment scheme for solids removal was deemed necessary as a precautionary measure to prevent basin sediments, possibly containing PCB's, from being incorporated into the aqueous layer by entrainment and being discharged to the municipal treatment plant. Consequently, the required pretreatment system consisted of rapid mixcoagulant addition, flocculation sedimentation, mixed media filtration and carbon adsorption.

Following installation of the aqueous layer pretreatment facilities, but prior to commencing full-scale operation, a pilot program testing performance of the treatment units at the rated flow capacities was conducted to determine optimum coagulation dosage; required pH control; filter bed life; amount of suspended solids in the influent, effluent, waste sludge and backwash waters; and total PCB concentration in an 8-hour composite sample of the effluent stream. During this test period, all effluent, backwash waste and sludge were discharged back to the stabilization basin so as not to contribute to environmental contamination. Upon commencing full-scale operations, the pretreatment facilities were operated to achieve a removal of not less than 90 percent total suspended solids and an effluent pH of between 8.0 and 8.5. Monitoring of the effluent discharged to the municipal treatment plant indicated that this pretreatment system was effective in preventing the discharge of PCBs into the municipal treatment system.

Aqueous layer pretreatment residual management ensured that operations resulting from supernatant disposal would not contribute to environmental contamination. Waste sludges and backwash waters from solids removal operations were discharged to a sludge settling tank. Immediately prior to filter backwashing, the sludge tank was dewatered by decanting the tank supernatant back into the basin. Settled sludge from the bottom of the settling tank was then pumped to a thickener unit. When the water level in the thickener exceeded a specified limit, supernatant was decanted to the basin. Sludge in the thickener was disposed of in the same manner as the basin sediments.

Sediment/Sludge Layer Excavation

Excavation of basin sediments was accomplished by scraping back the cement/bentonite cap off the stabilization fabric and forming a berm around the newly exposed area to be excavated. The basin sediment/sludge was excavated and stockpiled on a portion of the basin site to allow for dewatering, thereby limiting exposure to the boundaries of the basin site. The sediment dewatering piles were comvered to limit sediment volatilization during non-working hours or whenever it was deemed necessary due to temperature or wind conditions. Dewatering during excavation and backfill was accomplished by lowering sump pumps into the excavated area and removing groundwater and run-off draining back into the excavation area from sediment dewatering piles. These volumes were pumped and stored in a second sump area and were treated by the existing aqueous layer pretreatment facilities. Additionally, all rainfall coming in contact with any portion of the basin was routed to this sump for treatment.

Sediment/Sludge Layer Removal

Following dewatering, the basin sediments were placed in impermeable, reinforced containers. These watertight containers were filled, weighed and placed in an on-site storage area for loading onto specially equipped, hazardous waste vehicles for transportation to an approved, secure landfill site. The reinforced sediment containers were lifted by crane, and lowered into each trailer truck until the desired weight limit of approximately twenty (20) tons was secured by bolting to the truck upper body with a gasketed seal designed to prevent spillage in the event of an overturning accident. The transport vehicles were all equipped with sealed, lock down covers. The tailgate was bolted and sealed in the same manner to prevent any spillage in the event of an accident when in transit frm the loading site to disposal site. Convoys consisting of 8-10 trailers traveled to the disposal site accompanied by a radio equipped, truck carrying a load of sand, extra covering tarps and clean up tools. An experienced supervisor accompanied each loaded convoy; therefore, in the event of an accident appropriate emergency measures would be taken immediately. Upon arrival at the disposal site, only the tailgate required unbolting for tipping. The trailers were then rebolted, ready for travel back to the loading site.

Hard hats, steel-reinforced protective boots, safety goggles, neoprene rubber golves, neoprene rubber boots or overshoes, non-porous disposal coveralls and respirators with replacement filters to provide protection against organic vapors were required at all times while working at the excavation site. In addition, all items worn during the course of excavation were left at the site at the end of each day, and all soiled coveralls were disposed of by drumming.

To avoid environmental contamination during implementation, all equipment contacting the aqueous layer or sediment/sludge material was cleaned by flushing or swabbing contacted surfaces with an approved solvent prior to equipment removal from the site. The waste solvents were then disposed of in drummed containers. All equipment contacting basin material were stored in the segregated, stockpile area when not in use. No equipment was removed from the site without solvent cleaning. The aqueous layer treatment units and all auxiliary equipment were also cleaned at the conclusion of the project to remove all visible traces of sludge, resin, scum, or other residue. All material cleaned from the units, as well as solvents or other material used in the cleaning operation, were disposed of by packing into waste drums. The filter media was also drummed for disposal.

Implementation of the waste stabilization basin site closure was accomplished through the development and administration of three separate contracts. Contract No. 1 included all construction work necessary for the installation of a stabilization fabric followed by a 6-inch layer of prehydrated cement/bentonite slurry through the existing basin aqueous layer. The project also included the construction of storm water diversion facilities at areas around the basin. Contract No. 2 involved the instalation and maintenance of two (2), self-contained, 20 gpm treatment plants; WATER BOY Model WB-27, manufactured by Neptune Micro-FLOC, Inc., Corvalis, Oregon, for the treatment of waste stabilization supernatant. Contract No. 3 included the excavation of the basin sediment layer, placement of the material in containers, transport and disposal ina secure permitted landfill and subsequent backfilling in a prescribed manner with standard process, run of bank gravel, topsoiled and seeded.

SUMMARY

In summary, an effective source control and waste segregation program was implemented, and a pretreatment facility designed, constructed and is now in operation. Upon removing the waste stabilization basin from service, a characterization program was conducted which identified the existence of a contaminated sediment/sludge layer resulting from years of sediment deposition due to waste stabilization basin activity. This solids layer was overlain by a contaminated aqueous layer and had to be removed prior to removal of the sediment/sludge layer. O'Brien & Gere Engineers, Inc., identified environmentally-sound remedial options and upon evaluation of those alternatives. The selected approach was the excavation and disposal of these solids in an off-site secure landfill. The site was closed in June, 1981, thereby eliminating the waste stabilization basin as a source of environmental contamination.

CHAPTER 28

SITE SAFETY AND SAMPLING PLANS - THE FIRST STEP IN INVESTIGATING ABANDONED HAZARDOUS WASTE DISPOSAL SITES

John W. Edwards, Vernon M. Reid, and Paul B. MacRoberts
Black & Veatch, Engineers-Architects

ABSTRACT

The development of comprehensive site-specific safety and sampling plans is a prerequisite to field investigations of abandoned hazardous waste disposal sites. Preparation of safety plans requires consideration of natural hazards, as well as those posed by hazardous waste materials. Emergency plans are needed to ensure an automatic, immediate response by on-site personnel in the event of any of a number of possible accidents. The sites are divided into Hot, Decontamination, and Support Areas. Safety equipment, protective clothing, and safety procedures are prescribed for each area.

A sampling plan is needed to ensure that information collected during the investigation is pertinent, reproducible, accurate, representative, and accountable. The sampling plan details the steps required to have data of known and documented quality.

INTRODUCTION

Thee safest way to investigate an abandoned, uncontrolled hazardous waste disposal site is to do no field work. However, field investigations are needed to obtain information on the current environmental situations on and near abandoned sites. Given this reality, every reasonable precaution must be taken to protect the personnel performing the field work. This exposure to risks and hazardous substances must be minimized and unnecessary risks eliminated. Steps that can be taken to accomplish this include developing and implementing standard operating procedures, providing formal training of personal safety gear. This paper addresses site-specific standard operating procedures: site safety and sampling plans.

Separate safety and sampling plans should be prepared for each field investigation of abandoned, uncontrolled hazardous waste disposal sites. Just as the safety and sampling aspects of field

Management of Toxic and Hazardous Wastes, H. Bhatt, R. Sykes, T. Sweeney (Editors)

work are inseparable, the development and contents of the two plans are not mutually exclusive. The plans should not be elaborate; rather, they should be concise and written in plain, easily understood terminology and presented in a direct manner so that nothing is open to interpretation.

Black & Veatch has generic Safety and Sampling Manuals that present the firm's general policies and procedures for field investigation of hazardous waste sites. The site-specific plans are prepared following the guidance in these manuals and incorporate by reference the applicable general procedures in order to reduce the bulk of the plans. Development of the sampling plan is initiated prior to the safety plan because safety procedures are based on the type of work to be performed. However, after work on both plans has started, an iterative process takes place so that personnel safety and the purpose of the field investigation are not compromised. Because work on the sampling plan is started first, it will be discussed first.

SITE SAMPLING PLAN

Purpose of the Site Sampling Plan

Sampling and monitoring activities at hazardous waste sites produce data, information, and materials that may be used, among other things, as evidence in envorcement actions and legal proceedings. The evidence may be in several forms, including field logbooks, samples, photographs, and analytical results. Security and accountability of evidence must be maintained from the time it is collected. In addition, the evidence must be of known and documented quality. For these reasons, field investigations must be designed. The sampling plan establishes and presents this design.

The Site Sampling Plan details the course of the sampling and monitoring activities to be accomplished at a specific hazardous waste disposal site. The plan is based on specific objectives identified by previous site work or the project's scope of work. The plan is subjected to a quality assurance review prior to initiation of field activities. Field personnel are instructed to perform the sampling and monitoring activities in strict accordance with the plans to ensure that the collected information is accurate, reproducible, representative, and accountable.

Content of the Site Sampling Plan

For initial site entry and preliminary assessment type investigations when little is known about the site, a simple on-to-two page sampling plan is all that is needed. The site name, site location, field personnel, schedule, access conditions, and sampling philosophy are identified in such a limited plan. The sampling philosophy can be stated as simply as the following: take photographs of any drums or piles containing potentially hazardous wastes; collect soil samples from any on-site areas stained by suspected hazardous wastes or leachate; and collect samples of any observed surface water leaving the site.

For subsequent investigations to obtain data on the extent and level of contamination, such as EPA Superfund remedial investigations, a more detailed plan is needed. Our company's sampling plans are comprehensive, but as often as possible the reader is referred to the Sampling Manual, previous reports on the site, and other documents to keep the plan brief, concise, and readable. Our company's sampling plans are typically divided into seven sections; however, sections may be added or deleted to fit the specific site conditions.

Plan Section 1: Introduction

The introduction to the site Sampling Plan includes a physical description of the site, site history, and other information relevant to the objectives of the sampling and monitoring activities. The introduction typically contains the following:

a. Site location information such as description of site location, regional and local maps showing location, site owner's name and address, and site owner's involvement.
b. Site description information such as topography, hydrology, geology, groundwater hydrology, adjacent land use, area, internal access roads, and a site map showing pertinent features of these.
c. Site history information such as description of past site usage and operations involving hazardous substances; description of internal areas known or suspected to contain hazardous substances; description of on-site areas known or suspected to contain hazardous substances; synopsis of past investigations, monitoring data, accidents involving hazardous substances, and public complaints: and a site map showing pertinent features of these.

Plan Section 2: Objectives

The objectives of the sampling and monitoring activities are stated on an activity and purpose basis, e.g., collect groundwater samples from monitoring wells to check for contamination changes.

The establishment of objectives is an important step because the objectives control the direction and tenor of the field investigation. On one project involving a formerly-cladestine hazardous waste land disposal site, the objectives of (and hence the design of) the field investigation came from the interim results of a feasibility study. Remedial action alternatives had been developed and evaluated as far as existing data and assumptions permitted. The objectives of that investigation were essentially to verify the feasibility study assumptions. The objectives for an investigation at another site were directed at obtaining design data on the extent of contaminate-soil excavation needed during clean-up.

The formulation of objectives is one of the steps where unnecessary risks can be identified and eliminated. An investigation at a third site was originally planned to include off-site surface water and sediment, waste impoundment, and above-ground tank sampling. The off-site and waste impoundment sampling required relatively low levels of protection and were performed from the banks of ditches and impoundments, while the tank sampling required higher levels of protection and special entry equipment. The general objective of the investigation was to ascertain if this site was responsible for off-site contamination. Because tank sampling did not fit this general objective and would have been so different from other sampling, it was deemed an unnecessary risk an postponed to a later investigation.

Plan Section 3: Key Personnel

The names, organizations, and telephone numbers of certain key individuals are identified, such as the client contact person, the individual responsible for site access, the project manager, the project engineer, the field supervisor, and the public information liaison.

Plan Section 4: Field Activities

Sampling or monitoring locations for field activities are identified both on figures and in tables. On maps or sketches, sampling locations and their distances from physical features are indicated. The tabular presentation includes a brief description of each location. It is important, especially for long-term investigations involving repeated samplings at the same location, that the specified location is sampled and the sampling station description is consistently used in documentation. Evidence from an early EPA investigation conducted by others was thrown out of court because inconsistent documentation indicated the same sampling location was not used in repeated sampling efforts, whereas the exact same location actually was used.

Samples or monitoring data to be collected are identified by type of sample or data in another tabular presentation. The tabulation includes the sample location, sample type, sample identification code, analyses to be performed, and shipment classification. Such tables are an aid to field personnel in determining how to package and ship the samples in accordance with DOT regulations.

The methods to be used to collect different types of samples or monitoring data, preserve samples, and clean sampling equipment are presented. This is usually accomplished by referring to the Sampling Manual.

Arrangements that have been made, and the resulting field procedures regarding site access and access to other properties are described. A schedule for field activities is also presented.

Plan Section 5: Personnel and Equipment Requirements

The project teams needed to perform different field activities are described by indicating the number of personnel involved and their function title.

Sampling container requirements are identified in two tabular presentations. The first indicates the number of each type of sampling container required for different types of samples. The second indicates the total number needed, including allowances for breakage, of each type of sampling container. Lists are presented of general sampling equipment requirements and of sampling equipment required for different types of samples.

Plan Section 6: Custody and Documentation Requirements

Procedures for chain-of-custody, logbooks, photographs, and other sample documentation are described, usually by referring to the Sampling Manual.

Plan Section 7: Control of Contaminated Materials

Control methods are described for the handling of contaminated materials generated by the investigation, such as disposable sampling equipment, soil cuttings, and decontamination fluids. A statement is included on the storage or disposal obligations and responsibilities for such materials.

SITE SAFETY PLAN

Purpose and Basis of the Site Safety Plan

The Site Safety Plan identifies the known and potential hazards at the sitte and describes the safeguards to be implemented to protect personnel from those hazards. It details procedures to be used in the event of accidents and emergencies.

The generic Safety Manual serves as the basis for all Black & Veatch safety plans. This manual establishes basic safety policies and procedures for field team organization and training, medical monitoring work area controls, site operations, personnel and equipment decontamination, laboratory safety, and crisis situation responses.

Content of the Site Safety Plan

Black & Veatch's safety plans are typically divided into eight sections; however, as for sampling plans, sections may be added or deleted to fit specific site conditions.

Plan Section 1: Introduction and Objectives

This section provides information on the site including the following:

a. Background information such as site history, including hazardous materials produced or stored on the site; previous investigations or remedial actions; and health problems attributed to the site.
b. Hazardous materials information such as locations and types of hazardous wastes and contaminated media (soil, groundwater, surface water, air).
c. Investigation objectives.
d. Safegy Plan Objectives, including guidelines, information, and rules for the protection of the health and safety of project personnel.

Plan Section 2: Key Project Personnel

The project manager, project engineer, field supervisor, site safety officer, and client liaison are identified by name, organization, and telephone number in this section.

Plan Section 3: Emergency Services

Telephone numbers and street addresses are provided for local ambulance, hospital, poison control, police, and fire department services. These are confirmed prior to their listing in the plan. Also, these agencies are notified to alert them of the potential need for emergency services.

Plan Section 4: First Aid Equipment and Locations

The locations of first aid kits, eyewashes, personnel shower facilities, fire extinguishers, and other safety equipment are identified. The use of all equipment is addressed in the personnel training programs.

Plan Section 5: Site Hazards

Respiratory, dermal, and general physical hazards are documented. Respiratory hazards may include windblown dust laden with contaminants, organic vapors, or acid fumes. Substances may be present that affect the skin or enter the body through the skin. Physical hazards may include slips, trips, falls, cuts, snake bites, explosions, and the personal protection equipment itself. Personal protection equipment may impair sight, smell, hearing, taste, and feeling and may restrict movement and increase fatigue. A standard caveat in the plan is that the wearer may become overconfident in the protection afforded.

Plan Section 6: Site Layout

Sites are divided into three basic work areas: the Hot, Decontamination, and Support Areas. A description and map are provided as to the function and location of each area.

The on-site field investigations take place in the Hot Areas. The Hot Area may be subdivided for different levels of protection. There may be more than one Hot Area. Access to the Hot Area is restricted: only those individuals performing sampling and other specified activities are permitted.

The Decontamination Area is normally located on-site near the main entry point. An equipment washing station, a personnel washing station, emergency eyewashes, toilets, and a first-aid station are typically located in the Decontamination Area. Entry into and exit from the Hot Area for all personnel and equipment, except in emergency situations, is made through this area.

The field office, personnel staging area, equipment staging and storage areas, and rest area are contained in the support area. Entry into and exit from the Decontamination Area for all personnel and equipment, except in emergency situations, is made through this area.

Plan Section 7: Safety Equipment and Requirements

Personal protective equipment is identified for each Hot and Decontamination Area. Specific safety requirements are presented and field operating procedures are specified, e.g., air monitoring requirements.

Plan Section 8: Emergency Procedures

Emergency procedures for major and minor accidents in the Hot, Decontamination, and Support Areas are detailed.

A major incident is defined as one that potentially poses an immediate threat to life, limb, or health of one or more individuals, e.g., through explosion, fire, excessive concentrations of organic vapor, or equipment accident. Directions are outlined in the plan for the notification of other on-site personnel and the proper authorities in regard to the emergency and for the provision of aid to affected personnel. Emergency escape routes are delineated. A post-emergency personnel gathering location is specified so that head-counts can be taken.

Minor incidents are those that pose no immediate threat to life, limb, or health but do require corrective action be taken, e.g., sprain, cut, equipment malfunction. Appropriate response actions for minor incidents are described.

Implementation of the Plan

Prior to entering the site, all project personnel are required to read the Site Safety Plan and to sign an affidavit certifying that they have read, understood, and will abide by the Site Safety Plan. Before initiating site activities, a formal presentation of the Site Safety Pan is made to all project personnel to ensure a complete understanding of the plan. The plan is posted at conspicuous locations on the site, with at least one copy located in each of the tree work areas.

CONCLUSION

Safety of personnel is the utmost concern during the performance of sampling and monitoring activities on or near hazardous waste disposal sites. Interaction is needed between the safety and sampling planning efforts so that personnel safety is not compromised by field activities and the validity of the activity is not compromised.

Sampling and monitoring activities performed at hazardous waste disposal sites are potentially dangerous to the people involved, yet are performed because of project objectives. Sit sampling plans are developed and employed to ensure that a quality product will result. The safety and sampling plans are designed to protect the field personnel. These plans are only as effective as the people who implement them. Each member of the project team is responsible for the safe achievement of the quality product.

CHAPTER 29

REMEDIAL INVESTIGATION AND FEASIBILITY STUDY - TACOMA WATER SUPPLY WELLS COMMENCEMENT BAY AREA, TACOMA, WASHINGTON

Mark G. Snyder and Paul B. MacRoberts
Black & Veatch, Engineers-Architects

PROJECT DESCRIPTION

Contamination of groundwater drinking water supply wells in Tacoma, Washington was one segment of a hazardous waste contamination study in the Commencement Bay Area of Tacoma, Washington. Black & Veatch, Engineers-Architects, with geotechnical consultant services provided by Woodward-Clyde Consultants, implemented a remedial investigation at the Tacoma well 12A site for the U.S. Environmental Protection Agency [1]. This investigation was designed to support preliminary feasibility study work performed by Black & Veatch.

Major components of the remedial investigation and preliminary feasibility study included:

1. Installation of 13 groundwater monitoring wells ranging in depth from 58 to 199 feet.
2. Installation of four soil borings to a depth of 30 feet.
3. Collection and analysis of groundwater samples from the monitoring wells and existing public and private wells.
4. Description of geologic and hydrogeologic characteristics in support of identification of the extent and magnitude of aquifer contamination.
5. Development and screening of remedial action alternatives to mitigate contamination of the affected water supply wells.

BACKGROUND

The Tacoma City well 12A study area encompasses production well 12A and vicinity, as shown on Figure 1. Well 12A is one of 13 wells in the South Tacoma Well System, a 21 mgd firm yield groundwater supply providing an estimated 15 percent of the City of Tacoma's water supply [2]. Well 12A is 167 feet deep, has a

Management of Toxic and Hazardous Wastes, H. Bhatt, R. Sykes, T. Sweeney (Editors)

30-inch diameter screen installed between depths of 141 and 167 feet, and has a reported capacity of 4,000 gpm.

Chlorinated hydrocarbons were detected in groundwater samples from well 12A on July 24, August 21, August 27, and September 15, 1981. The range of concentrations for the detected contaminants is listed below:

1,1,2,2,-Tetrachloroethane	17 to 300 ug/l
1,2-Trans dichloroethylene	30 to 100 ug/l
Trichloroethylene	54 to 130 ug/l
Tetrachloroethylene	1.6 to 5.4 ug/l

Groundwater sampling programs were also initiated at wells 9A and 11A during the latter part of the summer of 1981. The contaminants which were detected in well 12A were also present in groundwater samples from well 9A but at much lower concentrations.

REMEDIAL INVESTIGATION

Approach

The Remedial Investigation was developed in a phased approach so that data obtained during initial phases could be used to site the monitoring wells to be installed in subsequent phases. The sites for the six Phase I wells were selected to give information on the direction and magnitude of groundwater movement, to provide information on stratigraphy and lithology to supplement the geologic model, and to provide water quality data for an evaluation of contaminant migration.

The screened interval for the Phase I wells was selected based on the geology and organic vapor analyzer (OVA) measurements of soil samples taken during the drilling program. Analysis of the OVA data from the first six wells indicated that the highest levels of contaminants were between elevations of 150 and 180 feet, mean sea level (msl).

The sites for the remaining six wells were selected one at a time as water quality data from the previous well became available. The screening interval for all five Phase II groundwater monitoring wells was in a high permeability zone where the highest contaminant concentrations were detected during the first phase of well installation. The water quality data for all groundwater samples were interpreted using simple groundwater and contaminant transport models. The purpose of the interpretation was to predict the possible locations of source areas so that subsequent wells could be located in the vicinity of a potential source area.

Conclusions

The primary conclusions resulting from the remedial investigation are listed below. Definitions of pertinent laboratory terminology and explanations of, as well as support, for the conclusions are embodied in subsequent sections of this paper.

1. The aquifer which currently supplies water for residential and industrial use for the City of Tacoma is contaminated.
2. Due to the expected proximity of the suspected primary source or sources of contamination in well 12A, contaminant concentrations will increase with time as the well is pumped.
3. Contaminant transport is predominantly in the horizontal direction. The approximate location of the zone is between the elevations of 150 and 180 feet.
4. The migration of contaminants in the vertical direction is impeded by low permeability layers of silts, silty or clayey sands, and "hardpan" which underlie the high permeability zone. The low permeability layers are discontinuous with zones of high permeability material extending through the low permeability layers.
5. Well 12A is screened at the bottom of what appears to be the contaminated high permeability zone and the upper portion of a deeper pervious zone.
6. The primary source or sources of contamination at wells 12A and 9A were not located during the remedial investigation.
7. The undisturbed pre-pumping or steady state groundwater gradient in the vicinity is in a west to east direction.
8. Analysis of available information and data indicates that high concentrations of contaminants are likely to be present in the aquifer north and east of the study well.
9. The boundaries of the contaminant plume are not well defined based upon data obtained during the investigation.

Air Quality Survey

The air quality in the vicinity of all monitoring wells and soil borings was monitored throughout the investigatin. A portable organic vapor analyzer (OVA) was used to monitor general air quality in the vicinity of field operations so that the level of respiratory and protective apparel could be specified.

Organic vapor levels were also obtained from extruded samples during the advancement of the borings. The soil samples were placed in glass jars for 15 to 20 minutes before OVA measurements were taken. The two soil samples from each well boring with the highest OVA measurements were sent to the EPA contract laboratory for analysis.

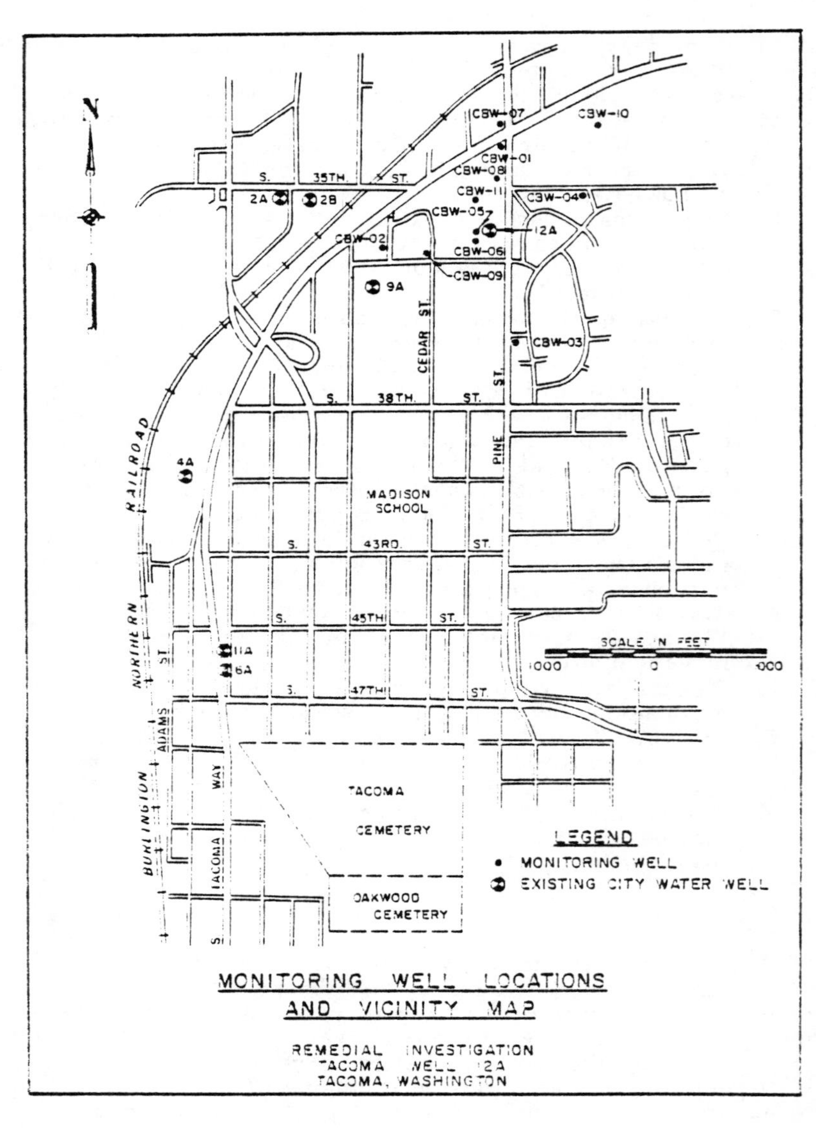

N
CBW-07
CBW-10
CBW-01
CBW-08
S. 35TH. ST.
CBW-11
CBW-04
2A
2B
CBW-05
12A
CBW-02
CBW-06
CBW-09
9A
CEDAR ST.
CBW-03
S. 38TH. ST.
PINE ST.
RAILROAD
4A
MADISON SCHOOL
S. 43RD. ST.
NORTHERN
S. 45TH ST.
SCALE IN FEET
11A
6A
S. 47TH ST.
ADAMS ST.
BURLINGTON
WAY
TACOMA
TACOMA CEMETERY
LEGEND
MONITORING WELL
EXISTING CITY WATER WELL
OAKWOOD CEMETERY
MONITORING WELL LOCATIONS AND VICINITY MAP
REMEDIAL INVESTIGATION
TACOMA WELL 12A
TACOMA, WASHINGTON

Figure 1.

Monitoring Well Installation

Eleven groundwater monitoring wells were installed during the field investigation (see Figure 1). Well locations were selected in conjunction with EPA Region 10 by Black & Veatch and Woodward-Clyde Consultants. The well installation activities were divided into two phases. Six wells were installed during Phase I and five during Phase II. The Phase I wells were installed from mid September to the first week in October, 1982. The Phase II wells were installed from mid-November to the first week of December, 1982.

The sites of the wells installed during Phase I were located in a quadrant arrangement to identify the general direction of contaminant migration. Monitoring wells CBW-01,02,03, and 04 were located approximately 1,000 feet north, west, south, and east of well 12A, respectively. Monitoring wells CBW-05 and 06 were also located approximately 20 feet west of well 12A. These wells were located approximately 10 feet apart and were screened in different layers. The purpose of these wells was to investigate the zone through which contaminant transport was believed to be occurring.

The depth of the layer or layers which had the highest contaminant concentrations was unkown prior to the drilling of the Phase I wells. The monitoring well-screening intervals were selected based on the site geology and the results of OVA measurements.

Wells CBW-05 and CBW-02 were installed first, and both were screened below the first encountered low permeability layer. Groundwater samples were obtained from the borings prior to completion as wells. The samples were taken from the high permeability zone above the low permeability layers. The groundwater samples from the borings were sent to the EPA Region 10 laboratory at Manchester, Washington for the analysis of volatile organics. Groundwater samples wre subsequently obtained from the completed wells, split, and submitted to the EPA Manchester laboratory for anlysis of volatile organics and to the EPA contract labs.

Monitoring well CBW-03 was installed to a depth of 199 feet. No "hardpan" or distinct low permeability zone was encountered during drilling. A groundwater sample was obtained prior to completion of the well from a high permeability zone at approximately the same elevation as the groundwater samples taken from well borings CBW-02 and 05. The well screen was installed in the bottom 20 feet of the boring.

Monitoring wells CBW-04 and CBW-06 were installed in the high permeability zone above low permeability layers. Groundwater samples were obtained and submitted for analysis of volatile organics. The wells were screened in the high permeability zone.

A review of the boring logs, water quality data from the EPA Manchester laboratory, and OVA measurements indicated that the zone with the highest contaminant concentrations was the high permeability zone in which wells CBW-04 and CBW-06 were screened at an elevation of approximately 150 to 180 feet. All subsequent wells were screened in this high permeability zone.

The sites of the monitoring wells installed during the second phase were located one at a time as volatile organics data from the Phase I and Phase II wells became available from the EPA Manchester laboratory. Monitoring well CBW-07 was located in an outwash channel to provide water quality data from the deeper pervious zone of the aquifer. A groundwater sample was obtained approximately 2 feet above the bottom of the boring and analyzed for volatile organics prior to well completion.

Based on the results of the volatile organics analyses from monitoring wells CBW-01 through CBW-07, monitoring well CBW-08 was located just south of a reported solvent-barrel-washing operation. This site was selected because the previous water quality data indicated that the suspected primary source or sources were located northeast of the well 12A site. Since the contaminant concentrations in the groundwater samples from borings CBW-01 and 07 were low, it was postulated that contaminant migration between the barrel-cleaning site and the outwash channel might be impeded because of a geologic structure. Well CBW-08 is located on the edge of the bluff.

The depth of well CBW-08 was selected so the well screen would be at approximately the same elevation as wells CBW-04, 06, 07, in a high permeability layer where the highest contaminant concentrations had been detected. Based on EPA analytical results of the volatile organics, the well screen was set in the high permeability layer immediately above the low permeability layer, and the well was completed.

The volatile organics analyses from well CBW-08 indicated relatively high values (500 ug/l for 1,1,2,2-tetrachloroethane). An analysis of the data collected to date indicated that a source was either very close to the west of or to the northeast of well CBW-08. The sites for wells CBW-09 and CBW-10 were located on either side of well CBW-08 to confirm the direction of contaminant migration. The planned screened intervals for both wells were at approximately the same elevation as the earlier installed monitoring wells CBW-04, 06, 07, and 08.

The boring for well CBW-09 was advanced to a depth of 161 feet, and a groundwater sample was obtained from a depth of approximately 158 feet. The results of the volatile organics analysis for this well were low by comparison to well CBW-08 (11 ug/L for 1,1,2,2-tetrachloroethane). This indicated that the area with the highest probability of containing the source was either very close to well CBW-08 or northeast of well CBW-08.

Permission to drill on private property for well CBW-10 was not granted until after well CBW-09 was completed. Therefore, well

CBW-11 was located and begun prior to well CBW-10. The boring for well CBW-11 was advanced to a depth of 169 feet and a groundwater sample was obtained from the boring at a depth of approximately 166 feet. The results of the volatile organics analyses confirmed the earlier indications that the area with the highest probability of containing the suspected primary source or sources was northeast of well CBW-08.

Monitoring well CBW-10 was then installed approximately 1,000 feet northeast of well CBW-08. The well screen was installed in what is believed to be the same high permeability layer indicated earlier.

Boring Construction

The borings for the 11 monitoring wells were advanced using an air rotary system. The boring was advanced with a 5-7/8-inch tri-cone roller bit. The hole was cased during the drilling process with a welded 6-inch-ID steel casing to stabilize the boring walls.

The borings were then completed as groundwater monitoring wells using 4-inch commercially manufactured PVC slotted (0.010-inch slot) well screen and threaded flush-coupled PVC riser pipe. The steel casing was partially removed from the hole and cut above the top of the PVC pipe, and a locking cap was installed.

Well Development

Each well was developed and flushed to remove possible contaminants introduced during installation and to remove suspended particulate from the screen area prior to sampling. All wells, with the exception of CBW-01, were developed and flushed with a submersible pump. The pump was lowered into the well attached to a steel cable linked to a portable sampling tower equipped with a winch. Monitoring well CBW-01 was developed and flushed using a point-source bailer. The volume of water removed from each well during development and flushing was approximately six to twelve well volumes until the effluent was clear.

Groundwater Samples from Well Borings

During the drilling of each well, one water sample was obtained and delivered to the EPA Manchester laboratory for analyses for volatile organics. This sample was used as an indicator to support siting of subsequent wells.

These samples were obtained by lowering a 3-inch diameter point-source bailer to the desired depth and slowly raising to the surface. One bailer was dedicated to each well.

Results of the Groundwater Samples from Well Borings

The reported concentrations of trichloroethylene, 1,1,2,2-tetrachloroethane and tetrachloroethylene, and 1,2-trans dichloroethylene for the majority of the groundwater samples from the borings at wells CBW-01 through CBW-11 were in excess of levels referenced in the water quality criteria [3].

A comparison of reported concentrations from the boring water samples and those from groundwater samples from the completed wells noted that concentrations were generally higher in the boring water samples, but the two sets of results were comparable. Also, contamination was reported in the boring water sample from well CBW-01, but later groundwater samples were reported free of contamination.

For these reasons and because the borings were not developed and flushed prior to sampling, results from analyses of boring water samples were considered to have significantly less importance than those from groundwater samples.

Groundwater Samples from Completed Wells

Groundwater samples were obtained from the 11 monitoring wells installed during this program and from existing public and private wells.

Each of the 11 monitoring wells was sampled from the bottom of the well using the point-source bailers. Each well was sampled within 24 hurs of flushing. Before sampling, the water level was measured with an electronic water-level indicator.

The reported concentrations of trichloroethylene, 1,1,2,2-tetrachloroethane and tetrachloroethylene, 1,2-trans dichloroethylene, dibromochloromethane, chloroform, napthalene, and pentachlorophenol for the groundwater samples from wells CBW-01 through 11, 9A, 12A, T10, and the two private wells were in excess of published water quality criteria [3].

Decontamination

Equipment was decontaminated throughout the investigation to control the movement of contaminants to clean areas, to prevent cross contamination of bore holes and samples, and to assist in maintaining the health and safety of personnel. The decontamination of drilling and sampling equipment, PVC pipe, screen, and well casing was performed prior to placement of each of the borings.

Geology of the Study Area

The geologic model for the study area includes three types of general hydrogeologic units:

- highly permeable, predominantly gravel intervals;
- less permeable sand and silt intervals; and
- low permeability layers with a high percentage of silt and clay.

The areal extent and approximate thickness of these three types of hydrogeologic units can be distinguished in the study area through the construction of schematic cross sections of the study area. Figures 2 and 3 show north-south and east-west schematic cross sections, respectively, through the study area. These cross sections were created by:

- designating predominantly gravel layers as the highly permeable hydrogeologic units;
- combining predominantly sand and silt (with some gravel) layers into the intermediate permeability hydrogeologic units; and
- designating the intervals with high silt and clay content as the low permeability hydrogeologic units.

Simulations of Aquifer Behavior

The behavior of the aquifer in response to pumping stress was simulated to examine the possible behavior of the contaminant plume as a result of the pumping-induced reversal of the undisturbed potentiometric head gradient. The numerical simulations were also performed to help identify the area with the highest probability of containing the contamination source or sources. Results from these simulations were instrumental in helping to explain the observed contaminant distribution and in identifying the probable source area.

Two pumping schemes were simulated:

1. The pumping of well 9A during the summer of 1982.
2. The pumping of wells 9A and 12A during the summer of 1981.

During and immediately following the pumping of both wells 9A and 12A during the summer of 1981, the reported concentrations of contaminants in well 12A increased rapidly during the early stages of pumping and then increased at a lesser rate thereafter. During this time, well 9A remained free of reported contamination.

The modelling approach assumed that the aquifer is homogeneous, isotropic, and infinite in areal extent.

The computer program used for simulation is based on an

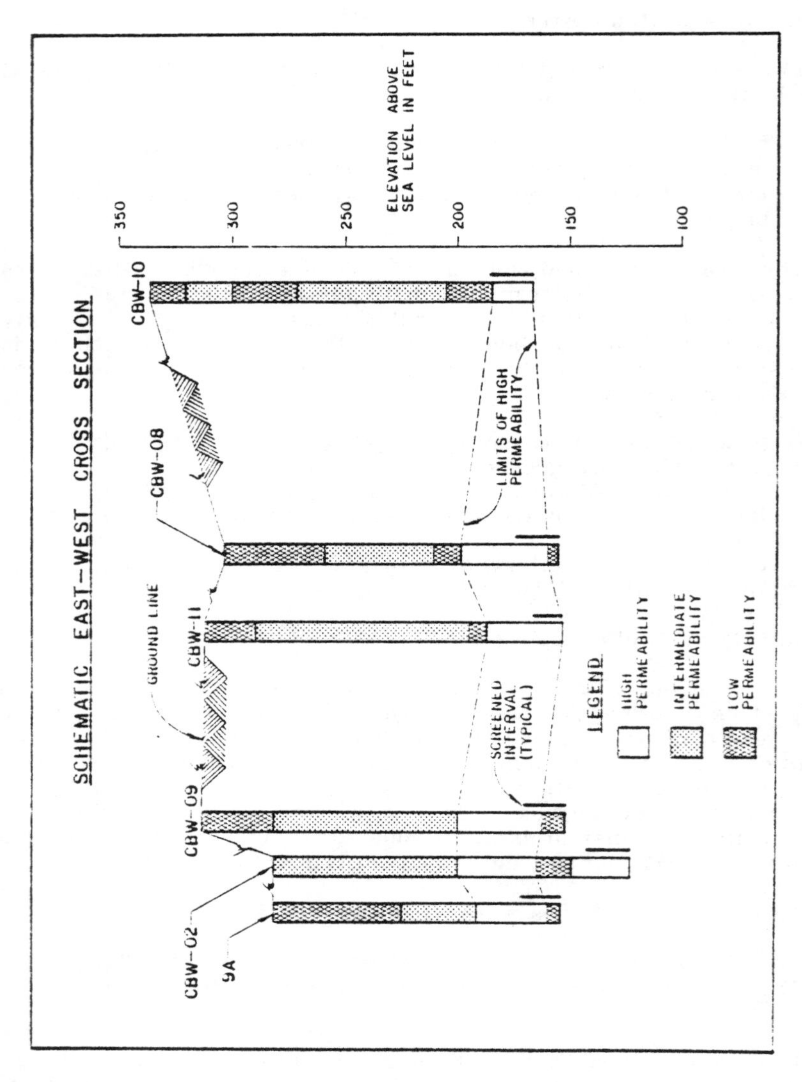

SCHEMATIC EAST–WEST CROSS SECTION
CBW-02
9A
CBW-09
GROUND LINE
CBW-11
CBW-08
CBW-10
SCREENED INTERVAL (TYPICAL)
LIMITS OF HIGH PERMEABILITY
LEGEND
HIGH PERMEABILITY
INTERMEDIATE PERMEABILITY
LOW PERMEABILITY
350
300
250
200
150
100
ELEVATION ABOVE SEA LEVEL IN FEET

FIGURE 2

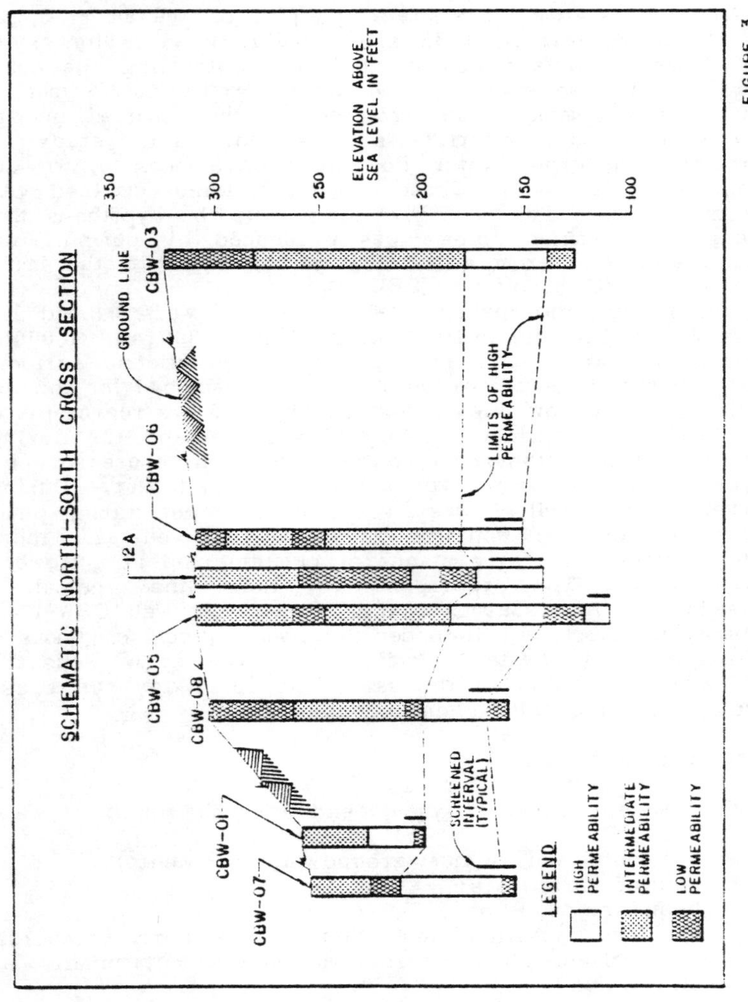
SCHEMATIC NORTH-SOUTH CROSS SECTION
CBW-07
CBW-01
CBW-08
CBW-05
12A
CBW-06
GROUND LINE
CBW-03
350
300
250
200
150
100
ELEVATION ABOVE SEA LEVEL IN FEET
SCREENED INTERVAL (TYPICAL)
LIMITS OF HIGH PERMEABILITY
LEGEND
HIGH PERMEABILITY
INTERMEDIATE PERMEABILITY
LOW PERMEABILITY

FIGURE 3

analytical solution to the Theis [3] equation describing groundwater flow. The model computes drawdowns in the aquifer due to pumping for any number of pumping and observation wells.

The results from the simulated pumping of well 9A at 3,300 gpm for 12 weeks during 1982 are on Figure 4. The simulated potentiometric surface was created by subtracting the computed drawdowns from an assumed static water elevation of 225 feet. This assumed static water level corresponds to the observed undisturbed steady-state potentiometric surface for the study area. Representative groundwater flow lines have been drawn on the potentiometric surface. These flow lines, when combined with the contaminant plume data on Figure 5, support the hypothesis that the observed contaminant plume was influenced by pumping-induced groundwater flow from a source area or areas in the immediate vicinity or northeast of well CBW-10.

The results from the simulated pumping of wells 9A and 12A for 12 weeks furing the summer of 1981 at 3,300 and 4,000 gpm, respectively, are on Figure 6. the potentiometric surface was created using the same procedure as that described for the well 9A simulation. the flow lines shown on Figure 6 are representative of the observed contaminant history at wells 9A and 12A during the summer of 1981 if the major contamination source area or areas is in the vicinity of CBW-10 or to the north and east. During the summer of 1981, well 9A was free of contamination during pumping, while the reported contamination level in well 12A increased rapidly during the early stages of pumping and at a slower pace thereafter. The representative flow line indicate that contamination in a source area or areas near well CBW-10 would probably not reach well 9A under the given pumping conditions. The approximate groundwater velocity in the vicinity of wells CBW-10 and 12A is probably in the same 1 to 10 ft/day range as that calculated for the well 9A simulation.

Contaminant Migration

Contaminant transport will depend primarily upon:

- horizontal and vertical groundwater gradients;
- aquifer parameters;
- boundary conditions;
- location, magnitude, and extent of the source or sources;
- chemical and physical properties of the contaminants; and
- site geology.

In areas which are far removed from teh source, contaminant transport will depend primarily on the groundwater gradients, aquifer parameters, and the boundary conditions. However, in areas which are close to the source, contaminant transport may be more strongly affected by the characteristics of the source, the properties of the contaminants, and the site geology. Climate may

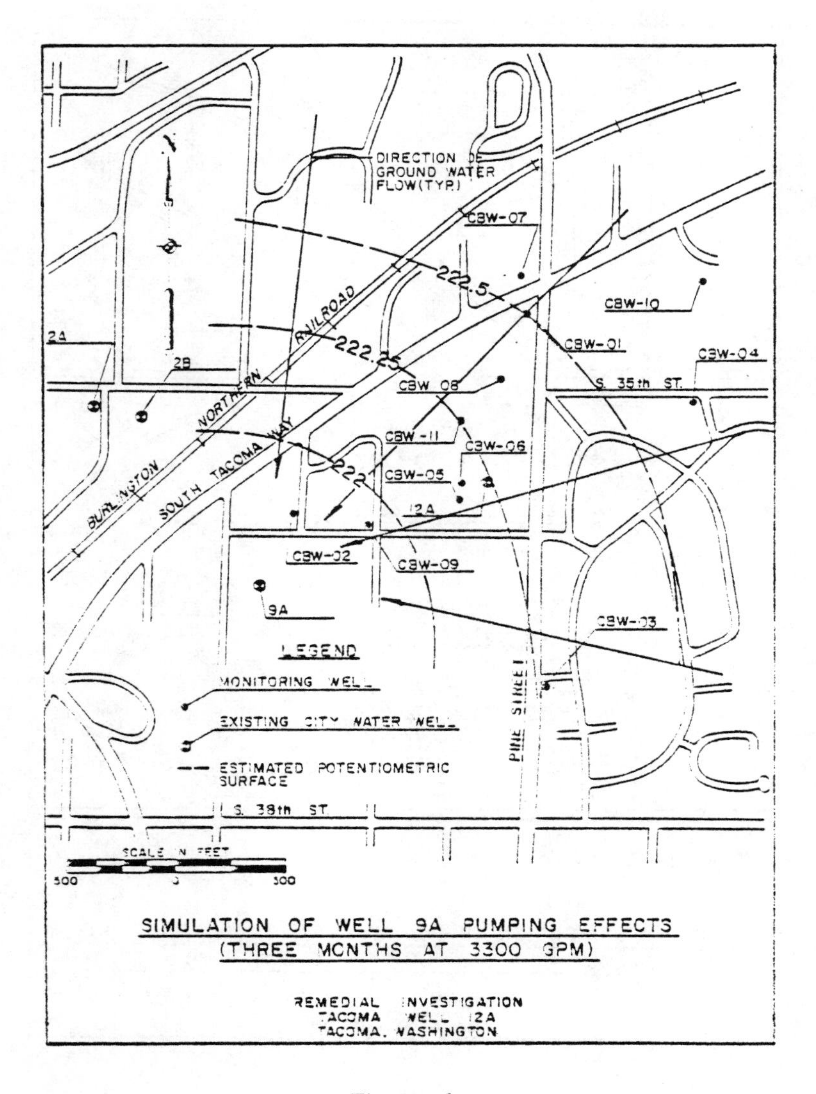
DIRECTION OF GROUND WATER FLOW(TYP.)
CBW-07
222.5
CBW-10
RAILROAD
222.25
CBW-01
CBW-04
2A
2B
NORTHERN
CBW 08
S. 35th ST.
CBW-11
CBW-06
BURLINGTON
SOUTH TACOMA WAY
222
CBW-05
12A
CBW-02
CBW-09
9A
CBW-03
LEGEND
MONITORING WELL
EXISTING CITY WATER WELL
ESTIMATED POTENTIOMETRIC SURFACE
PINE STREET
S. 38th ST.
SCALE IN FEET
300
0
300
SIMULATION OF WELL 9A PUMPING EFFECTS
(THREE MONTHS AT 3300 GPM)
REMEDIAL INVESTIGATION
TACOMA WELL 12A
TACOMA, WASHINGTON

Figure 4.

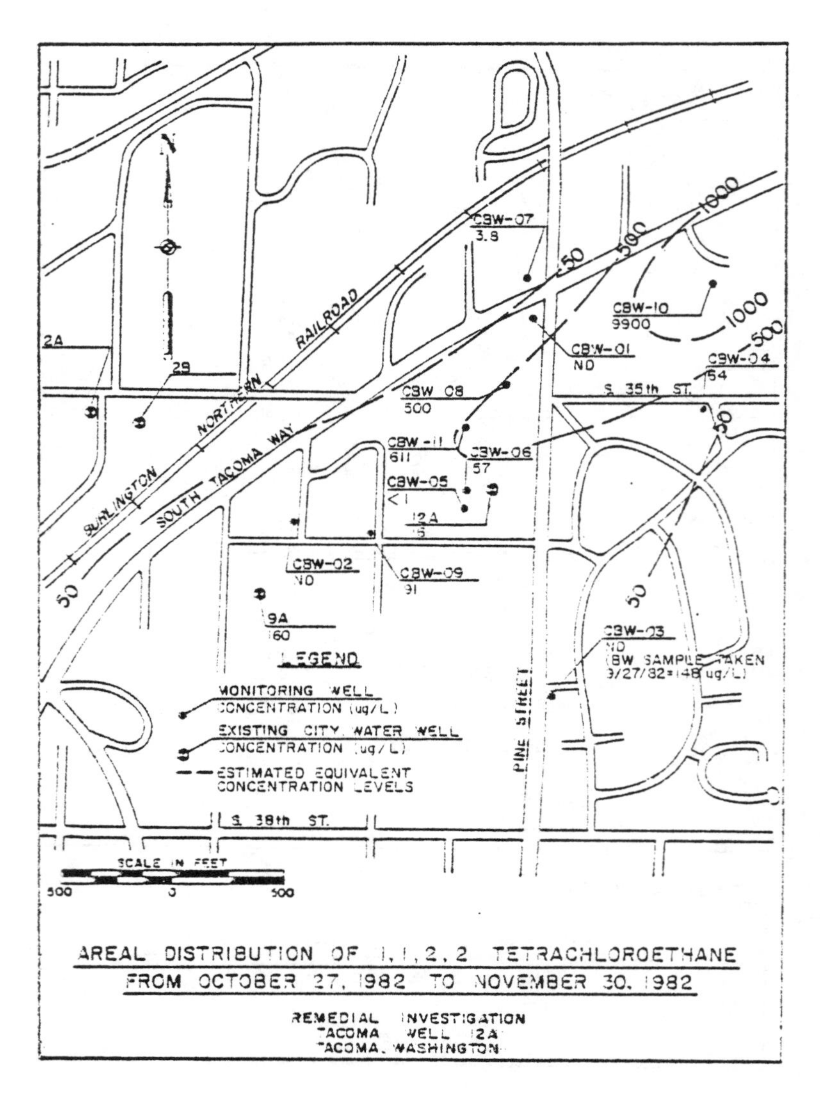

CBW-07
3.8
CBW-10
9900
CBW-01
ND
CBW-04
54
CBW 08
500
CBW-11
611
CBW-06
57
CBW-05
<1
12A
CBW-02
ND
CBW-09
91
9A
160
CBW-03
ND
(BW SAMPLE TAKEN 9/27/82=148 ug/L)
2A
29
1000
500
50
BURLINGTON NORTHERN RAILROAD
SOUTH TACOMA WAY
S. 35th ST.
S. 38th ST.
PINE STREET
LEGEND
MONITORING WELL CONCENTRATION (ug/L)
EXISTING CITY WATER WELL CONCENTRATION (ug/L)
ESTIMATED EQUIVALENT CONCENTRATION LEVELS
SCALE IN FEET
500 0 500
AREAL DISTRIBUTION OF 1,1,2,2 TETRACHLOROETHANE FROM OCTOBER 27, 1982 TO NOVEMBER 30, 1982
REMEDIAL INVESTIGATION
TACOMA WELL 12A
TACOMA, WASHINGTON

Figure 5.

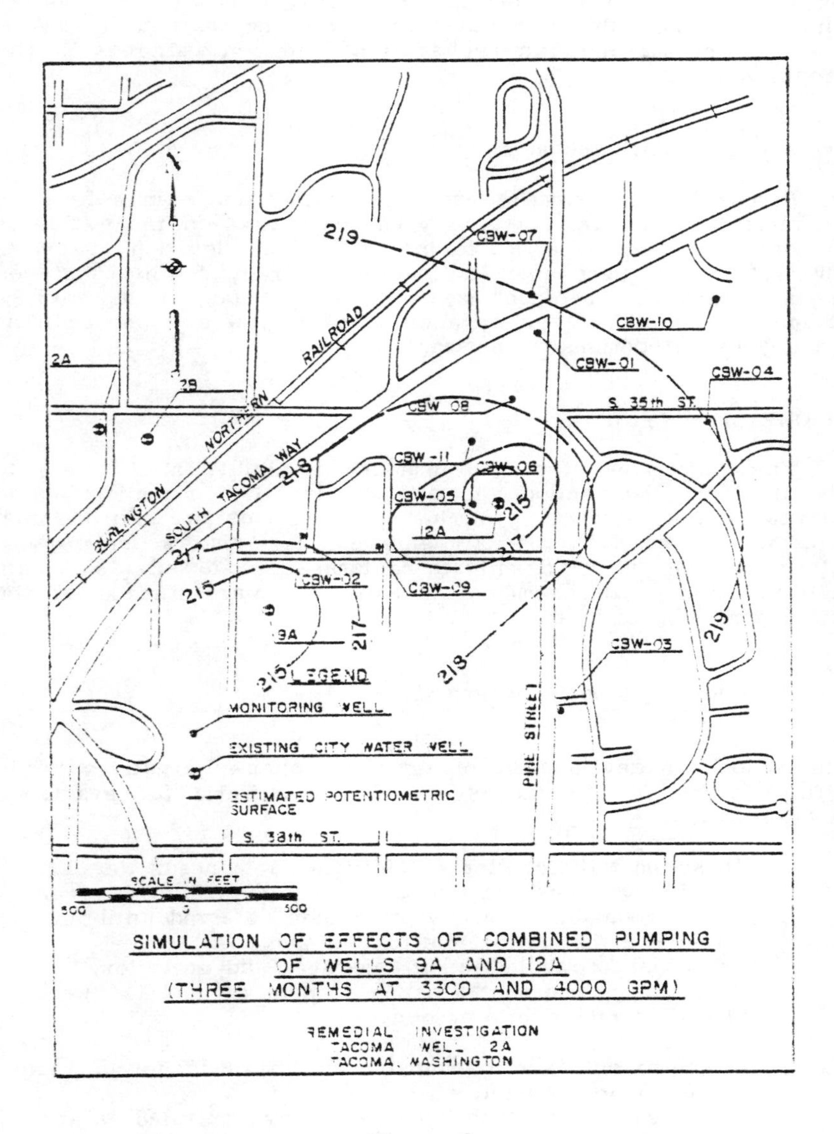

CBW-07
CBW-10
CBW-01
CBW-04
S. 35th ST.
CBW 08
CBW-11
CBW-06
CBW-05
12A
2A
2B
BURLINGTON NORTHERN RAILROAD
SOUTH TACOMA WAY
CBW-02
CBW-09
9A
CBW-03
PINE STREET
219
218
217
215
LEGEND
MONITORING WELL
EXISTING CITY WATER WELL
ESTIMATED POTENTIOMETRIC SURFACE
S. 38th ST.
SCALE IN FEET
500
0
500
SIMULATION OF EFFECTS OF COMBINED PUMPING
OF WELLS 9A AND 12A
(THREE MONTHS AT 3300 AND 4000 GPM)
REMEDIAL INVESTIGATION
TACOMA WELL 2A
TACOMA, WASHINGTON

Figure 6.

also play an important role in the source area. Percolation of rainwater through the unsaturated soil above the water table may be the primary method for recharge of the contaminants in the groundwater.

Primary Transport Mechanism

Results from soil samples and from groundwater samples from the soil borings and wells strongly indicate that contamination is present in a highly permeable interval at an elevation of 150 to 180 feet. This permeable interval is generally confined between two less permeable "hardpan" layers or zones. Due to the presence of low permeability layers, groundwater flow and contaminant transport is predominantly horizontal.

FEASIBILITY STUDY

The objectives of the preliminary feasibility study were to identify remedial action objectives, identify remedial action alternatives, conduct a preliminary assessment of environmental impacts, and develop order-of-magnitude cost estimates for remedial alternatives which would mitigate contamination at well 12A. Data collected during the remedial investigation were utilized in the initial screening of alternatives.

Identification of Remedial Alternatives

The scope for development of remedial action alternatives began with selection of remedial objectives and criteria for evaluation of alternatives.

1. To support these efforts, additional data needs were identified. These data needs were reviewed to determine readily available data and other data deferred until the remedial investigation was further along.
2. Preliminary remedial action objectives and goals for mitigation of the contamination at wells 9A and 12A were identified taking into account:

   - the extent to which substances pose a danger to public health or the environment; and
   - the extent to which substances have migrated or are contained by natural barriers.
3. Preliminary technical, environmental, and economic criteria for the evaluation of alternatives were identified. Criteria considered included:

- reliability;
- implementability;
- technical effectiveness/efficiency;
- environmetal concerns;
- safety requirements;
- operation and maintenance requirements;
- present worth cost analysis;
- regulatory requirements;
- public acceptance; and
- timeliness.

Based upon the identified criteria and objectives, remedial alternatives were identified utilizing the following process:

1. Appropriate technologies were identified for the development of remedial alternatives, e.g., control, containment, handling, disposal, and treatment.
2. Remedial action alternatives were defined which incorporate remedial technologies identified above and which address the objectives and criteria developed earlier. Alternatives for onsite, offsite, and non-cleanup actions were developed, as appropriate, as well as a no-action alternative.
3. The identified site alternatives were screened in accordance with the NCP (Federal Register: March 12, 1982) by applying the following criteria:

   - order-of-magnitude cost;
   - significant adverse environmental effects;
   - adequate control and/or effectiveness;
   - accepted engineering practice and technical feasibility; and
   - timeliness.

The initial screening of alternatives produced two basic alternatives which could mitigate contamination at well 12A without specifically identifying the contaminant source or sources. Thus, the alternatives were limited to control and/or treatment of contaminated groundwater so that well 12A and other water supply wells in the vicinity could be protected from further contamination and could potentially provide potable water for the city. The two alternatives were:

1. Groundwater pumping or other hydraulic controls.
2. Onsite treatment of groundwater utilizing air stripping and/or carbon absorption.

Alternative Selection

Well 12A was reviewed for its effectiveness as a barrier to contaminant migration towards other water supply wells (e.g., 9A and 11A). It was determined that well 12A could be pumped at approximately 2,000 gpm adn that the water could be treated utilizing a forced-draft aeration tower to reduce the concentrations of calibrated organic solvents in the well discharge. This treatment, in addition to nearly constant operation of well 12A at minimum of 2,000 gpm, was selected as the most feasible interim alternative until enforcement actions could pinpoint specific sources for remedial action.

Current Project Status

The U.S. Environmental Protection Agency is pursuing the installation of the forced-draft aeration system at well 12A coupled with efforts to locate sources of contaminant which may contribute to contamination of the well 12A aquifer.

ACKNOWLEDGMENTS

Essential information and assistance in support of this project were provided to Black & Veatch by the following entities:

- U.S. Environmental Protection Agency (EPA), Region 10;
- EPA Region 10 Laboratory, Manchester, Washington;
- City of Tacoma, Water Division;
- City of Tacoma, Sewer Utility Division;
- City of Tacoma, Department of Public Works;
- Tacoma-Pierce County Health Department (TPCHD);
- State of Washington, Department of Ecology (DOE).

Their cooperation during this work was appreciated.

REFERENCES

1. Black & Veatch, Engineers-Architects, Woodward-Clyde Consultants, February, 1983, Remedial Investigation, Tacoma Well 12A, Tacoma, Washington, U.S. Environmental Protection Agency, Cincinnati, Ohio.
2. Economic and Engineering Services, Inc., December 1980, Tacoma Water System Plan, Volume 1, Water System Plan, Water Division, Tacoma Department of Public Utilities.
3. U.S. Environmental Protection Agency, Water Quality Criteria Documents, November 28, 1980, Federal Register, Part V, Vol. 45, No. 231.
4. Theis, C.V., 1935, The Relation between the Lowering of Piezometric Surface and the Rate and Duration of Discharge of a well Using Groundwater Storage, Trans, Am. Geophys. Union, 16th Ann. Meeting, pt. 2.

CHAPTER 30

SOIL INVESTIGATION AT THE RE-SOLVE, INC., HAZARDOUS WASTE SITE

Tom A. Pedersen
Camp, Dresser & McKee Inc.

INTRODUCTION

The Re-Solve, Inc. hazardous waste site located in Dartmouth, Massachusetts, is listed on the U.S. Environmental Protection Agency's (USEPA) interim list of 115 top-priority disposal sites and on the proposed national priority list of 419 hazardous waste sites promulgated December 30, 1982. As such, the site is eligible for funds under the Comprehensive Environmental Response, Compensation and Liability Act of 1980 (CERCLA), known as the "Superfund".

The Remedial Action Master Plan (RAMP) prepared for the site provided an assessment of the existing conditions and identified a number of remedial actions to be taken at the site[1]. One of the remedial actions recommended in the RAMP was the preparation of a detailed site soil characterization study to be used in developing the final source control program. This paper presents a discussion of the approach used and the results of the investigations undertaken to determine the degree and extent of soil contamination at the Re-Solve Inc. hazardous waste site.

BACKGROUND

The 6 acre Re-Solve, Inc. hazardous waste site is an abandoned solvent reclamation facility located in a rural area of Bristol County, Massachusetts, approximately 10 miles northeast of New Bedford. During the early 1950s the site was an active sand and gravel pit. From 1955 to 1980 a number of buildings and distillation towers were located at the site. These facilities were used in the separation of reusable products such as lacquer thinners and degreasing agents from dry cleaning and industrial process wastes [2]. Historic aerial photographic evidence suggests that drums, tanks and lagoons were present on the site by 1961 [3]. The site was closed and abandoned in 1980 and all buildings and distillation towers on the site were razed and removed in 1981. Containerized wastes were also removed from the site in 1981.

Management of Toxic and Hazardous Wastes, H. Bhatt, R. Sykes, T. Sweeney (Editors)

The waste sludges generated at the facility were disposed in a number of lagoons on the site and historic aerial photographs indicate that disposal of waste oils on the soil surface may also have occurred [3]. As a result of past disposal practices, contamination of soil, groundwater, surface waters and sediments has been documented at the site [4]. In addition contaminants have migrated off site and have been encountered in shallow groundwater monitoring wells and in the Copicut River immediately adjacent to the site.

MATERIALS AND METHODS

The information available at the outset of site investigations indicated that a number of potential sources of contaminatin remained on site. These sources of contamination include four waste lagoons, a cooling water pond which has been filled with clean sands and an alleged oil spread-area.

The first step in investigating the soils on the site was an initial inspection to evaluate surficial soil conditions during which limited surficial probing with a soil auger was performed. The soil conditions observed provided information on the extent of surficial disturbance. As a result of the initial probing six surficial soil samples were obtained within the oil spreading area and submitted for analyses. The intial walkover also allowed for selection of areas in which to excavate the soil test pits.

Four soil test pits (Figure 1) were excavated at the Re-Solve, Inc. site to allow for the preparation of detailed morphological descriptions and for obtaining soil samples. Soil profile descriptions included horizonation, estimated soil texture, structure and color. Soil samples were obtained from the test pit profiles and for analysis.

The test pits were excavated to depths of approximately 6 feet and were 4 to 5 feet wide and 15 feet long. Photoionization detector (HNu) readings were made during excavation and subsequently in the test pits to identify zones of organic contamination. A hand bucket auger was used to bore to the water table in each test pit and obtain soil samples at depths greater than 6 feet.

Morphological descriptions of the exposed soil profiles were prepared and soil samples were obtained from prominent horizons. Soil texture and color were used to evaluate potential for contaminant migration within the profiles. Evidence of contaminant migration observed during soil investigation included oil in groundwater and discoloration of mineral soils.

One soil test pit (TP-1) was excavated to the west of the lagoons. The intent of locating the test pit in this area was to evaluate any changes in morphology resulting from construction of the lagoons and evaluate the presence of any seepage planes.

Although the surficial soil probing did not reveal the

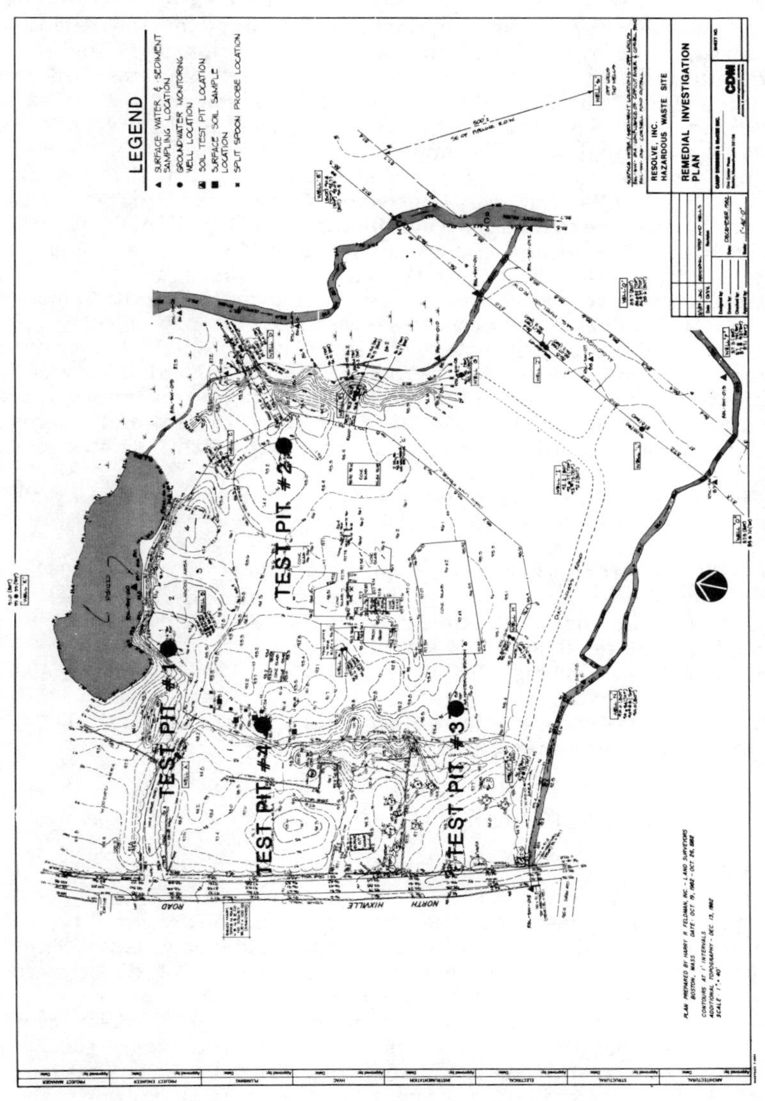

Figure 1. Soil Test Pit Locations

presence of any contaminants in the vicinity of the filled cooling water pond, a test pit (TP-2) was excavated in this area. This test pit was excavated at the western boundary of the filled pond to determine the location of the pond bottom prior to filling.

Within the suspected oil spreading area two soil test pits were excavated (TP-3 and TP-4). The surface soil in this area was discolored and the test pits were excavated to evaluate soil morphology in relation to percolation of wastes.

Soil samples obtained at the Re-Solve, Inc. site were analyzed through the USEPA National Contractor Laboratory Program (NCLP). The analytical services were coordinated by VIAR (Alandria, Virginia) as management consultant to USEPA's Sample Management Office (SMO). All NCLP analyses were performed by private laboratories of proven ability which have won competitive contract awards. Laboratory performance was assured through ongoing quality assurance evaluations conducted by the Environmental Monitoring and Support Laboratory/Las Vegas (EMSL/LV). EMSL/LV was also responsible for developing all methods, standards and protocols used by contractor laboratories. Final data review and evaluation was conducted by the NCLP support staff with assistance from EMSL/LV.

The soil samples were analyzed for the following parameters using the methodology cited:

- EP Toxicity (SW 846-7.0; 40 CFR 261.24)(SW 846-8.24)
- Polynuclear Aromatic Hydrocarbon Scan (SW 846-8.30)
- PCB/Chlorinated Pesticide Analysis (SW 846-8.08)
- Trace Metal Scans (SW 846-8.49)
- Cyanide Analyses (SW 846-8.55)

The discussion presented in this paper is limited to soil polychlorinated biphenyl (PCB) contamination.

RESULTS AND DISCUSSION

Test Pit 1

Morphological features observed in Test Pit 1 are typical of those found in soils subject to anthropogenic activities. The lack of significant structural development indicates that disturbance of these soils occurred recently (within 10-30 years). The disturbance of these soils is associated with the construction of the lagoons immediately to the east of the test pit. Fill consisting of trash, rusted cans and miscellaneous refuse was observed in the excavation indicating that this area was used for disposal of trash prior to or during construction of the lagoons. Table I provides a description of the soil profile encountered in Test Pit 1 and Figure 2 is a depth function of PCB content of soil horizons.

The surface horizon at Test Pit 1 was coarse textured and permeable. Immediately beneath this horizon were two distinct horizons (8-15 in. and 15-32 in.) which had a marked textural change. The texture of the horizons was estimated by visual appearance and feel when rubbed between gloved hands. Differentiation between fines (i.e., clay and silt separates) by this technique was difficult, and based on color changes it is inferred that textural change may be due to presence of organic sludges. The colors within these horizons ranged from pinks to greens. These colors were observed in sludge samples obtained from lagoons located adjacent to the test pit.

With the absence of clay lens or other features that would indicate the illuvation of clay from the overlying horizon, it is postulated that the morphological changes observed in these horizons are due to seepage of sludges from the lagoons into the soil materials. Although some PCBs were encountered in the surficial horizon at this test pit, the most significant concentrations were encountered in the horizon. This due to concentration of PCBs on soil materials as sludges migrate through the soil.

Insignificant levels of PCBs were encountered in Test Pit 1 below a depth of 32 inches. These horizons were coarse textured and did not exhibit color changes observed at shallower depths. Probing of the lagoon adjacent to Test Pit 1 indicated that it was 6 to 7 feet deep. The absence of a seepage plane in the soil profile indicates that the lagoon bottoms and sides at depth are sealed.

It has been alleged that burning of the lagoon contents took place on a number of occasions. This could have resulted in sealing the lower portions of the lagoon. Sealing of the lower portions of the lagoon could also have resulted as the heavier waste material settled out, causing the lighter oils and wastes to rise. Migration of the lighter and less viscous materials could therefore be restricted to the upper portions of the lagoons.

Test Pit 2

Test Pit 2 was excavated in the eastern portion of the filled cooling pond. The excavation was made in a low spot (elevation) of the filled pond. The morphological description of the soil is presented in Table II and Figure 3 contains a depth function of PCB content within the soil profile.

The soil exposed in this excavation was coarse textured fill material. Horizonation differentiation was due to observed differences in placement of fill lifts within the cooling pond. The horizons had variable thicknesses and sloped inwards from the edge of the filled cooling pond towards its center due to the nature of placement of materials.

A minor horizon comprised of brown silty clay and roots was

Table I. Soil Profile Description - Test Pit 1

Depth (inches)	Color (moist)	Texture*	Structure	Comments
0-8	10YR6/3 pale brown	cobbly sandy loam	weak subangular blocky	test pit surface about 2 feet above lagoon sludge
8-15	mixed colors, greens, pinks, black, white	sand and silty clay loam	structureless, single-grain, some blocky clay	many roots, fill
15-32	5YR7/3 pink	silty clay	very fine blocky	scattered pockets grayish white clay, black solid tar, light green clay
32-33	black	decomposing organic matter and refuse	--	no sample taken
33-58	10YR6/4 light yellowish brown	cobbly sandy loam	--	roots concentrated at 54-58 in.
62-64		rusting iron	--	no sample taken
66-88	10YR5/6 yellowish brown	gravelly loamy sand	structureless, massive	sampled at 66-68 in.
102-108	10YR5/2 grayish brown	coarse sandy clay loam	structureless, massive	water table at 102 in.

*Estimated.

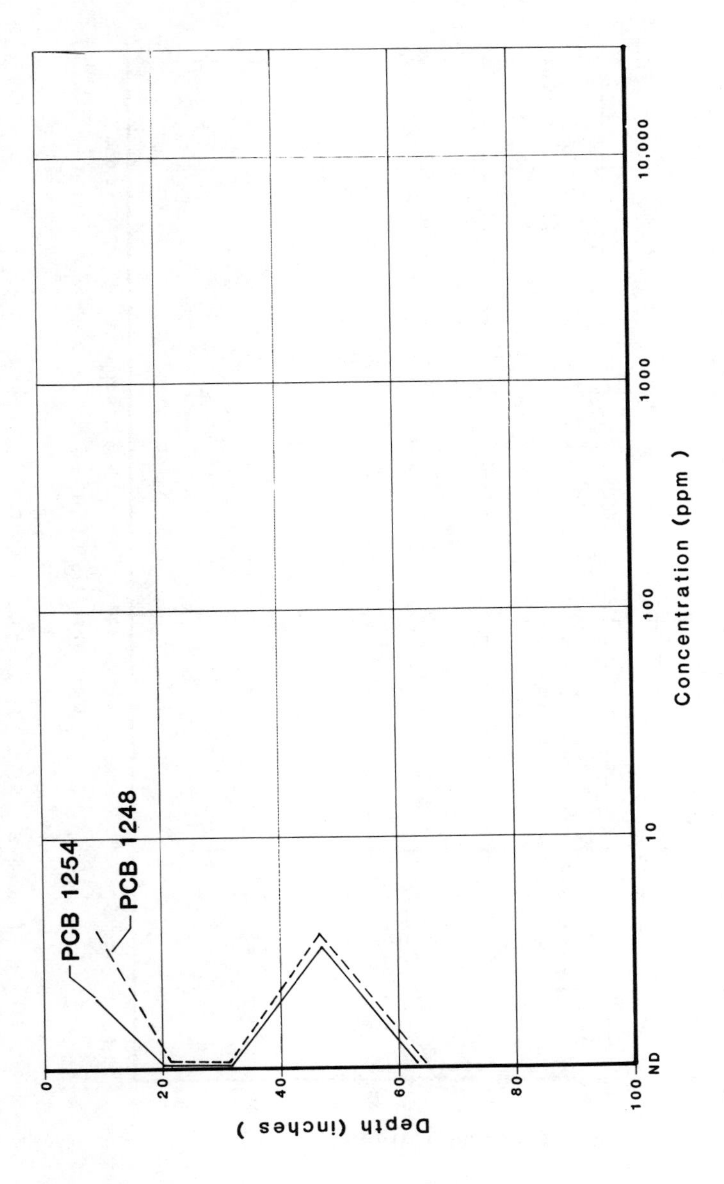

Figure 2. Test Pit 1 - Soil/PCB Depth Function

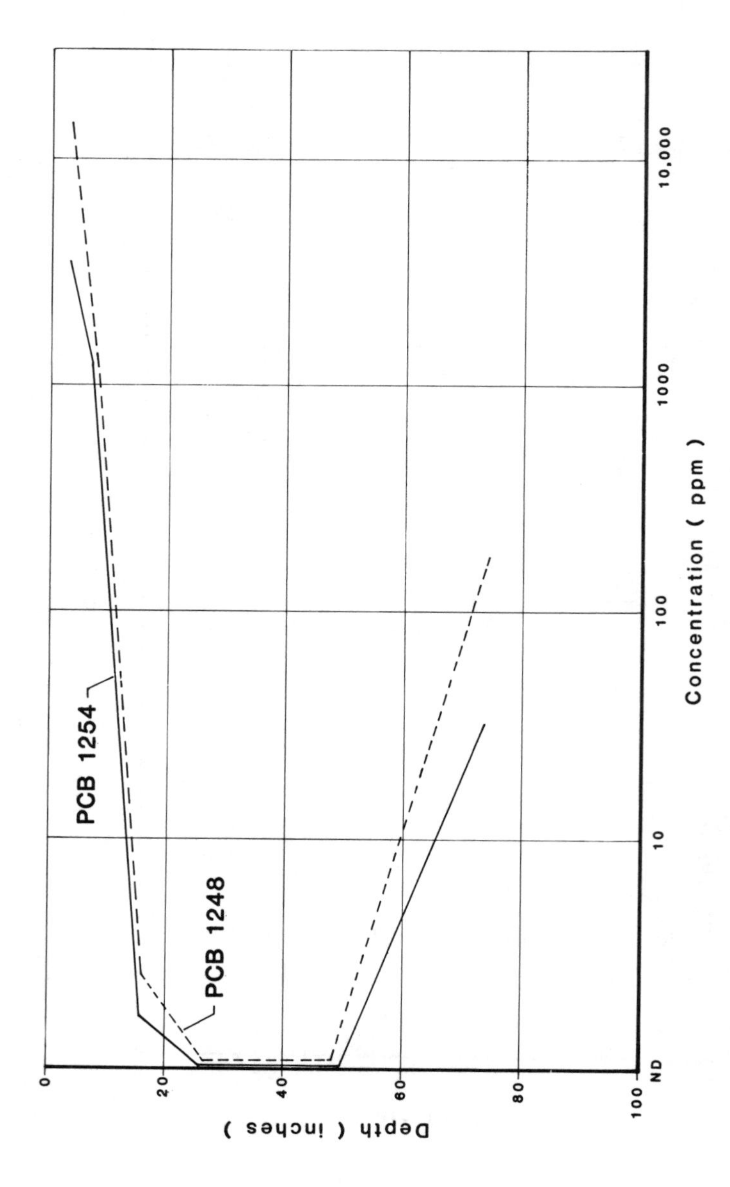

Figure 3. Test Pit 2 - Soil/PCB Depth Function

Table II. Soil Profile Description - Test Pit 2

Depth (inches)	Color (moist)	Texture*	Structure	Comments
0-2	10YR8/3 very pale brown	very fine sand	structureless, single-grain	fine gravel lenses
2-16	10YR7/3 very pale brown	gravelly fine sand	structureless, single-grain	horizon thickness ranges from 2-22 in. east to west
16-24	10YR5/4 yellowish brown	loamy sand	structureless, single-grain	1/4 in. lenses of brown (10YR5/3) silty clay, roots
24-38	10YR6/3 pale brown	coarse gravelly sand	structureless, single-grain	some cobbles
48-50	10YR5/6 yellowish brown	loamy sand	weak subangular blocky	(bottom of pit)
50-76	10YR5/1 gray	coarse sandy loam	--	water table at 74 in.

*Estimated.

encountered at a depth of approximately 16 in. This horizon is the zone of accumulation of organic materials (leaves, etc.) at the limits of the cooling pond during its use. The surface soil contained PCB concentrations greater than those of any subsurface horizon. Analysis of replicate samples from this horizon showed considerable variation in PCB content. Presence of PCBs on the soil surface may be due to deposition of finer particles such as silt which have been eroded from other areas on the site. PCB content of the subsurface horizon between 2 and 48 in. was comparatively low. However, at 48 to 50 in., the PCB content of the soil increased slightly to approximately 7 ppm. This increase is correlated with a slightly advanced structural development in this horizon where there is a higher percentage of fines that have a greater affinity for adsorption of PCBs than sands. The water table was encountered within this test pit at 74 in. and no PCBs were encountered in soil material in or below the water table (October, 1982).

Test Pits 3 and 4

Test Pits 3 and 4 were excavated in the area of the Re-Solve, Inc. site suspected of being used for landfarming of oily waste and sludges. Prior to excavation of the test pit a number of surficial soil samples were obtained and submitted for analyses. The PCB contents of these samples are listed in Table III. There were also high levels of heavy metals, especially lead as high as 6 percent, and volatile organics (1,000 ppm range).

Table III. Surficial soil Sample PCB Content (ppm)

Sample	PCB 1242/1016	PCB 1254
SL01	13,000	2,000
SL02	13,000	7,300
SL03	12,000	7,800
SL04	27,000	25,000
SL05	15,000	6,100
SL06*	11,000	5,500

*Duplicate of SL05.

The PCB content of the surficial soils is extremely high, ranging from 15,000 to 52,000 ppm. These high levels of contaminations, as well as the physical appearance of the soil, confirm the allegation that this area was used for disposal of waste oils.

The soil profiles exposed in Test Pits 3 and 4 reveal a soil that lacks any significant structural development (Tables IV and V). The soil appears to be fill because morphologically it has no features observed in contiguous soil road cuts. The weak platy structure observed from 9 to 23 in. in Test Pit 5 is likely the result of vehicle traffic in this area.

The surface horizons at both Test Pits 3 and 4 have been modified by plowing, or some other mixing operations to incorporate the disposed oils. In Test Pit 4 discrete tarlike materials and oozing oils were observed at the 0 to 8 in. depth.

The weakly developed platy horizon in Test Pit 3 decreases the amount of PCBs which have migrated into the subsoil (Figure 4). The continuous lens of fine gravel present at a depth of 23 in. also serves as a hydraulic restriction for vertical migration of contaminants under non-saturated conditions. Relatively low PCB

Table IV. Soil Profile Description - Test Pit 3

Depth (inches)	Color (moist)	Texture*	Structure	Comments
0-6	10YR4/4 dark yellowish brown	very gravelly	structureless, massive	
6-9	10YR7/2 light gray	fine sand	structureless, single-grain	
9-23	10YR7/4 very pale brown	very fine sand	weak thin platy	silt varves
23-29	10YR8/2 white	gravelly sand	structureless, single-grain	lenses of fine gravel
29-72	10YR7/2 light gray	very fine sand	structureless, massive	coarse gravel lenses at 41 in.
72-76	10YR7/6 yellow	fine loamy sand	--	water table at 72 in.

*Estimated.

contents were encountered in this and underlying horizons above the water table. PCB content of the soil within the water table was approximately 200 ppm. An oil sheen was observed on the water seeping into the test pit, which indicates that contaminants have migrated to this level.

Within Test Pit 4 the PCB content of the soils remains fairly uniform from the surface to a depth of 36 in. (8,890 to 17,650 ppm) (Figure 5). The PCB content of the soil decreases to less than 5 ppm within the loamy sand horizon encountered from 36 to 62 inches. Above this horizon is a coarse sand and gravel which would not be expected to retain more PCBs than the underlying soil based on textural considerations. However, this abrupt textural change can prevent vertical migration of contaminants.

The water table was observed in this profile at a depth of 68 in. (October, 1982). Within this horizon the PCB content of the soil totaled 4,331 ppm. The features of this horizon indicate that

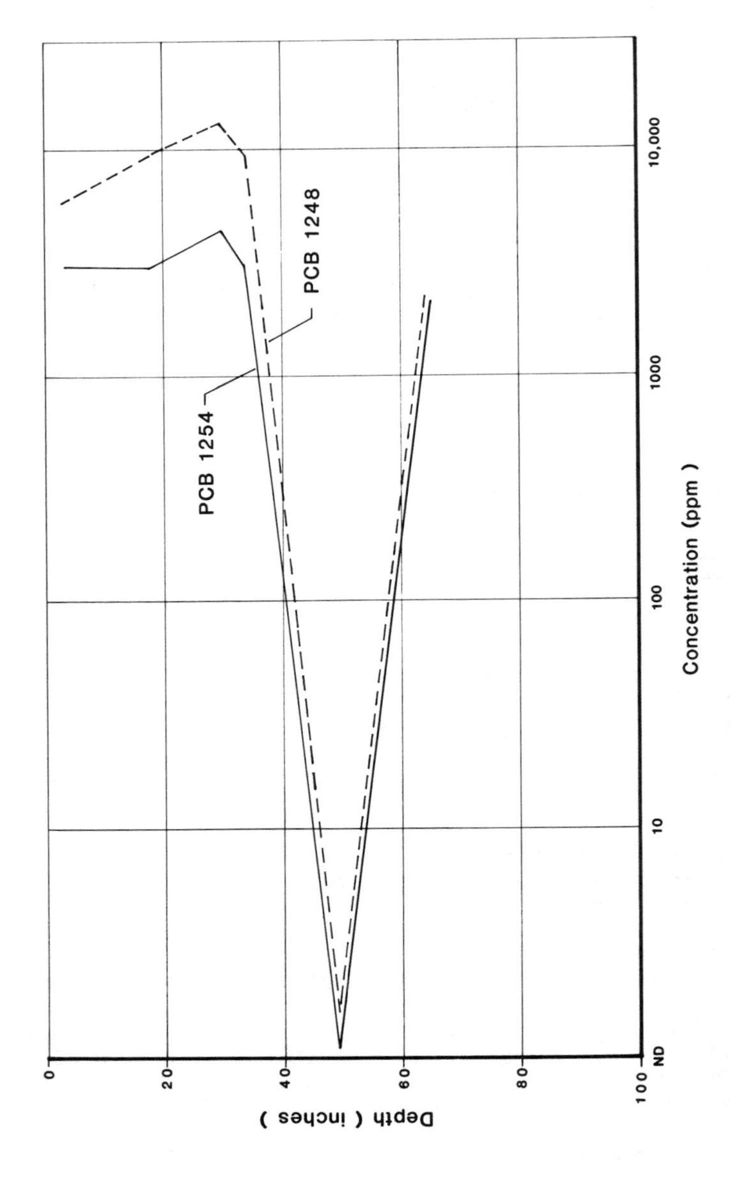

Figure 4. Test Pit 3 - Soil/PCB Depth Function

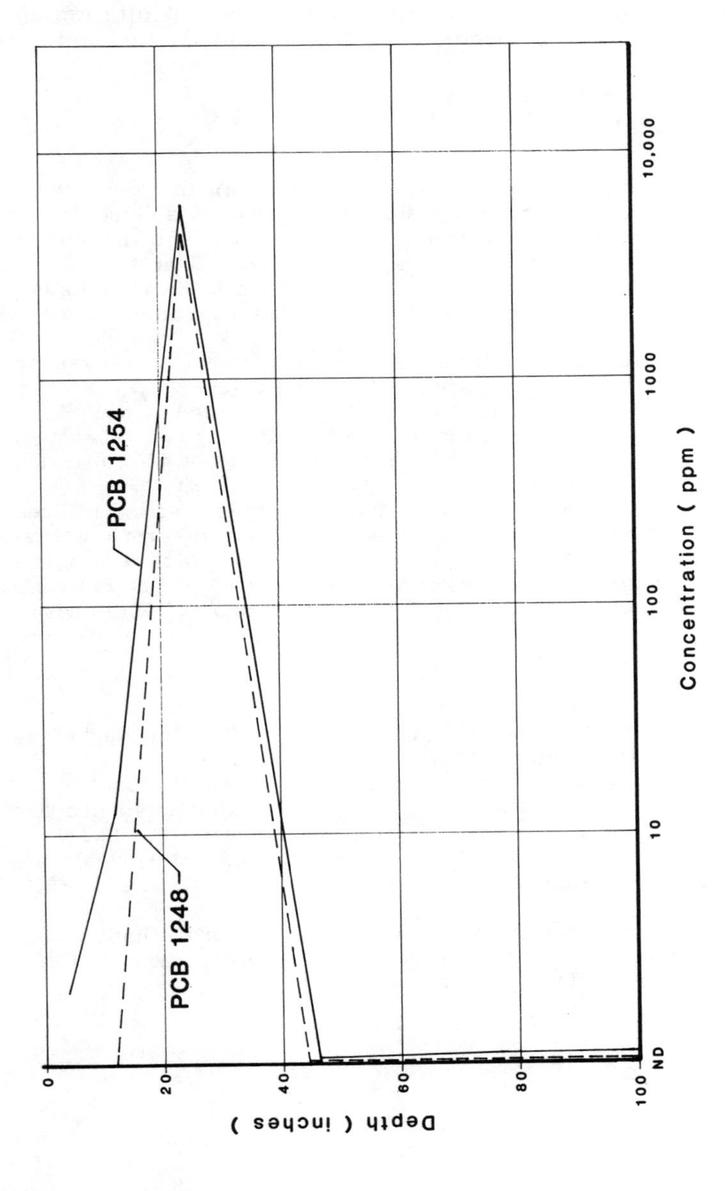

Figure 5. Test Pit 4 - Soil/PCB Depth Function

the water table is present at this depth for a significant portion of each year. An oil sheen was also observed on this water table in Test Pit 4.

SUMMARY

Four (4) soil tests pits were excavated at the Re-Solve, Inc. hazardous waste site to ascertain the degree of soil contamination at the site and to evaluate the extent of and potential for continued migration of contaminants off-site. The test pits were located in areas where waste materials were known to have been deposited on the soil surface or to evaluate migration of wastes in the soil.

The results of the investigations confirmed the presence of contaminants, especially PCBs, at elevated levels in the surficial soils on the site. Contaminants have migrated from the waste lagoons on the site to contiguous soils and disposal of oily wastes on the soil surface has led to the migration of contaminants vertically in the soil profile.

The results of the soil investigations were utilized in conjunction with soil boring logs and groundwater analysis to develop a series of source control measures which are currently being implemented by the Massachusetts Department of Environmental Quality Engineering, USEPA and US Army Corps of Engineers.

REFERENCES

1. CDM, 1982. Remedial Action Master Plan for Re-Solve, Inc., Dartmouth, MA. June 15, 1982.
2. Massachusetts Department of Environmental Quality Engineering. "Re-Solve, Inc. - Remedial Response Fact Sheet for Allocation of Funds" (February 16, 1982).
3. Crook, D., Environmental Photographic Interpretation Data. Re-Solve Inc., Dartmouth, MA. USEPA TS-PIC-82024 (1982).
4. Massachusetts Department of Environmental Quality Engineering. Data Summary - Re-Solve, Inc., North Dartmouth (May 3, 1982).

SECTION VII

RISK ASSESSMENT

CHAPTER 31

ENVIRONMENTAL RISK ASSESSMENT

Lynne M. Miller
Risk Science International

INTRODUCTION

Most companies that manufacture, handle or dispose of chemicals or petroleum products have the potential to cause environmental impairment, especially by long-term or gradual release of materials into the environment. This impairment creates potential liabilities that result from numerous regulations as well as from common law. One way to identify these potential environmental liabilities and exposures is through environmental risk assessment. The primary focus of an environmental risk assessment is to evaluate the potential for off-site gradual impairment arising from a company's operations. Such an assessment reviews the status of the firm's environmental risk exposure, both as a snapshot of the present operations and as a review of past operations. In addition to being a useful internal tool for corporate planning, the assessment can also be used in obtaining environmental impairment liability (EIL) insurance, which provides coverage for gradual impairment that results in third-party bodily injury or property damage.

OBJECTIVES OF RISK ASSESSMENT

There are three major objectives in performing an environmental risk assessment: 1) risk identification; 2) risk evaluation; and 3) risk reduction. To identify risk, numerous criteria are used to determine the individual factors contributing to the potential for off-site exposure. The interaction of these individual risk factors is then used to evaluate the overall potential for gradual environmental impairment. After the risks have been evaluated, recommendations are made to reduce the potential exposures. The disciplines necessary to meet all three objectives include environmental engineering, toxicology, ecology, environmental science, and chemistry. Assigning a risk assessment project solely to an environmental engineer could result in inadequate assessment of certain critical elements in a nonsudden evaluation, such as

Management of Toxic and Hazardous Wastes, H. Bhatt, R. Sykes, T. Sweeney (Editors)

toxicology and susceptibility of target populations.

A first step in the risk assessment involves an on-site inspection of selected locations of the company being evaluated. Locations are selected that are representative of the company's overall operations or that are believed to present an increased risk of environmental impairment. Factors used to select sites include, where available: nature of the operations; size and age of the facility; surrounding environment; present and past handling of chemicals or other potentially hazardous materials; existence of prior environmental claims, complaints or litigation; and status of regulatory compliance and environmental permits.

RISK IDENTIFICATION

The identification of environmental risk at selected locations goes far beyond the relatively simple question of technical compliance with mandated environmental regulations. In many areas of environmental and health risk, there are no effective "standards" that serve as protective safety demarcation lines, inside of which it can be argued that no damage can be caused and hence there can be no liability exposure. The discrete standards, where they exist, should be regarded as analagous to the 55 mph national speed limit. A driver can still be found responsible for injury even when he has strictly obeyed the speed limit. Compliance is therefore a minimum requirement rather than guarantee of zero liability.

Risk Science International's risk assessment methodology is based on the interrelationship of several key risk factors. Our approach is not based on a checklist or a model, because we believe that could lead to a superficial examination of inherent uncertainties at individual locations or to overlooking information that requires careful professional scrutiny and judgemet as to its risk potential. The environmental risk assessment takes into account four factors for each location being assessed:

1. Environmental routes.
2. Target populations.
3. Facility operations and practices.
4. Characteristics of materials.

Environmental Routes

This factor describes the pathways by which materials may move off the premises of the facility. Materials can move off-site by three major pathways: surface water, groundwater and air. soil contamination is not considered as a separate route, because on-site soil contamination can affect off-site groundwaters or surface waters, and would be considered under those routes.

Evaluating the possibility of surface water contamination involves understanding such factors as the nature of nearby surface waters (i.e., lakes, estuaries, streams, ocean); physical and geographic characteristics of the surface water (i.e., distance from facility, flow rates, depth); and the nature of indirect and direct discharges into surface water. The possibility of groundwater contamination is evaluated by considering the nature and use of the groundwater aquifer(s); depth of groundwater; bedrock depth and permeability; soil permeability; and net precipitation. Assessing the possibility of air contamination involves considering such factors as known and expected stack emissions; fugitive emissions; mean wind speed and direction; and topography.

Target Populations

This factor identifies the human and non-human populations that could be affected by materials moving off the premises through the environmental routes discussed above. Characteristics of the neighborhood are reviewed to evaluate potential adverse effects, such as size of residential human population, source of water supply, nature of human population, distance to residential and non-residential areas (i.e., food crops, livestock, fish) and ecological habitat, as an indicator of species composition and diversity.

Facility Operations and Practices

Whether a facility could result in gradual environmental impairment depends to a large extent on how it is maintained and operated. Factors such as the company management system, management attitudes, training programs, environmental compliance status, present and past handling and disposal of chemicals, past violations, prior claims history and facility design as it may affect pollution containment are considered.

Characteristics of Materials

This factor evaluates the potential impacts on the exposed populations by considering the intrinsic hazards of the materials that could be released. These materials could originate from numerous sources at the facility, including handling, treatment, storage and disposal. If certain materials at the facility are adequately contained, or appear unlikely to travel off-site, their inherent hazards are of less concern. For individual chemicals of concern, the following areas are evaluated: carcinogenicity, mutagenicity, teratogenicity, other chronic effects, persistence and bioaccumulation.

RISK EVALUATION

The next step in the risk assessment involves integrating the four factors to arrive at an overall assessment. Each factor is separately ranked as to its potential for contributing to the risk of environmental impairment by scoring it as low, low-to-moderate, moderate, moderate-to-high, or high. The overall assessment is then ranked according to the same procedure, taking into account the interaction of the four factors. For example, a company that manufactures highly toxic chemicals (i.e., "high" potential risk under Characteristics of Materials) would not present an overall high risk of environmental impairment if the facility is well designed and managed (i.e., low or moderate risk under Facility Operations and Practices), even if there were pathways of potential exposure and potentially exposed populations. How all four factors interact is the key in assigning an overall risk rating.

RISK REDUCTION

Once the major areas of risk have been identified and evaluated, a list of priority recommendations are made to reduce the potential for environmental impairment. Of course, the specific recommendations depend on the nature of the risk, but may include the need for monitoring, soil and water testing, better containment, records search, specific toxicity testing, improvements in mangement practices, etc.

USES OF THE ASSESSMENT

The risk assessment is of great importance to internal company management in anticipating potential problems of a nonsudden nature. The risk assessments will provide a list of priority recommendations of ways to reduce risk exposures, including appropriate data gathering in those cases where ignorance and uncertainty about a situation could hve serious consequences. It will assist in assigning priorities to expenditures for environmental controls to reduce potential exposures. Because the document establishes a baseline of information, it has been used in dealing with the public or regulatory agencies when allegations are made of environmental impairment. An environmental risk assessment is also an important tool for companies considering a merger or acquisition. In the absence of such an assessment, the new owner could unknowingly assume obligations for substantial third party liability or remedial clean-up costs.

Environmental risk assessmenta are usually required by underwriters of "nonsudden pollution" insurance or Environmental Impairment Liability (EIL) insurance. Because this coverage is relatively new, underwriters do not have an actuarial data base on which to base premiums. For the underwriting market, the environmental risk assessment serves as a valuable baseline for the existing situation at the time the applications was filed. The assessment is an important tool for establishing rates, deductibles, limits of coverage, exclusions, and other terms. Also this baseline, coupled with the manner in which the applicant implements the report recommendations, can give the underwriter a fair picture of the client's good faith attempts to minimize the overall risk. The assessment can also give the underwriter a mechanism that can be used to improve the terms of the insurance at renewal time, e.g. by offering improved premiums or policy terms.

The assessment may be the only source of information about the types of past practices that may result in future claims. The risk assessment must be prepared carefully in order to serve as important evidence if and when the time comes for adjusting any claims against the policy. The report will represent the technical status of the client's facility at the time of the insurance issuance, and assuming no substantial physical changes have been made, will represent the status at the time the claim is filed.

Some insurers have in-house capabilities for risk assessments, but most do not, and therefore require that a potential insured contract directly with an outside firm. These underwriters have a list of approved firms to perform risk assessments. If a company is interested both in exploring EIL insurance and having an environmental risk assessment performed, it is important that the firm performing the risk assessment be acceptable to the insurers. A company should also consider whether the risk assessment firm carries adequate limits of professional liability coverage, such as errors and omissions insurance. It is crucial that the company have an opportunity to review the assessment before it is sent to underwriters.

In summary, an environmental risk assessment is a qualitative review of key factors, including environmental routes, target populations, facility operations and practices, and characteristics of materials. The assessment is useful both to corporate planners as well as underwriters offering Environmental Impairment Liability insurance. The insurance industry relies on risk assessments for EIL insurance, because underwriters can only successfully provide this coverage if they can adequately determine the risk involved.

SECTION VIII

LEGAL CONSIDERATIONS

CHAPTER 32

MANUFACTURERS' WARRANTIES ON HAZARDOUS WASTE DISPOSAL EQUIPMENT

Stanley A. Reigel
Morrison, Hecker, Curtis, Kuder & Parrish

INTRODUCTION

Practically everyone is familiar with the ancient legal maxim "caveat emptor" or "let the buyer beware". This maxim accurately characterizes the relationship that existed between buyer and seller during the early days of the industrialization of America. As Kropp puts it: "There was something American (like Will Rogers and apple pie) about the horse trade in which each party sought to take advantage of the other" [1]. Today, however, caveat emptor has been replaced by the Uniform Commercial Codes' system of warranties, disclaimers, and remedies [2].

The Uniform Commercial Code ("UCC") has been adopted as the law of commerce in 49 of the 50 states (Louisiana excepted) and for all intents and purposes it is the same law in each state. The UCC is divided into 11 articles; Article 2 contains the law of sales of "goods" [3]. Hazardous waste disposal equipment falls within the UCC definition of "goods" [4]. In this paper the terms "equipment" or "hazardous waste disposal equipment" will be used instead of the term "goods."

When a buyer purchases hazardous waste disposal equipment from a seller, the buyer may be entitled to the protection of two separate and distinct types of warranties respecting the quality and performance of the equipment: 1) express warranties; and 2) implied warranties.

EXPRESS WARRANTIES

Express warranties by the seller as to the quality and performance of hazardous waste disposal equipment are created as follows [5]:

1. Any affirmation of fact or promise made by the seller to the buyer which relates to the equipment and becomes part of the basis of the bargain creates an express warranty that the equipment shall conform to the affirmation or

Management of Toxic and Hazardous Wastes, H. Bhatt, R. Sykes, T. Sweeney (Editors)

promise;

2. Any description of the equipment which is made part of the basis of the bargain creates an express warranty that the equipment shall conform to the description; and
3. Any sample or model which is made part of the basis of the bargain creates an express warranty that the whole of the equipment shall conform to the sample or model [6].

No particular formalities or "magic words" such as "guarantee" or "warrant" are necessary to the creation of an express warranty [6]. Thus, technical specifications, blueprints, promotional brochures, quotations, proposals and the like transmitted by the seller to the buyer may all contain express warranties of quality of performance.

Notice that for an express warranty to exist the statement, promise, affirmation of fact, etc. must be the "basis of the bargain." Although this phrase has no precise legal meaning it is generally thought that the promise, statement, etc. should, At least, have been one of the factors leading the buyer to purchase the hazardous waste disposal equipment [7].

Typical express warranties that might be extended on hazardous waste disposal equipment include:

- Compliance with applicable state and federal laws and regulations.
- Destruction efficiency and maximum emission to atmosphere for hazardous waste incinerators.
- Pressure and leakage tolerances for gas cylinders.
- Lack of reactivity with specific hazardous waste for containers.

IMPLIED WARRANTIES

Although express warranties must be affirmatively given by the seller to the buyer before they exist, the UCC provides for two automatic warranties which exist unless the seller specifically disclaims them. These are the implied warranties of "merchantability" and "fitness for a particular purpose".

Merchantability

Unless excluded or modified, a warranty that the hazardous waste disposal equipment shall be "merchantable" is implied in a contract for sale if the seller is a "merchant" [8].

All sellers are not "merchants" under the UCC. A "merchant" means a person who deals in goods of the kind being sold or otherwise by his occupation holds himself out as having knowledge or skill peculiar to the goods [9]. Thus, if an environmental engineer sells his car to a used car dealer, the environmental engineer probably is not a "merchant"; but when the used car dealer resells it to a member of the public he does so as a "merchant".

What does it mean to say that hazardous waste disposal equipment must be "merchantable?" The UCC recites six requirements which boil down to the proposition that in order to be merchantable, the equipment must be of average quality such as would ordinarily pass without objection from seller to buyer [10].

Fitness for a Particular Purpose

In addition to the implied warranty of merchantability, the buyer may be entitled to the implied warranty of "fitness for particular purpose". The "fitness" warranty arises when the seller, at the time of contracting, has reason to know of any particular purpose for which the equipment is required and that the buyer is relying on the seller's skill or judgement to furnish suitable equipment for that purpose [11]. Thus, if a buyer were to send a purchase order to a seller for one hundred 55 gallon steel drums, no "fitness" warranty would arise. However, if the buyer were to ask the seller to quote suitable containers for the permanent storage of 6,000 gallons of acidic waste containing 10% HCL the fitness warranty would arise upon purchase by the buyer of the seller's recommended containers.

DISCLAIMERS OF WARRANTY

The UCC permits sellers to disclaim the implied warranties of merchantability and fitness for a particular purpose [12]. To exclude or modify the implied warranty of merchantability, the language must mention "merchantability" and in case the exclusion or modification is in writing (which it need not be), it must be conspicuous. The exclude or modify the implied warranty of fitness, the exclusion must be in writing and must be conspicuous. There are three general exceptions to these methods of excluding implied warranties:

1. Unless the circumstances indicate otherwise, all implied warranties are excluded by expressions like "as is", "with all faults" or other language which is common understanding calls the buyer's attention to the exclusion of warranties and makes plain that there is no implied warranty; and
2. When the buyer, before entering into the contract, has examined the equipment or the sample or model as fully as desired, or has refused to examine the equipment, there is no implied warranty with regard to defects which an examination would have revealed; and
3. An implied warranty can also be excluded or modified by common trade usage or by the past practices of the buyer and seller in their dealings with each other [3].

It sometimes happens that a seller's proposal for the sale for hazardous waste disposal equipment, e.g. a hazardous waste incinerator, contains conflicting language regarding warranties. For instance, a proposal might state that the hazardous waste incinerator offered for sale will be capable of burning 1000 pounds per day of a particular spent solvent. This same proposal might also state in fine print on a separate printed "terms and conditions" page: "Seller makes no warranties express or implied, including without limitation, the implied warranties of MERCHANTABILITY AND FITNESS FOR A PARTICULAR PURPOSE." In such a case is the incinerator warranted to burn 1000 pounds per day or isn't it? In all probability an express warranty that the incinerator will burn 1000 pounds per day has been extended notwithstanding the disclaimer. Words which tend to create an express warranty and words tending to negate or limit a warranty are construed, wherever reasonable, as consistent with each other. But insofar as such construction is unreasonable the words of negation or limitation are inoperative and the extension of the warranty prevails [14].

DAMAGES FOR BREACH OF WARRANTY

If the seller breaches an express or implied warranty, the buyer is entitled to damages. Moreover, if the warranty has been breached there is virtually no defense available to the seler. For instance, if the seller warrants that a hazardous waste incinerator will burn 1000 pounds per day of spent solvent and the incinerator is incapable of this performance, it will avail the seller nothing to argue that he designed the incinerator in accordance with good engineering practice and did everything in his power to attempt to make the incinerator perform as warranted.

The theory of damages in contract law (a sales transaction is a contract between buyer and seller) is to put the non-breaching party in a good a position as he would have been in had the breaching party performed his obligation, at least insofar as money is able to do this.

In general, the measure of damages for breach of warranty is the difference between the value of the equipment accepted and the value the equipment would have had if it had been as warrantied [15]. In addition to these "direct" damages, the buyer may also be entitled to "incidental" and "consequential" damages. Incidental damages are those miscellaneous expenses incurred by the buyer which result from the seller's breach, such as cost of inspection; reciept, transportation, care and custody of the goods; long distance telephone calls; etc. [16]. Consequential damages resulting from the seller's breach include losses resulting from the buyer's requirements and needs of which the seller had reason to know and which the buyer could not reasonably prevent after the breach occurred. Consequential damages also include injury to persons or property "proximately" resulting from the seller's breach [17]. "Proximate" is a word of legal art generally meaning that the causal connection between seller's breach and buyer's loss is not unreasonable or speculative. To give an extreme example, suppose buyer purchased a hazardous waste incinerator from seller which was expressly warranted to be capable of burning 1000 pounds per day of spent solvent and further suppose that the seller breached this warranty. Because of seller's breach, buyer and seller schedule an engineering meeting, to be held at seller's offices, during which an engineering solution to the problem (i.e., a "fix") will be discussed. Buyer dies in an automobile accident enroute. In these circumstances, buyer's wife would not be able to sue seller for buyer's death. Although there may be a very tenuous cause and effect relationship between seller's breach and buyer's untimely demise no court would hold that seller's breach was the proximate cause. This chain of events was totally unforeseeable at the time buyer and seller entered into their contract. Moreover, there are probably many intervening causes such as wet and slippery roads, driver negligence, etc. which caused the automobile accident.

If buyer and seller wish, they may "liquidate" damages at the inception of the sales contract. In this event buyer and seller will pre-agree on a reasonable amount which will be paid by seller to buyer in the event a warranty is breached. The amount must be a reasonable estimate of the anticipated or actual harm which would be caused by the breach and may take into consideration the difficulties of proving the amount of loss and the inconvenience or nonfeasibility of obtaining any other adequate remedy. An unreasonable large liquidated damage provision will be voided as a penalty [18]. Remember, it is the function of the law of contract damages to make the nonbreaching party whole, not to penalize the breaching party.

The buyer and seller may also agree that in lieu of damages for breach of warranty, the buyer will be limited to some other remedy altogether, such as return of the equipment repayment of the price or repair and replacement of the equipment. Buyer and seller may also agree to limit or exclusion would be "unconscionable" [19]. It is beyond the scope of this paper to discuss the concept of "Unconscionability."

EXAMPLE

A concrete example will serve to illustrate some of the above concepts. Suppose Seller extends the following warranty to Buyer regarding the sale of a hazardous waste incinerator:

> Seller warrants that the system as quoted in this proposal will:
>
> 1. Provide 99.99% destruction and removal efficiency (DRE) for each principal organic hazardous constituent (POHC)

defined by the facility permit when burning no more than 100 gph of spent solvent as specified in Buyers request for bid dated June 6, 1983.

2. Provide 99% HCL removal from flue gas if waste contains more than 5% CL when burning no more than 100 gph of spent solvent as specified in Buyers request for bid dated June 6, 1983.
3. Emit no more than 180 mg. of particulate per day standard cubic meter of flue gas (0.98 grains per standard cubic foot).

The foregoing express warranties are in lieu of all and any other warranties, express or implied, including the implied warranties of MERCHANTABILITY AND FITNESS FOR A PARTICULAR PURPOSE. This warranty is based on the system being operated at the operating requirements specified in any permit issued by the Environmental Protection Agency. This warranty is also based on the system being operated properly and in accordance with Seller's instructions. This warranty will be in effect until a properly qualified independent testing company has performed an emmision test which shows compliance or the governing agency has approved the system, or 6 months from the date of initial operation, whichever occurs first. Seller has the right to be notified of an emission testing schedule 1 week prior to actual testing, to be present during the testing, and to review the test data and results. Should the system fail to perform as warrantied, buyer will promptly notify Seller in writing of such failure and Seller will have a reasonable time to make any repairs or modifications it desires to bring the system into compliance. If Seller is unable within a reasonable time to make the system perform as warrantied, or is unwilling to do so, then buyer's sole and exclusive remedy for breach of the above warranty shall be the refund to buyer by Seller of an amount of money not to exceed 20% of the purchase price of the system (not including field erection, insulation, wiring and piping) less the monies already spent by Seller to attempt to repair the system.

With respect to the above warranty notice the following:

1. It is an express warranty;
2. It disclaims the implied warranties;
3. There are conditions to the validity of the warranty;
    a. The system must be used to burn "no more than 100 gph of spent solvent as specified in Buyers request for bid dated June 6, 1983;
    b. The system must be operated according to the EPA permit;
    c. The systems must be operated "properly and in accordance with Seller's instruction";
4. The warranty has a limited duration, i.e., 6 months,

acceptance, or successful completion of performance testing, whichever comes first;

5. Buyer is obligated to notify Seller in writing of any breach;
6. In the event of breach, damages have been liquidated to 20% of the purchase price less certain deductions;
7. The remedy set forth is the sole and exclusive remedy, thus direct, incidental and consequential damages will not be recoverable.

CONCLUSION

Warranties are the buyers best protection that he is getting what he pays for; they should not be bargained away.

REFERENCES

1. Knopp, E.L. "Vendor Guarantees -- After the New Wears Off," paper presented at 75th Annual Meeting of the Air Pollution Control Association, Philadelphia, Pennsylvania, 1981.
2. Uniform Commercial Code, The American Law Institute, National Conference of Commissioners on Uniform State Laws, 1978.
3. Id. at Sic 2-102.
4. Id. at Sic 2-105(1).
5. Id. at Sic 2-313(1).
6. Id. at Sic 2-313(2).
7. Bender's UCC Service, Duesenberg and King, Sales and ulk Transfers Sic 6.01, note 2, Matthew Bender & Co.
8. Uniform Commericial Code, The American Law Institute, National Conference of Commissioners on Uniform State Laws, 1978, Sic 2-3141(1).
9. Id. at Sic 2-104(1).
10. Id. at Sic 2-314(2).
11. Id. at Sic 2-315.
12. Id. at Sic 2-316(2).
13. Id. at Sic 2-316(3).
14. Id. at Sic 2-316(1).
15. Id. at Sic 2-714(2).
16. Id. at Sic 2-715(1).
17. Id. at Sic 2-715(2).
18. Id. at Sic 2-718(1).
19. Id. at Sic 2-719.

CHAPTER 33

FEDERAL AND STATE ENFORCEMENT OF HAZARDOUS WASTE LAWS

William W. Falsgraf
Baker & Hostetler

INTRODUCTION

When the historians turn their attention to the dominant political and social issues of the 80's preeminent among them will be the treatment, storage and disposal of the enormous volume of hazardous chemical wastes generated as by-products of our consumer oriented society. As is the case with all significant political movements, the first step is the raising of the public consciousness to an appreciation of the enormity of the problem and the consequences of neglecting remedial action designed to alleviate it. In the matter of hazardous wastes this process has been going on for some time with incidents such as the PCB contamination of the Hudson River and the Kepone discharges into the James River providing fuel for the growing fire of public outrage. However, the single most important consciousness raising event was the discovery of the leaking chemicals and their effect on the health of the residents of the Love Canal area in Niagara Falls, New York. This event provided the national news media with a dramatic example of the serious consequences which can result from releases of toxic wastes and they lost no time in reporting the details of this disaster. The impact of the chemical releases upon innocent property owners was so shocking that the "Love Canal" became synonymous in the minds of Americans with the dangers posed by land disposal of toxic wastes.

In the wake of these, and other less celebrated, but equally dangerous, releases of toxic wastes, Congress enacted two statutes the purpose of which was to address the multiple problems of handling, transportation, storage, disposal and cleanup of hazardous chemical wastes. The first effort took the form of the Resource Conservation and Recovery Act of 1976 ("RCRA" - 42 USC Sic 6901 _et seq_.) and the second, the Comprehensive Environmental Response, Compensation and Liability Act of 1980 ("CERCLA" - 42 USC Sic 9601 _et seq_.). While four years separated the enactment of RCRA and CERCLA the implementing regulations relating to RCRA were not published until mid 1980 and so the impact of these two

Management of Toxic and Hazardous Wastes, H. Bhatt, R. Sykes, T. Sweeney (Editors)

statutes came at roughly the same point in time.

The thrust of RCRA and CERCLA is entirely compatible, the former establishing the mechanism for regulation current and future handling and disposal of hazardous waste and the latter dealing with releases from existing sites which in some cases are inactive or even abandoned. In short, the concept was to provide a cradle-to-grave regulatory system which would allow the government to track and control the handling of hazardous waste from the time of generation to the point of ultimate disposal and further to provide response authority in the event of a release of the waste at any stage of its transporation or storage.

It is beyond the scope of this paper to delve into the intricacies of these statutes. Suffice it to say that they are at once comprehensive and arcane. Compliance is difficult, time consuming and expensive and the potential for severe financial loss as a result of violations by generators, transporters and disposers is enormous. Obviously such statutes are not self-executing and without sweeping enforcement authority in the EPA the intent of congress would in all probability be frustrated. It is the purpose of this paper to examine the enforcement provisions of both RCRA and CERCLA, the application of such provisions in specific cases and finally the defenses and protective steps available to those who may be subjected to enforcement activities under these acts.

ENFORCEMENT PROVISIONS OF RCRA AND CERCLA

The enforcement mechanism of the RCRA is contained in Section 3008 of the Act and it is by all odds a formidable deterrent to violation. When the administrator of EPA finds that a person is in violation of any requirement of the Act, he may issue an order requiring compliance or may seek an injunction from a court for the same purpose. In the event the violator fails to take corrective action within the time specified, he shall be liable for civil penalties of up to $25,000 per day of civil penalties and injunctive relief, the Act provides for severe criminal sanctions for those who knowingly handle hazardous waste without a permit or falsify required reports or records. criminal penalties include fines of up to $50,000 per day of violation or up to two years in prison or both. Further, Section 3008 provides for additional criminal penalties if a violation of the Act is known by the perpetrator to have placed another person in imminent danger of death or serious bodily injury. The penalties for this type of violation run as high as $250,000 for individuals, $1,000,000 for organizations and up to five years in prison.

One other section of the RCRA deserves mention in the context of a discussion of enforcement provisions and that is the so-called "Imminent Hazard" provision in Section 7003. This section simply authorizes the administrator to seek an injunction to restrain the handling of hazardous wastes in ways which present an imminent and substantial endangerment to health or the environment. The administrator may also issue his own orders under this section, violations of which can lead to fines of up to $5,000 per day.

The reason that Section 7003 is worthy of note is that most of the RCRA enforcement actions to date have been brought under this section. Apparently the rationale of the EPA and the Justice Department in proceeding in this fashion is that the burden of proving imminent and substantial endangerment under Section 7003 is easier to bear than proving a violation of the RCRA under Section 3008. Another consideration may have been the lack of implementing regulations and a viable permit program which rendered Section 3008 useless until their effective date in late 1980.

Whatever the reason, the fact is that Section 7003 is where the RCRA enforcement action has been and the response from the courts has been mixed. While some courts have interpreted Section 7003 as providing a broad, substantive grant of enforcement authority to the EPA, several other courts have found that Section 7003 is merely jurisdictional in nature, restricted in its application and useless as an omnibus enforcement tool. If this latter view is adopted by the appellate courts and the Supreme Court the utility of this section in enforcement actions will have been seriously diminished.

In contrast to RCRA, CERCLA is remedial as opposed to regulatory in nature. CERCLA started out as an oil spill bill which incidentally provided for the clean-up of hazardous substances. As the precursors of CERCLA were debated and reworked by various Congressional committees, the oil spill aspect of the legislation was deleted and in the waning days of the 96th Congress a watered-down compromise version of CERCLA was passed. Its object was to provide the federal government with the authority and funds to clean up chemical spills and hazardous waste disposal sites from which there is a release or threatened release of hazardous substances.

The CERCLA provides that any person in charge of a facility or vessel and who has knowledge of a release of hazardous substances therefrom shall immediately notify the National Response Center of such release. Failure to notify is punishable by fines of up to $10,000 or a jail term of up to one year or both.

CERCLA also authroizes the EPA to proceed with the clean-up of various high priority hazardous waste sites as well as current spills and to finance these clean-up programs with the so-called "Superfund." The fund in turn is financed by a tax on crude oil production and importation and a tax on the sale by the manufacturer of various chemicals such as benzene, toluene, chlorine and mercury which are either hazardous themselves or are an integral component of various hazardous substances.

The CERCLA also provides a complex system for the imposition of liability on generators, transporters and site owners for the costs of clean up and any longer term remedial action. If found liable these persons can be required to reimburse the superfund for all costs associated with the clean-up an remedial action at a site or a spill location. In addition those parties who are liable for such costs and who have failed to provide removal or remedial action upon the order of the government may be liable for punitive damages in treble the amount of the actual costs incurred by the fund.

ENFORCEMENT ACTIONS UNDER RCRA AND CERCLA

Ohio has been the scene of several of the earliest and largest enforcement actions under CERCLA. These actions have given rise to substantial interest, activity and consternation on the part of generators and transporters of hazardous substances. There are two major CERCLA cases pending in Ohio now, with several more in the initial stages of preparation for litigation. In all of these cases the state or federal government has obtained records from the disposal site. The government then uses these records to develop a list of contributors to the site. These generators and transporters are then notified in writing of their potential liability and are invited to arrange and pay for the site clean-up and ultimately any remedial action which may be necessary. They are advised that if they fail to respond in a very short period of time, usually a month or six weeks, the government will proceed with the clean-up and then will press claims against the potentially responsible parties for the costs. The class of potentially liable persons is typically very large, often numbering in the hundreds, which makes a coherent group response within the prescribed time frame impossible. What usually happens is that the government proceeds with the clean-up and while that is proceeding negotiates settlements with the various contributors either on a group or individual basis or both. Those who feel they are not liable and therefore refuse to settle usually end up as defendants in a lawsuit in which the government seeks reimbursement for the clean-up costs.

While no one seriously questions the propriety or necessity for a clean-up of hazardous waste sites which are threatening human health, there is considerable controversy over who should bear the burden of the costs of such clean-up. The CERCLA imposes liability for the costs of clean-up on generators, transporters and owners alike. The government takes the position that this liability extends to persons who generated or transported hazardous substances which ended up at the site in question at a time before the enactment of CERCLA in December of 1980. This position has the effect of applying the Act retroactively to a group of persons who were behaving in a perfectly legal way when they sent their material off for disposal but who are now faced with potential liability measured in multimillions of dollars. In view of the government position, it is no wonder that CERCLA and more specifically its liability section has spawned a rash of litigation across the country with no end in sight.

Within the past month the writer has argued the question of whether Congress intended that CERCLA be applied retractively to a transporter before a U.S. District Judge in the case of Ohio vs. Donald Georgeoff, et al. Case No. C81-1961(N.D. of Ohio). The court has the question under advisement and will rule shortly. That ruling will be the first judicial determination of this particular issue in the country. It certainly will not be the last court to consider this issue and in all probability this and other CERCLA issues will find their way up to the Supreme Court for ultimate resolution.

RCRA enforcement has also been pursued in this state but not with quite the precedent-setting impact that has attended the CERCLA actions. For example in U.S. vs. Diamond Shamrock Corporation, Case No. 80-1857 (N.D. Ohio, May 29, 1981) Judge Krupansky held that Section 7003 of RCRA provided ample authority for the court to order Diamond Shamrock to clean up its inactive liquid waste disposal site near Painesville. Interestingly another district judge found just the opposite in the recent North Carolina case of U.S. vs. Waste Industries, Inc., 18 ERC 1521 (E.D. No. Carolina, January 3, 1983). These two cases are merely representative of the diversity of interpretations being placed upon Section 7003 by the courts. This being the case, the entire question of the scope and nature of Section 7003 will have to await determination by the appellate courts and perhaps ultimately the supreme Court.

AVOIDANCE OF INVOLVEMENT IN ENFORCEMENT ACTIONS

While there are defenses available to a defendant in a RCRA or CERCLA enforcement action, the assertion of such defenses will have to be made in a litigation context which means hiring counsel and incurring substantial expenses even if the defenses are successful. Secondly, the availability of the defenses, except for purely legal challenges to the statute or regulations, depend upon events which have already taken place and are therefore beyond the control of company personnel. The point is that one should only rely upon defenses such as acts of God or third party acts or omissions as a last resort. By far the best and cheapest approach is to establish internal procedures and systems which will minimize exposure to enforcement actions in the first instance.

Avoidance of enforcement actions under RCRA is easier than those brought pursuant to CERCLA because RCRA is prospective in nature and, albeit complicated, does provide detailed rules and regulations compliance with which will preclude enforcement actions. The single most important step in avoiding RCRA and CERCLA enforcement is to determine whether your operation is generating substances which are subject to control pursuant to these statutes. This may seem to be belaboring the obvious but in my experience many operations do not bother to analyze their waste streams or run material balances to determine what is actually being discharged as wastes.

Having determined the nature and volume of the various waste streams, a periodic review and analysis thereof with attendant recordkeeping should be instituted. THese records may serve to free you from potential enforcement involvement in that they will confirm that your operation is not a potential source of a hazardous substance which has been released into the environment.

If your operation does generate hazardous substances make certain that it is stored in accordance with the RCRA regulations while on site and that it is delivered to a licensed and responsible hauler and ultimate disposer. The mere possession of a license or permit, particularly an interim status permit, is not necessarily conclusive evidence of responsibility. Independent investigation of both transporter and disposer could disclose practices which could involve a generator in enforcement actions and subject him to substantial liability.

The manifest and recordkeeping requirements of RCRA must, of course, be scrupulously followed. Not only is it illegal not to do so, but also these records will provide you with a first line of defense if there is a problem.

Finally, it is helpful to obtain an indemnity agreement from your hauler and/or disposer. While CERCLA does not permit the shifting of liability by means of an indemnity agreement it does not prohibit such an agreement. To the extent that you have a financially responsible indemnitor, you will have insulated your operation from potential liability which can be very significant.

CONCLUSION

Hazardous waste disposal and the clean-up of abandoned sites and spills is a top priority item in this country and absent a war or similar cataclysm, will continue to be at least for the remainder of this decade. Congress has already responded to voter concerns about the handling and disposal of hazardous substances with RCRA and CERCLA and most states, including Ohio, have done likewise. However, this is not the end! We are just beginning to see the emergence of toxic tort cases such as the asbestos and black-lung claims. Such claims, together with government enforcement actions, are certain to occupy the time of more and more industrial technicians. Hopefully this paper has served to alert the participants in the Third Annual Ohio Environmental Engineering Conference to some of the dangers that lie ahead and some methods for avoiding these pitfalls. Those hazardous waste generators and transporters who are not so aware and persist in conducting business as usual are bound to become enmeshed in a tangled net of enforcement activity and/or private damage claims which could lead to economic losses which could well be fatal to their entire enterprise.

CHAPTER 34

GENERATOR LIABILITY UNDER SUPERFUND

Richard T. Sargeant
Eastman & Smith

INTRODUCTION

After a lengthy debate, the Comprehensive Environmental Response, Compensation and Liability Act of 1980 (hereinafter "CERCLA" or "Superfund") was hurriedly written and passed. It was intended to provide funding and a mechanism for cleaning up "leaking" inactive hazardous waste sites. Despite the fact that the idea of such a fund had been debated for years, the statute that was hurriedly passed in December of 1980 was a last-minute compromise passed by a lame duck Congress just prior to the final recess in December of 1980. Consequently, portions of the statute are not particularly well drafted, and the legislative history is incomplete. Indeed, there were ambituities purposefully left in the statute by Congress because it was unable to make certain necessary policy decisions regarding liability. Issues concerning whether responsible parties under CERCLA will be subject to strict and joint and several liability are examples of these unresolved issues.

While CERCLA's scope is not limited to abandoned hazardous waste sites, it was this problem that caused Congress to pass CERCLA. It was perceived that there a large potential for harm from abandoned, but leaking facilities and that "something should be done." Implicit in this position was the conclusion that the then existing environmental statutes and common law were inadequate to provide the mechanisms and resources needed to clean-up those sites. Therefore CERCLA is different from most environmental statues because its thrust is not regulation of ongoing activities (although there is some of that in the statute), but rather taxation and assignment of liability.

ASSIGNMENT OF LIABILITY

Predictably, the difficult decisions in drafting such a statute concerned who would have to pay for the clean-up of the sites and the damage caused by them. Traditionally in this

Management of Toxic and Hazardous Wastes, H. Bhatt, R. Sykes, T. Sweeney (Editors)

country an injured person has a right to recover from persons who were at fault and who proximately caused the injury. The problem with abandoned and leaking waste sites is at least twofold. First, it is frequently difficult to prove who, other than the owner or operator, is at fault. Secondly, the problems caused by some of these sites are frequently extremely expensive to remedy, and the owners or operators are frequently judgement-proof.

Congress was concerned about the situation where an abandoned site presented a problem, but the owner-operator lacked the resources to remedy the situation, i.e., finding a responsible party with enough money to pay for cleaning up the site. If Congress had done nothing, the burden of the hazards at certain inactive hazardous waste facilities would have remained upon the enviornment and potential victims. Congress had other choices, one of which was to spend general revenues and thus spread the liabilty over all taxpayers. A third choice was to deliniate persons with resources and require them to pay for the clean-up of the sites even though they may not be at fault or have caused the problem.

The first option was fair in the sense that it did not create liability for persons who were not at fault, but was frequently ineffective because it did not provide the money necessary to clean-up the problem. The second option was less fair, but very effective. The third was likewise frequently unfair, and is somewhat less effective than the second option. Congress chose the third option.

The Resource Conservation and Recovery Act (hereinafter RCRA) required persons generating hazardous wastes to internalize the costs of the proper disposal of those wastes into the cost of the product manufactured or the service provided. That is a fair requirement so long as it is applied prospectively. However, pursuant to CERCLA, a generator who may have paid a licensed and legitimate recycling or disposal facility to properly recycle or dispose of its wastes in accordance with then-approved methods, may nonetheless be responsible for the costs of remedying the effects of not only the disposal of its own wastes, but also other persons' wastes at a particular site.

The taxes and liability in CERCLA have been construed as user fees. Under this analysis, the business which creates the hazardous waste and which uses the environment to dispose of those wastes, is required to pay the costs associated with that use. However, user fees are normally not imposed on activity which took place prior to the adoption of the fees. Generators who arranged for the disposal of hazardous wastes according to government-approved methods by government-licensed recyclers or disposers are, in some instances, not being sued by those same government for the remedial measures and damages caused by the above described activity. CERCLA does not appear to limit how far back the government can go to find a generator with resources adequate for the government's purposes.

Retroactive application of the liability provision in CERCLA may violate the U.S. Constitution, Article I, Section IX, which prohibits *ex post facto* laws. *See also, Green v. U.S.,* 376 U.S. 149, 160 (1963).

SOURCES OF LIABILITY WITHIN CERCLA

A. Section 103

1. Releases: 103(a) and 103(b)

CERClA requires persons to report releases of hazardous substances to the National Response Center. Persons required to report under this section are persons who are in charge of a facility and who learn of a release of a reportable quantity of hazardous substances.

The definition of "hazardous substances" is broad and includes: (1) substances designated to pursuant to Section 311 of the Clean Water Act, (2) hazardous waste listed or having the characteristics identified in regulations adopted pursuant to 3001 or RCRA, (3) pollutants identified as toxic or priority pollutants under Section 307(a) of the Clean Water Act, (4) substances identified as hazardous under Section 112 of the Clean Air Act, (5) imminently hazardous substances designated such under Section 7 of the Toxic Substances Control Act and (6) substances specifically designated by the United States Environmental Protection Agency (hereinafter "EPA") as hazardous substances under the authority found in CERCLA.

"Reportable quantities" of these hazardous substances are to be determined by the EPA. The definition of "release" is broad, and includes: "any spilling, leaking, emitting, discharging, injecting, escaping, leaching, dumping, or disposing into the environment" [para. 101(22)]. Federally permitted releases, such as those pursuant to para. 402 of the Clean Water Act, are excluded from the requirements of the section.

Generally, the persons who are required to report under Section 103 (a) and (b) are those persons who are in charge of the facility from which the hazardous substance is released. Consequently, the responsibility to report will be that of an owner or operation. While generators frequently are also owners or operators of hazardous waste facilities, the liability depends upon the person's authority over the facility and not the source of the wastes.

However, the effect of a report of a release is to precipitate the requirement that the released substances be cleaned up and that future releases be prevented. While the government will normally look first to the owner-operator, if, for whatever reason, the owner-operator is unable or unwilling to remedy the problem, the government may then turn to generators who may have "arranged" for the disposal of their wastes at the facility, and who thus may be "responsible parties".

Subsection 103(c) required persons to report or notify the EPA of the existence of hazardous wastes facilities. The notification was to be submitted by June 9, 1981, and was to indicate: (1) the amount and type of hazardous substances at each facility, and (2) any known or potential releases of said substances. The requirements pertain primarily to persons who owned or operated the facility at the time of the disposal or at the time or reporting, or who controlled activities at such a facility immediately prior to its abandonment. Although generators frequently also own or operate hazard substance treatment, storage or disposal facilities, the reporting obligation is triggered by the status of owner or operator rather than of generator. Consequently the obligations set forth in this subsection are largely beyond the scope of this chapter.

Section 103 provides a fifty year retention requirement for records pertaining to the location, title or condition of hazardous waste facilities, and the identity, character, quantity, origin or condition of hazard substances deposited in hazardous waste facilities. It is a criminal offense to knowingly destroy, mutilate, erase, depose of, conceal or otherwise render unavailable or unreadable or falsify any such records. Persons required to retain these records are the persons who are subject to the reporting obligations in Section 103(c) noted above.

B. Section 104

In cases where hazardous substances are released or where there is a substantial threat of such a release into the environment which may present an imminent and substantial danger to the public health or welfare, Section 104 of CERCLA provides authority to the President to act to remove or arrange for the removal of, or to provide remedial action relating to said hazardous substance. The President may also take other response measures which he deems to be necessary to protect the public health or welfare or the environment.

The above actions may not be taken if the President or his designee (the administrator of the EPA and the Coast Guard) determine that the removal and remedial action will be properly done by the owner or operator of the facility. Any actions taken must be consistent with the National Contingency Plan. With certain exceptions, the government's response pursuant to para. 104(c) (1) may not continue after one million dollars ($1,000,000.00) has been obligated or six months has elapsed from the date of initial response or threatened release.

Subsection 104(e)(1) requires persons who store, treat, or dispose of hazardous wastes (and where necessary generators of transporters) to furnish information relating to the hazardous substances and to permit, upon request, the EPA to have access to and to copy records relating to such substances. Likewise the EPA may enter (at reasonable times) any establishment where such hazardous substances have been generated, stored, treated or disposed of, to inspect and obtain samples.

The EPA may not provide any remedial actions pursuant to para. 104 unless the state agrees to: (1) provide future maintenance of the removal and remedial actions, (2) assure the availability of a hazardous waste disposal facility acceptable to the EPa, and (3) pay or assure payment of 10% of the cost of all remedial action, or a minimum of 50% in response to a release at a facility that was owned at the time of disposal by the state or a political subdivision thereof.

C. Section 106

Section 106 of the Act is the imminent hazard section of CERCLA, and is analogous to 7003 in RCRA. Section 106 allows the EPA to respond to what it determines to be an imminent and substantial endangerment to the public health and welfare or the environment because of an actual or threatened release of a hazardous substance from faciliity.

The EPA has the option of bringing an action in a federal district court to abate the danger or threat, or to issue "such orders as may be necessary to protect the public health and welfare and the environment". However, few of these orders have been issued; the EPA normally sought an injunction in matters asserted to involve imminent and substantial endangerment. The statute requires the EPA to establish and to publish guidelines for using these imminent hazard provisions.

The penalty for willful violation of or refusal to comply with an order issued by the administrator pursuant to 106 includes a fine of up to $5,000.00 per say of noncompliance. Pursuant to para.107, failure to properly provide removal or remedial action in response to such an order may subject the responsible party to treble damages.

"Imminent" has been interpreted to require something less than an "immediate" threat. Action may be brought under the provision if conditions may present such a hazard. Proof of actual harm is not required, nor need the harm be truly imminent. The court in United States vs. Vertac Chemical Corp., 489 F. Supp. 870 (E.D. Ark., 1980), constued similar language in para. 7003 of RCRA and determined that it was sufficient if the situation "gives rise to a reaonsable medical concern for the public health." The court balanced the low probability that the harm would occur against the very serious and dire consequences of such an occurrence. See, however, the standard used in United States vs. Solvents Recovery Service of New England, 496 F. Supp.

1127, 1143, note 29 (D. Conn., 1980 which states:

> "The defendants may rest assured that this court does not view section 7003 as a general "clean-up statute".
>
> * * *
>
> "Only proof of health or environmental emergency for which the defendants are responsible would justify some of the releif sought by the United States in this action."

In footnote 29, the court added:

> "... situations which do not present true emergencies are better health with through the more comprehensive, if more cumbersome, provisions of RCRA and the ERA regulations promulgated thereunder than in an action under 7003."

See also para. 300(i) of The Safe Drinking Water Act in Title 42 United States Code:

> "Nor is the emergency authority to be used in cases where the risk of harm is remote in time, completely speculative in nature, or de minis in degree.

See also Mott, Liability for Clean-Up of Inactive Hazardous Waste Disposal Sites, Hazardous Waste Litigation, Practicing Law Institute, 1981.

In U.S. v. Wade, No. 79-1426, (E.D. Pa., Sept. 7, 1982), the district court prohibited the EPA from using para. 106 (and para. 7003 of RCRA) to force non-negligent past generators to pay in advance for remedial measures. The court noted that the defendant generators were not accused or currently dumping waste at the site or of acting negligently in previous disposals at the site.

The government has sought to recover from the off-site generators for expenses incurred or to be incurred, in planning and cleaning up the waste site. The court refused to accept the argument that non-negligent past off-site generators were "contributing to" the problem as that term is used in para. 7003 of RCRA.

The district court acknowledged that para. 107 of CERCLA provided a mechanism whereby the government can recover the clean-up costs of from off-site generators whose wastes have caused a pollution problem. The court said that, had the government undertaken the clean-up and then proceeded against the defendants pursuant to para. 107. statutory authority to support recovery would have been clear. However, the court found no evidence that Congress intended past off-site generators to be liable under the emergency injunction relief authority of para.

106 merely because such generators appeared to be liable under the statutory scheme established by para. 107. The court also found the government's demand for what was essentially money damages not be authorized under para. 106 of CERCLA or para. 7003 of RCRA.

The court in Wade relied in part on the district court's decision in United States vs. Price, 523 F. Supp. 1055, 1067 requested injunction was"... designed essentially to remedy the ill effects of past actions, not to restrain ongoing statutory violations." Also: "A plaintiff cannot transfer a claim for damages into an equitable action by asking for an injunction that orders the payment of money."

The United States government appealed the lower court's decision denying preliminary relief pursuant to 106 and 7003. The appellate court in United States vs. Price, 688 F.2d 204 (3rd Cir., 1982) affirmed the result in the lower court, but criticized some of the reasoning. It determined that the issue of who should bear the cost of abating a hazard and obtaining alternate water supply was not properly decided pursuant to an application for preliminary injunction. The Third Circuit disagreed, however, with the district court's somewhat narrow interpretation of the remedies available the government pursuant to para. 7003 of RCRA and para. 106 of preventive in nature rather than compensatory.

An issue likely to be litigated under para. 106 of CERCLA is whether the section imposes retroactive liability for conduct begun and ended prior to the effective date of CERCLA. Similar arguments were made pursuant to para. 7003 of RCRA in United States vs. Solvents Recovery Service of New England, supra. There the court determined that: (1) federal common law of nuisance applied, and (2) the determining factor was the offensive conditions themselves, not the activity that gave rise to those conditions or whether there was continuing disposal. The court indicated that the leaching was ongoing, and thereby avoided determining whether RCRA could be applied retroactively by deciding that retroactive application was not involved in the particular case.

The Court in United States vs. Midwest Solvent Recovery, Inc., 484 F.2d 138 (N.D., Indiana, 1980) construed para. 7003 as jurisdictional only and required the government to provide not only endangerment, but also that in the absence of preliminary injunctive relief, irrepairable harm will result. The government did so in that case, and the requested injunctive relief was granted.

D. Section 107

Section 107 describes the liability of responsible parties under CERCLA. The section identifies a responsible or liable party as:

> "...any person who by contract, agreement, or otherwise arranged for disposal or treatment, or arranged with a transporter for transport for disposal or treatment, of hazardous substances owned or possessed by such person, by any other party or entity, at any facility owned or operated by another party and entity and containing such hazardous substances ... from which there is a release or a threatened release which caused the incurrence of response costs, of a hazardous substance." (emphasis added).

The government must consequently prove that a putative resonsible party "arranged" for the disposal of substances that were hazardous and which were owned or possessed by such person. The arrangements must be for the hazardous substances to be disposed or treated in a facility owned by another person, and one which contains such substances, i.e., the substance once belonging to the alleged responsible party. The government therefore must provide that the "arranging" party's wastes are in the particular facility from which there is a threatened release.

However, there is some question with respect to whether the language of the statute requires that the substance threatening to be released and causing the incurrence of response costs be the particular hazardous substance formerly belonging to the putative responsible party. The language of the statute can be read to require only a release of a hazardous substance from the particular site with which the generator's hazardous wastes are still present. However, since this interpretation essentially imposes liability for associating or doing business with someone whose actions have resulted in a hazard, it is unlikely to be adopted.

Pursuant to the language of 107(a)(3), a generator who did not choose the particular disposal site should not be liable. If generator arranged for its wastes to be disposed of in site X, and the transporter transported the substances to site Y, the generator may argue that the it did not "arrange" for the disposal of its waste at the facility from which there has been a release.

The statutory limitations of liability are so high as to not warrant much discussion. The maximum liability for stationary facilities other than vessels, motor vehicles, aircraft or pipelines is fifty million dollars ($50,000,000.00) for damages to natural resources. There is no limit on the total liability for reimbursement of response costs.

Additionally, subsection 107(c) (2) provides that there is no limit to a responsible party's liability for response costs or damages if: (1) the release or threat of release was the result of willfull misconduct or willful negligence (within the privity or knowledge of such person), (2) the primary cause of the release was a violation (within the privity or knowledge of such person) of applicable safety, construction or operating standards or regulations, or (3) such person fails or refuses to provide all reasonable cooperation and assistance requested by a responsible public official in connection with response activities under the national contingency plan. "All reasonable cooperation and assistance" is not defined or described by the statute; likewise, "a responsible public official" is not defined.

Additionally, if a person fails "without sufficient cause" to properly effect the removal or remedial action pursuant to an order issued pursuant to para 104 or 106 of the statute, said person may be liable for punitive damages in an amount of at least the amount of any costs incurred by the fund as the result of such failure to take proper action, and up to three times said amount.

In an attempt to define what the Congress left undefined, i.e., "without sufficient cause", one comentator said:

> " A colloquy on the Senate floor between Stafford and Helms, however, helps provide some guidance. This discussion makes clear that sufficient cause includes such things as (1) the person who is subject to the President's order was not a party responsible for the release of the hazardous substance, (2) substantial facts or questions gave a party good reason to believe himself not be a responsible party, or (3) the party subject to the order was not a substantial contributor to the release or threat of a release. Finally, sufficient cause for not complying with an order will cover a party if at the time he did not have the necessary financial or technical resources to comply or if no technical means for compliance was available." Evans and Frost, "Changes In Spill Liability Occasioned by the Passage of Superfund", 1982 Hazardous Spills Conference.

The willful conduct or willful negligence specified in the statute need not be that of the defendant generator or his agents, it may also be binding upon a generator if it is done by someone in privity with him (or his agents). As an example, a generator may contract with the transporter for the disposal of waste at a responsible and secure disposal site. However, if the transporter dumps the waste in a stream, the generator is arguably responsible. (However in the instance, as noted above, the generator could reasonably argue that it did not "arrange" for the disposed of the substances at that manner). If the company that the generator paid to recycle its wastes chooses not to recycle, this section of CERCLA may be read to nonetheless impose liability on the generator.

Responsible parties are liable for the following items or damages:

"1. Any costs of removal or remedial action incurred by the United States Government or a State not inconsistent with the national contingency plan;

2. any other necessary cost of response incurred by any other person consistent with the national contingency plan;

3. damages for injury to, destruction of or loss of natural resources including the reasonable costs of assessing such injury, destruction or loss resulting from such a release."

Section 107 does not provide for liability for anticipatory response costs; instead it provides for liability for costs already incurred or damages already suffered. The government may not rely on 107 for authority to being suit for future damages or response costs to be incurred later.

The court in Wade, supra discussed what was remedial relief vis-a-vis prospective relief. The court in Wade refused to allow the government to circumvent the requirement that the costs be already incurred by styling the matter as one based on solely para. 106 injunctive relief.

This argument, however, is double-edged. If the government does the work, it will typically cost much more than if it had been done by private parties. This provides an incentive to settle before the government incurs response costs. If the generator's liability is likely, even though the amount is not established, it is likely that the most cost-effective method of responding to the situation is to do it privately. (The use of a performance bond, such as was used in the Seymour Recycling settlement, can bridge the gap between private party's estimate which normally will include generous "safety factors".

In addition to being "incurred", the response costs for remedial measures must not be inconsistene with the terms of the national contingency plan. (If the response costs are incurred by persons other than the federal or state govenments, they must be consistent with the national contingency plan).

Responsible parties are liable for damages for injury to or destruction of or loss of natural resources resulting from a release of hazardous substances. The definition of natural resources, found in para. 101(6) requires that any such natural resource have some special relationship to the federal, state or local governments, thus preserving an argument to the putative responsible party that wildlife or other natural resources are not "natural resources" as used in para.107 unless they have this special status or relationship to the government or were located on public property.

CERCLA does not provide a mechanism for third parties to recover for personal injuries or medical expenses. Although earlier drafts of the statute included such provisions, the promoters of the concept did not have the votes to make it part of the statute. Liability for personal injuries from exposure to releases from hazard waste facilities is unchanged. The normal common law burdens and principals of causation with respect to these types of injuries are substantial, but are necessary to tie the injury to the entity alleged to be responsible for it.

While parties may not contract away liability under CERCLA through the use of indemnity or hold-harmless agreement, para 107(e) provides that parties may execute and enforce hold harmless or indemnity agreements among themselves. In addition, subsection 107(e) (2) provides that causes of action that a person otherwise might have in subrogation or under a guaranty agreement are not invalidated by CERCLA.

DEFENSES

It will not be possible to confidently describe the scope of off-site generators' liability under Superfund until the resolution by the courts of issues such as: (1) whether the act provides for strict liability, (2) whether the act provides for joint and several liability, and (3) generally, whether liability will attach to a statutory "responsible party" in situations where the party did not cause the conditions which occasioned the remedial actions or damages.

A. Strict Liability

Congress did not specify whether strict liability will be the standard of liability under CERCLA, or, if it is, in what form. Strict liability elimintes the requirement that a plaintiff provide fault or negligence. Section 101(32) of CERCLA provides that the liability under that statute is to be the same as the standard for liability under the para. 311 of the Clean Water Act. For cases construing para. 311 as prvoding for liability without fault for clean-up costs, see Burgess vs. M/V Tomano, 564 F. 2d 964, 981-982 (1st Cir., 1977) 39 435 U.S. 9 (1978), and U.S. vs. Bear Marine Services, 509 F. Supp. 710, 714 (E.D., La., 1980). For contrary authority see In Chicago, Milwaukee, St. Paul & Pacific Railroad C. vs. United States, 575 F.2d 839 (CT. Cl, 1978), and United States vs. M/V Big sam, 505 F. Supp. 1029 (E.D., La., 1981).

While the House Bill originally contained a provision providing for the application of strict liability, it was deleted in Senator Stafford's compromise. Although the original provision requiring the use of strict liability was deleted, para. 302(d) of the act provides the CERCLA shall not be interpreted as modifying the obligation or liabilities of persons under federal or state law, including common law, with respect to release of hazardous substnces. The subsection provides specifically that:

> "The provisions of this Act shall not be considered, interpreted or construed in any way as reflecting a determination, in part or whole, or policy regarding the inapplicability of strict liability, or strict liability doctrines, to activities relating to hazardous substances, pollutants, or contaminants or other such activities."

However, strict liability is not synonymous with absolute liability or market share liability. Generators may reasonably argue that plaintiff still must prove a causal connection of link between the injuries or damages and the activity of the defendant.

B. Defenses Listed in para. 107(b)

The defenses expressly set forth in the statute are narrow. Subsection 107(b) provides that there shall be no liability for a person otherwise liable who can establish by a preponderence of the evidence that the release of a hazardous substance and the damages resulting therefrom were caused soley by: (1) an act of God, (2) an act of war, or (3) an omission of a third party other than: (a) an employee or agent of the defendant or (b) one whose act or omission occurs in connection with a contractual relationship.

The third defense above cannot be used unless the defendant establishes that he exercised due care with respect to the hazardous substances, and that he took precautions against foreseeable acts or omissions of such third party (and the consequences of those acts or omissions). Upon its face the statute appears to create liability for any person who hazardous substances are found at the facility in question regardless of whether those hazardous wastes were actually released or whether those hazardous wastes in any way caused the need for remedial measures or the damages.

C. Causation

A generator may choose to defend an action brought pursuant to CERCLA by arguing that its hazardous substances are not the cause of the "problem". There are at least three levels of involvement. The first is the situation where a particular generator's hazardous substances went to the hazardous waste facility from which there is presently a release of hazardous substances, but the government is unable to provide that the particular generator's wastes are presently at the facility. In this instance, the generator may argue that it is not liable since the government cannot provide that the facility contains "such" hazardous substances as required by para. 107(a) (3) of

CERCLA. In addition, the government could not prove that the particular generator's hazardous waste caused the damages or the need for remedial action.

The second level is where the generator sent waste to the hazardous waste facility and it is present at the facility, but there is not evidence that the particular generator's wastes are leaching from the site. While the government may be expected to argue that the statute provides for liability under these circumstances, the generator may argue that its wastes did not cause the damage or the need for remedial actions. An argument by the government that a generator is liable merely because it dealt with a party who has caused a "problem" is flawed, and is unlikely to be endorsed by the courts.

The third level is the situation where the generator sent hazardous substances to the facility, those substances remain at the facility, and those substances are leaching, but are not the substances that are creating the need for particular response costs. This generator also has the argument that the response costs are not caused by its waste. There should be not liability for the costs of cleaning up other generator's waste. In situations such as this, the liability for damage is reasonably separable and apportionable. Liability should be assigned to those persons whose actions and wastes caused the need for the remedial measures.

D. Joint and Several Liability

The earlier versions of CERCLA provided for joint and several liability. References to joint and several liability were deleted in the final bill because the votes were not there to pass it. There are statements which were placed into the record by Congressman who were attempting to establish that existing common law principles of liability already required joint and several liability with respect to abandoned hazardous waste sites. However, other Congressmen put into the record statements to the effect that the deletion of the references to joint and several liability in the act should be interpreted by the courts to reflect Congress intention that liability under CERCLA should not be joint and several. The conflicting statements by individual Congressmen prove only that Congress deferred the question of joint and several liability to the courts to decide based upon the existing common law. Liability under the act under CERCLA will not be joint and several unless the courts construe the particualr situation in a particular case to require it.

Senator Helms entered a statement into the record to the effect that CERCLA provided that:

"The Government can sue a defendant under the bill only

for those costs or damages that it can prove were caused by the defendant's conduct.

Representative Florio entered into the record a letter from Justice Department to the effect that:

> "It is the Department's view that common law provides for joint and several liability where the act or omission of two or more persons results in an indivisible injury."

In both of these interpretations is the requirement that before a generator may be held liable, it must have caused at least part of the injury. In the situation where the government cannot prove that the particular generator's hazardous waste remains at the site or is causing any of the problem, there should be not liability.

From a fairness point of view, in many applications, the concept of joint and several liability in cases involving hazardous inactive waste sites has little to recommend it. Unless the individual involvement of each generator can be established, and unless the injury is truly inseparable, joint and several liability under CERCLA would result in the government transferring society's problem onto others who were not the cause of the problem. The government is thus narrowing the risk to parties who may not be at fault, and who were not the cause of the problem.

Arquably EPA has acknowledged that appointment is appropriate instead of joint and several liability by offering to settle with generators in the ChemDyne and Seymour Recycling matters on the basis of the volumes contributed. However, this willingness may only evidence the reality that no one contributer would agree to pay for the entire clean-up when others were also obviously involved.

EPA's official position is that CERCLA and the common law provides for joint and several liability for generators who arranged for waste to be sent to a problem site. The notice letters and subsequent negotiations have included threats by the government to impose on off-site generators (including those only minimally involved with the site) the liability for the entire cost of cleaning up the site. (EPA does recognize and acknowledge a de minimus or "one barrel" exception to the claimed joint and several liability).

The issue of whether or not joint and several liability will apply to generators will at all likelihood be settled through interpretation of or expansion of the common law rather than any discovery of any such requirement within the language of CERCLA.

Contribution

If joint and several liability is imposed for some or all generators, the issue of whether there is a right to contribution between joint tortfeasors must be resolved. As a general rule, the common law does not provide a right of contribution between joint tortfeasors. However, most states have adopted statutes which expressly provide for contribution, e.g., Ohio Revised Code para. 3207.31-2307.32.

It is also true that there is a no common law right of contribution when joint and several liability is imposed pursuant to federal statute unless the statute provides for the right or provides the federal courts with the authority to create common law on the subject. Texas Industries, Inc. vs. Radcliff Materials, Inc., 101 s. C. 1061 (1981).

However, certain generators who may be liable under CERCLA are more attractive candidates for contribution than were anti-trust defendants. The punitive damage provisions in the anti-trust laws include an element of fault and conscious improper action. By contrast, generators who may find themselves strictly and jointly and severally liable under CERCLA will frequently not even have caused the injury or damages for which the government has sought reimbursement. There will be instances where persons are found liable for very substantial amounts in the absence of any fault or any meaningful control over actions of those who actually caused the damaged. Indeed, the amounts in question may be ruionous for some small or medium-sized companies, and will be material even for the largest corporations.

If they discover no other method, the courts may reasonably find in the legislative history of CERCLA the intention of Congress to have the federal courts create common law on the subject of generators' right to contribution in these cases. The courts will find a method of allowing these defendants the right to seek contribution:

> "There is obvious lack of sense and justice in a rule which permits the entire burden of a loss, for which two defendants were equally, unintentionally responsible, to be shouldered onto one alone, according to the accident of a successful levy of execution, the existence of liability insurance, the plaintiff's whim or spite, or his collusion with the other wrongdoer, while the latter goes scot free." Prosser, Law of Torts, 4th Edition, para. 50 (West, 1971)

Indeed, Prosser strongly favors apportionment of damages saying:

> "In general, it may be said that entire liability will be imposed only where there is no reasonable alternative." Id. at para. 52

Apportionment

Section 433(a) of the Restatement (Second) of Torts is captioned "Apportionment of Harm to Causes" and provides that damages or harm are to be apportioned among two or more causes where there are distinct harms or there is a reasonable basis for determining the contribution of each cause to a single harm. Thus, if the damages are not truly inseparable, the Restatement provides for apportionment between the persons causing them.

> "...among multiple generators who can prove the quantum and type of waste there is a 'reasonable basis' for apportionment provided the rule is a relaxed one. Similarly, in cases of successive ownership or use, there would seem to be a basis for apportionment." Rodburg, "Apportionment of Damages in Hazardous Waste Litigation", page 197, Hazardous Waste Litigation, 1982, (Practicing Law Institute, 1982), page 183.

That author continued on page 205:

> "The absence of such a joint and several liability provision in Superfund means that congress has not come to grips with any such public policy issue. To the extent that it has been left to the courts, therefore, the courts should be guided by traditional common law principles of apportionment. Accordingly there is no a prior reasoning to find joint and several liability, but no reason not to in a proper case."

The justification for joint and several liability under comon law is that the actions or omissions of two or more persons cause or result in an indivisible injury. If the injury is not truly indivisible, joint and several liability should not be imposed.

There is a distinction in the common law between the situation when two causes concur to cause an injury, and the situation where there is either a breach of a joint duty by several defendants or several defendants act in concert thereby producing harm of a indivisible nature. Rodberg, supra, at 191, cites the Comments to Restatement Second of Torts, para. 433(A) to the effect that when tortfeasors act separately, and when the harm may be divided reasonably (e.g., by degree), the injury should be apportioned among the tortfeasors.

The main objection to apportionment is the claimed difficulty in calculating the respective shares. Prosser, however, notes:

> "The difficulty of any complete and exact proof in assessing such separate damages has received frequent mention in all these cases, but it has not been regarded as sufficient justification for entire liability." Id. at para. 52

Prosser notes:

> "Nuisance cases, in particular, have tended to result in apportionment of the damages, largely because of the interference with the plaintiff's use of his land as tended to be severable in terms of quantity, percentage, or degree. Thus, defendants who independently pollute the same stream, or who flood the plaintiff's land from separate sources, are liable only severally for the damages individually caused, and the same is true as to nuisances due to noise, or pollution of the air." Id.

In the same section Prosser notes that that the requisite proof does not have to be exact, and cites examples where courts have been very liberal in permitting a jury to "award damages where the uncertainty as to their extent arises from the nature of the wrong itself, for which the defendant, not the plaintiff is responsible." Prosser noted that requirements of proof in those cases are relaxed and the courts have not required exact evidence; general evidence with respect to proportions will suffice.

A primary difference between apportionment and contribution is the risk to the plaintiff that one or more of the liable parties will not have the resources to remedy the problem.

Contribution in some jurisdictions is calculated in accordance with the degree of liability or participation in the negligentt conduct, e.g., comparative negligence. In situations where defendants were all negligent or in some way at fault, it may be equitable to assign some risk of overpayment to them. However, in situations where the defendants are strictly liable in the absence of any fault, the argument that the risk should remain with them is not persuasive.

The Rodberg article made the further interesting distinction between apportionment and contribution. Apportionment divides the damages by calculating the portion of the injury caused by each respective defendant. Contribution either divides the damages pro rate by the number of solvent defendants or in accordance with the relative degree of fault of the defendants. The example used was that of two parties, one of whom sent 200 drums to an abandoned site and the other 1,800 drums. Apportionment could result in a 10/90 split. contribution may result in a 50/50 split, since both defendants are equally "at fault" under strict liability. Id. at 200.

While it is not yet known how the courts will decide the issue of joint and several liability under CERCLA, they should consider not only the public's interest in acquiring money to clean-up abandoned waste sites (to a major degree this has already been done through a tax on crude oil and chemical feedstocks), but also society's interest in the equitable resolution or division of responsibility.

CHAPTER 35

ENVIRONMENTAL LAW AND CONTRACTOR LIABILITY

J. Wray Blattner, Esq.
Smith & Schnacke, L.P.A.
Edward A. Hogan, Esq.
Lowenstein, Sandler, Brochin,
Kohl, Fisher, Boylan & Meanor

INTRODUCTION

By now, all of you are aware of the scope of the potential environmental problems surrounding the management of hazardous wastes. The statistics are off-quoted and speak for themselves: Approximately 47.5 million metric tons of hazardous wastes are generated annually in the United States. In 1979, the United States Environmental Protection Agency estimated that approximately 90 percent of all hazardous wastes ever generated had been disposed of improperly. Today the Agency is actively investigating over 7,000 of these sites, over 1,000 of which are believed to pose significant and immediate threats to human health and the environment.

What you may not be aware of is the extent to which recent developments in the law have substantially broadened the scope of potential liability for those who handle hazardous wastes. The principal statutes in this area are the Resource Conservation and Recovery Act of 1976 (RCRA) and the Comprehensive Environmental Response, Compensation and Liability Act of 1980 (commonly known as Superfund). There are also a number of other federal statutes, including the Hazardous Materials Transportation Act, the Clean Water Act, and the Toxic Substances Control Act that regulate the handling of toxic and hazardous wastes, and which impose severe penalties for their violation. Additionally, many states have statutory programs that substantially adopt the basic elements of the federal program. The passage of these laws has greatly enlarged the potential liability for mishandling hazardous waste. The exact scope of statutory liability for improper handling of hazardous wastes, particularly where an imminent hazard is alleged to exist, is not yet clear. Courts are, at the same time, expanding the scope of common law tort liability for injuries and damage caused by hazardous wastes and lowering some of the traditional procedural barriers, which in the past have limited the

Management of Toxic and Hazardous Wastes, H. Bhatt, R. Sykes, T. Sweeney (Editors)

ability of a plaintiff to obtain relief.

In short, this is a rapidly evolving area of the laws, and the potential liabilities of those generating, handling, or disposing of hazardous wastes, and those constructing waste facilities or engaging in waste site cleanups are likely to increase at a similar pace.

This presentation will examine the potential liabilities confronting contractors in the area of hazardous waste management, with a focus on the undertaking of hazardous waste "cleanup" operations at hazardous waste sites. I will examine four primary sources of liability: RCRA, Superfund, the common law, and contractually assumed liability.

II. RCRA AND SUPERFUND

RCRA established a "cradle to grave" regulatory approach to the management of hazardous wastes, and represented Congress' first attempt to address the problems of hazardous waste disposal. RCRA authorized EPA to adopt standards and practices governing generators, transporters, and disposers of hazardous waste. While, as will be discussed a little later, courts have applied RCRA in the context of liabilities for harm arising from abandoned waste sites, the law was designed to address ongoing and future waste management. In order to adequately respond to the dangers posed by abandoned and uncontrolled hazardous waste sites, in 1980 Congress enacted Superfund.

Superfund established a $1.6 billion hazardous substance response trust fund to finance prompt government cleanup of abandoned hazardous waste dump sites.

The fund is financed by taxes levied over a five year period on the oil and chemical industry and by smaller contributions from general revenues. The fund provides a reservoir of finances for the cleanup of waste sites targeted by the "National Contingency Plan." The Act also established the liability of generators, transporters, and disposers of hazardous waste for "releases" to the envieronment and authorized fund representatives to sue responsible parties for reimbursement of cleanup costs incurred by the Government at each disposal site.

The National Contingency Plan establishes procedures and standards for responding to hazardous waste releases, including methods for discovering, investigating and cleaning up waste sites. Recently the Plan has received a great deal of attention as the Agency has proposed a statutorily required list of approximately 400 of the waste sites which have been deemed to pose the most serious dangers.

In reality, the fund is not so "super"; it is of such a limited size that its reserves would be quickly exhausted without the liability-assessment provisions of the Act. Because many disposal site owners and operators and parties who transported the materials to the site are either insolvent or impossible to locate, the Government's success at replenishing the fund depends upon finding responsible parties who generated wastes to hold liable,. As an alternative to pursuing the cleanup and then seeking reimbursement, EPA has made demands upon potentially responsible parties to voluntarily effect a cleanup of the alleged hazardous waste site.

III. THE ADMINISTRATION OF SUPERFUND CLEANUPS

Before turning to potential liabilities that may be incurred in Superfund cleanup operations, I'd like to briefly identify the governmental entities which are administering funds for site cleanups. These entities contract with engineering and contracting firms for the design and implementation of hazardous waste removal activities.

The first and most obvious entity is the U.S. Environmental Protection Agency itself. Many of the initial emergency cleanup actions under Superfund were administered directly by EPA. It should be noted however that those actions do not constitute a large number of sites -- only the most threatening ones.

In February 1982, EPA and the Army Corps of Engineers reached an agreement under which the Corps will manage Superfund contracts awarded by EPA for design and construction work at abandoned hazardous waste sites. The agreement is quite vague and allows the Corps to refuse to take on the management of any Superfund contract if it decides that the remedial action selected by EPA is not, in the Corp's view, reasonable. Under this agreement, EPA will manage all activities at the site before the Corps accepts the project and all remedial program activities not performed by the Corps. In addition to managing the design and construction of remedial action assigned by EPA, the Corps will provide technical assistance to the Agency in support of response actions. It should be noted that virtually all of the actual cleanup and remedial work is expected to be contracted out to private concerns.

Before EPA awards design and construction projects, the Agency must perform feasibility studies and prepare remedial action plans. This work, initially done by the Agency, will also be handled by private firms. On September 30, 1982, EPA awarded two prime zone contracts for all technical support work performed under Superfund. The first contract, covering USEPA's Regions I and IV, was awarded to NUS Corporation of Maryland and its three subcontractors DynaMac Corp., Brown & Root, and IT Enviroscience. The second prime zone contract went to CH2M Hill of Virginia and its subcontractor Ecology & Environment. Each of these four-year contracts are for approximately $80 million.

While these companies will be preparing feasibility studies and remedial action plans, it is expected that some sites, those posing imminent hazards, will require interim removal action. EPA has recognized this and specified that over the four years of each contract, $30 million of immediate removal work must be bid out to other subcontractors; geologists, earth science specialists, fence builders, earth movers, haulers, and the like.

The third entity that will be spending federal Superfund dollars will be the states. A number of states under grants and contracts with USEPA are administering Superfund cleanup actions.

The fourth entity that may be dealing directly with contractors will be groups of potentially responsible parties who are negotiating voluntary cleanup operations with the Agency.

Obviously, hazardous waste site cleanup operations are fought with such risks as the danger of human exposure to toxics, fire or explosion, as well as the risk of a release of previously contained wastes. These potentialities must be borne in mind throughout the planning and implementation of any hazardous waste cleanup effort. The occurrence of any of these incidents creates the possibility of significant liabilities.

IV. LIABILITY UNDER SUPERFUND

Liability under Superfund is far reaching. Where a facility experiences a release or threatened release of a hazardous substance the following persons may be liable for cleanup and remedial action costs:

1. The owner and operator of a facility;
2. Anyone who at the time of disposal of any hazardous substance owned or operated the facility;
3. Anyone who "arranged" for disposal or treatment, or "arranged" with a hauler for transport for disposal or treatment of hazardous substances, at a facility owned or operated by someone else; and,
4. Anyone who accepted a hazardous substance for transport to a disposal or treatment facility selected by that transporter.

Liability for these costs is strict, in the sense that no showing of fault or negligence is required.

Very high dollar limits on liability are provided under the Act. For most facilities, the limit is the "total of all costs of response plus $50 million for any damages under this title." Those limits do not apply if willful misconduct or willful negligence is shown.

It is important to note that, given the trend towards expanding environmental liability, it is conceivable that a contractor who designed and constructed a waste facility could be held to have "arranged" for disposal or treatment. Similarly, a contractor engaged in a waste site cleanup operation could be viewed as a facility "operator." If so, then such persons, as well as a contractor who hauled wastes to a facility site, faces potential liability under Superfund.

There is no limit under the statute to how far back in time an act by a former waste generator, owner, operator, or transporter who contributes to a current release or threatened release may have occurred in order for him to be liable. Similarly, there is no guarantee that the treatment or disposal of a substance or a waste _today_ will not render the generatory, owner, operation, or transporter liable many years in the future for extensive cleanup and remedial costs if there is at that _later_ date a release or threat of a release.

EPA and the Justice Department have taken the position that liability under Superfund, based upon the statutory language and general common law principles, is joint and several. Under this theory, the Government may proceed against any one or a group of potentially liable parties for the total cost of the cleanup. If such liability were established, the only remedy of the defendant(s) would be to seek contribution from other responsible parties, if they can be found and to the extent allowed by state law.

I should point out that Section 107(d) of Superfund protects persons from liability for damages arising from Government sanctioned cleanup oeprations. Note, however, that this protections only applies as to potential liability under Superfund; it does not protect a contractor against potential liability under RCRA, common law, or contractual claims. Nor would Section 107(d) protect a contractor from Superfund liability for a release which occurred during a strictly private, non-National Contingency Plan clean effort or where the contractor is included as a responsible party because he designed and constructed the facility or hauled the wastes to the site.

Superfund also creates a release notification requirement. Under Section 103(b) (3), any person "in charge" of a facility from which a "reportable quantity" of hazardous substance is released must immediately notify the National Response Center. A contractor running a cleanup operation could be such a person "in charge." Failure to notify carries a maximum $10,000 fine and a possible one year imprisonment.

With regard to third party claims, contractors hired under Superfund should be concerned about the extent to which the government can indemnify them the their liability in official Superfund cleanup actions. Since the Summer of 1981, a standard clause has been included in all EPA Superfund cleanup contracts stating that the government agrees to cover third party claims against Superfund contractors up to the level of funds available in unobligated balances in the applicable appropriation for the Superfund program. Under this clause, third party claims must be paid out of appropriated funds for the Superfund program that are available at the time of a final court judgement or settlement with EPA. What this means is that there is a possiblity that if a claim is made at the end of a fiscal year and there are not enough funds available to cover this claim, the contractor would be left holding the bag. The Government apparently views this liability as a cost of contract performance. This partial indemnification may not be available under State or responsible party-generated cleanup contracts.

V. LIABILITY UNDER RCRA

We'll now look at the potential for liability under RCRA. As I mentioned earlier, RCRA has been applied in the context of abandoned hazardous waste sites. In fact, the chief legal theory underlying the federal government's initial enforcement actions against inactive disposal sites was the "imminent hazard" provision of RCRA, Section 7003. That provision provides that where hazardous wastes may present an "imminent and substantial endangerment" to health or the environment, the Administrator may initiate an action to immediately restrain any person "contributing" to the endangerment. While the courts are divided on this question, three federal district courts have recently held that Section 7003 does in fact apply to imminent and substantial endangerments arising from inactive sites. See, U.S. v. Reilly Tar & Chemical Corp., 546 F. Supp. 110 (D. Minn. 1982), U.S. v. Price, 525 F. Supp. 1055. (D. N.J. 1981), 496 F. Supp. 1127 (D. Conn. 1980).

Note that Section 7003 creates potential liability for any person "contributing" to an imminent and substantial endangerment. Accordingly, a contractor involved in the design and construction of either an abandoned facility or an ongoing waste management facility could be deemed to have "contributed" to the waste management which is now alleged to pose an imminent and substantial endangerment.

Section 7003 penalties may reach $5,000 per day of violation.

EPA has clarified its hazardous waste regulations to indicate that a person engaged in treatment or containment activites undertaken in immediate response to discharges of hazardous wate is not subject to the RCRA standards for owners and operators of hazardous waste treatment, storage, and disposal facilities (rp C.F.R. part 264 and 265) (48 FEd. Reg. 2508, January 29, 1983). However, a contractor involved in an extended remedial cleanup action could very well be considered a "generator" of hazardous waste under RCRA.

EPA has laready indicated that it views the term "generator" broadly. In 1980, the Agency amended its RCRA regulations to modify the definition of generator so that it covers persons who remove hazardous wastes from product or raw material storage tanks, transport vehicles or vessels, or manufacturing process units in which the hazardous waste is generated. Because, in EPA's view, both the owner/operator of the process unit or storage tank as well as the contractor who may, for instance, need to empty its contents to clean, repair, or replace it contribute to the generation of a hazardous waste and because none of the parties stands out in all cases as the predoninant contributor, EPA has concluded that all the parties stands out in all cases as the predominant contributor, EPA has concluded that all the parties should be jointly and severally liable as generators. (45 Fed. Reg. 72026, October 30, 1980).

If, in the course of a cleanup operation, the contractor hauls hazardous wastes, he becomes a transporter; if he disposes of the waste at his own facility, or if the waste is stored or treated prior to disposal, the contractor becomes subject to the full gamut of RCRA standards and obligations.

Failure to comply with the RCRA regulations carries a potential civil penalty of not more than $25,000 for each day of continued non-compliance.

I. OTHER SOURCES OF LIABILITY

Companies whose activities involve hazardous materials have understandably focused their attention on RCRA and Superfund. Those laws seek to make companies liable for and financially responsible to the public for damages caused by their activities. Apart from these statutorily imposed liabilities though, anyone in the hazardous waste management business must recognize that there are well established common law theories that can form the basis for recovery by plaintiffs for personal injury and property damage. The following are common law theories for which there is a substantial body of existing stte law upon which lawsuits could, and in fact are being based.

1. Negligence
   Negligence is "conduct which falls below the standard established by law for the protection of others against unreasonable risk of harm." The law has long recognized that if a person discharges pollutants negligently, and as a result someone else suffers personal injury or property damage, a cause of action may be maintained for the damages caused as a result.
2. Trespass
   Trespass involves interference with a person's possessory interest in land. Most states recognize that one who pollutes the environment so as to cause physical damage to another's property is liable for the resulting damages in a trespass action.
3. Nuisance
   A "private" nuisance is an unreasonable interference with another's use and enjoyment of his land, or related personal or property interest. A "public" nuisance is one which involves interference with a general public right. A civil cause of action may be maintained based on either type of nuisance.

4. Strict Liability; Ultra-Hazarardous Activity in environmental-related damage litigation, there is a growing trend to hold parties strictly liable for the consequences of their actions involving hazardous materials. Strict liability, i.e. liability without fault, is related in concept to strict product liability. Just as a manufacturer may be held strictly liable for injuries caused by a defective product under that theory, firms which manage hazardous materials similarly may be held strictly liable if those materials escape and cause injury.

Most recently, additonal theories of "toxic tort liability" are being advanced with the aim of easing the plaintiff's burden. These doctrines include those of "alternative liability" and "enterprise liability."

"Alternative Liability" may be applied in situations where two or more defendants acted in a way that may have caused injury to the plaintiff, but it is not possible to tell which of their actions in fact was the cause. Today, this theory is being applied in the environmental damage context.

"Enterprise liability" address the situation where an industry wide practice may be harmful. If it can be established that an entire group breached a duty to the plaintiff, as a result of which he was injured, and through no fault of his own he is unable to identify which member or members of the group actually caused the injury, the entire group may be jointly and severally liable. This enterprise group could include the contractor who designed or constructed a waste site or who participated in a waste site cleanup action.

VII. CONTRACTUALLY ASSUMED LIABILITY

Finally, I'd like to speak briefly about a fourth source of contractor liability; contractually imposed liability.

In Superfund cleanup actions, it is probably in the interest of most potentially responsible parties to voluntarily agree to site cleanup, since in this way they can have some actually say in minimizing the expenses of cleaning up the site as well as avoiding the possibility of having penalties for non-cooperation imposed. The negotiated settlement process probably begins with EPA making demands upon potentially responsible parties (i.e. the generators of the hazardous wastes found at an abandoned waste site) to voluntarily participate in cleanup efforts. In situations where there are numerous potentially responsible parties, they may assemble a committee and request bids for remedying the site. Through a negotiation process, an agreement may be reached with EPA and, where necessary, the affected state. Any contractual obligations between the contracting parties can be expected to vary considerably and similarly can be expected to contain a great deal of variation in the distribution of liabilities, indemnifications and the like.

In the Seymour Recycling Corporation situation, 24 generators, responsible for 51 percent of the site's hazardous wastes, entered a court-approved Settlement Agreement with the Government whereby the collectively paid $7.7 million, the settling parties were relieved of all future liability for what could turn out to be extensive subsurface damage, and consequently, expensive remedial work.

Chemical Waste Management was chosen to serve as the prime contractor in the Seymour cleanup operation, and actually became a party to the consent Agreement. Thus, Chemical Waste Management is not only obligated by contract with the generators to clean up the site but the performance of that obligation is enforceable through a contempt of court measure.

In exchange for the $7.7 million contract, CWM has assumed a significant amount of responsibility. Under the Consent Agreement, CWM has agreed to "assume and does assume any and all liability arising out of its acts of omissions in the performance of the work or its failure to satisfactorily and fully perform and complete the project." CWM is not released from liability arising out of or relating to the transportation, treatment, handling, disposal, storage, or release or threatened releases of materials at, to, from, or near the Seymour site resulting from its performance. Additionally, CWM must obtain a $15 million performance bond, all applicable federal, state, and local permits, maintain careful records of its operations, and is liable for $1,000 a day penalties for delay in completion of its tasks.

This type of arrangement, whereby the clean-up contractor, for a fixed sum, assumes, with Governmental approval, all statutory, comon law and contractual obligations and risks, is attractive to responsible parties: they can "cash in" and run. This is particularly attractive when there are a large number of responsible parties, none of who wish to engage in protracted supervisory activities concerning the cleanup.

VIII. CONCLUSION

In summary, the potential liabilities of a contractor range the full span from ancient common law doctrines to the most recent federal and state statutory and regulatory enactments. A vast amount of remedial work needs to be performed at the heretofore unattended waste disposal sites throughout the nation. The extent of a contractor's potential liability will depend not only on statutory and regulatory pronouncements, the specifics of the waste site and the range of services the contractor agrees to provide, but also with the contracting entity with which it deals. Before agreeing to participate in any cleanup operation, a contractor must carefully examine all the potential pitfalls.

INDEX

Activated carbon, 330-331
Acute, 30
Adsorption, bone char, 85
Air emissions, 250-251
Alkyl benzene sulfonate, 177
Alpha, 13
Aquifer, 121, 316, 376, 383
Asphaltic sludge, 177, 180
Austrian Federal Waste Exchange, 247
Bacteria, anaerobic, 140
Bacterial, coliform, 126, 133
Barrier design, 78
Bartlett's test, 110
Benzyl alcohol, 139
Beta, 13
Bone char, 85
Bulk liquids, 319
Cement/bentonite cap, 357
CERCLA, 256, 395, 424, 417
Chauvenet's criterion, 102
Chebyshev'e inequality, 111
Chemical analysis, 122, 131
Chemical exposure badge, 34
Chloroform, 381
Chromium, 205, 243
Chronic, 30
Citizen involvement, 48
Clay inerology, 262
Clean Air Act, 419
Clean Water Act, 419, 437
Colloidal chemistry, 262
Community aspects, 53
Community planning, 54, 59-60
Compensation measures, 71-72
Composited wastes, 166
Comprehensive Environmental Response, Compensation and Liability Act, 255
confidence limits, 110
Contaminant migration, 181, 376, 386
Criminal sanctions, 409
Cyclone separator, 172
Decontamination, 34, 371
Dewatering, 75, 81
Dibromochloromethane, 381

1,1,2,2 Dichloroethylene, 381
Disopropylmethylphosphonate, 78
Dicyclopentadiene, 78
Dredging, 321
Economic analysis, 174
Electroflotation, 163, 166
Electroplating, 193, 194, 196
Enforcement, 409
Ethylene oxide, 329
Facility, siting, 13, 20, 47, 48
Financial responsibility, 254-256
Fluoride, 85, 126, 132
German National CHamber of Commerce, 246, 247, 248
Groundwater barrier, 75
Groundwater cleanup, 329
Groundwater monitoring, 121, 134, 153, 313, 373, 377
Groundwater movement, 375
Grout curtains, 254
Halogenated organics, 317
Hazardous Materials Transportation Act, 437
Hazardous substance migration, 366
Hazardous waste treatment, 43
Heat stress, 33
Heavy metal, 163, 170, 342, 345
Hexachlorobutadiene, 139
Hydraulic conductivity, 95
Hydrology, 366
Illinois, 319
Iminnent hazard, 409, 439
Impoundment, 252
Incentives, 53
Incinerator, 50
Indemity agreement, 414
Ingestion, 33
Ketones, 138
Land disposal facilities, 49
Land disposal regulations, 249
Land use, 54
Landfills, 55, 249
Leachate, 131, 182, 315, 365
Liability, alternative, 442
Liability, enterprise 442
Liability, joint, 431, 434, 441
Liability, several 431, 434, 441
Liability, strict 429
Lime softening, 85
Liners, clay, 315
Liners, plastic, 315
Manufacturing equipment, 241
Matching funds, 9

Mercury, 126, 133, 243
Methylene chloride, 139, 346
Michigan, 330
Mitigative measures, 54, 68
Naphthalene, 381
National Contingency Plan, 341, 438
National Libray of Medicine, 37
National Oil and Hazardous Substances Contingency Plan, 349
National Primary Drinking Water Standards, 342
New York State Industrial Materials Recycling Act, 217
NIOSH, 30
Nitrobenzene, 139
Nordic Waste Exchange, 246
Northeast Industrial Waste Exchange, 232
Ohio Environmental Council, 50
Opposition tactics, 64
Organic compounds, 121
Organic lubricants, 138
Organic thickeners, 138
Organic vapors, 370, 371
OSHA, 35
PCB, 156, 231, 342, 345, 348, 398-400
Percutaneous absorption, 31
Perimeter dikes, 189
Permits, 44, 48, 234
Pesticides, 121, 133
Phenol, 126, 133, 329
Photochemical decomposition, 140
Pilot plant, 167
Pretreatment, 165, 358
Priority list, 7
Process modifications, 353
Process substitutions, 193
Product reformulation, 244
Production operations, 241
Protective clothing, 31, 32
Psychological stress, 33
Public health, 29
Public hearing, 29
Public involvement, 62-64
Public participation, 15, 50
Public relations, 338
Purge wells, 330, 334
Radioactive, 126
Radioactive wastes, 164
Radium, 135
RCRA, 1-6, 41, 47, 58, 193, 234, 253, 314, 316, 437, 442
Recharge, 75, 81
Recovery, 193
RECRA - See RCRA

Recycling, 47, 55, 241
Regulations, 47
Remedial action, 29, 181, 182, 190, 347, 348, 350, 390
Remedial action program, 341
Remedial action master plan, 342, 395
Re-solve, Inc. Hazardous Waste Site, 199, 341, 395
Resouce Conservation and REcovery Act - See RCRA
Respirators, 31
Respiratory hazards, 370
Retrofitting, 251, 254
Reverse osmosis, 194, 199, 207
Sampling, 123-129, 364-365
Sediments, 320, 367
Site safety, 363, 369
Site selection, 62, 298
Siting boards, 16
Siting procedures, 4, 16, 60-61
social issues, 64
Sodium, 133
Sodium sulfate, 177
Solid Waste Disposal Act 1965, 57
Solid Waste Facilities, 54
Solvent extraction, 353
Statistical evaluation, 95
Sulfonic acid, 177
Sulfuric acid, 177
Superfund, 7-11, 35, 313, 365, 395, 417, 438, 440
Supernatant removal, 184
Tacoma, Washington, 373
Tetrachloroethene, 347, 375, 381
Tetrahydrofuran, 138
TOC, 126, 133
Toluene, 345
TOX, 126, 133
Toxic Substances Control Act, 240, 419, 437
Trichloroethylene, 233, 346
USEPA, 58, 131, 270, 313, 395, 419, 438
Volatile organic, 143, 156, 345, 379
Vinylidene chloride, 329
Warranties, manufacturers, 415
Waste characterization, 165-166
Waste exchange, 231, 245, 247
Waste impoundment, 367
Waste reduction, 55
Waterway containment, 319
Well drilling, 130
Well location, 138
Well monitoring, 377-378
Well sampling, 122, 199
White oil, 177
"W" test, 110
Zoning, 48, 55